THE MUSHROOM

Edible and Otherwise
Its Habitat and its Time of Growth

With
Photographic Illustrations
of
Nearly All the Common Species

M. E. HARD, M. A.

Superintendent of Public Instruction

Kirkwood, Mo.

MJP Publishers

Chennai   New Delhi   Tirunelveli

To My Wife

Whose thorough knowledge of plant life,
and whose patience in preserving fungal specimens—sometimes
beautiful but often odorous—scattered
from the back porch to the author's library,
whose eyes, quick to detect structural differences,
and whose kindly and patient help have been a constant
benediction, this work's inscribed.

ISBN 978-81-8094-172-6

Printed and bound in India

MJP 122

MJP Publishers

New No. 5, Muthu Kalathy Street
Triplicane,
Chennai 600 005

MJP's imprint: 2013

INTRODUCTION

I would agree with those who might maintain that no Introduction is needed for this book on mushrooms. Nevertheless a word may not be out of place for the inception of the work is out of the ordinary. Mr. Hard did not decide that a book on this subject was needed and then set about studying these interesting plants. He has observed them, collected them, induced many friends to join in eating those which proved to be palatable and delicious—really meddled for years with the various kinds which are edible and otherwise, and then recently he has decided to publish a book on his favorite subject. The interesting occupation of photographing the mushrooms and the toadstools doubtless has contributed largely to the determination culminating in the materialization of the treatise.

If I have correctly apprehended the origin and the contributing causes, we would expect this book to be different from the other books on mushrooms—not of course in scope and purpose; but the instruction and suggestions given, the descriptions and general remarks offered, the wide range of forms depicted in word and picture, the whole make up of the book in fact, will appeal to the people at large rather than the college student in particular. The author does not write for the specially educated few, but for the mass of intelligent people—those who read and study, but who observe more; those who are inclined to commune with nature as she displays herself in the glens and glades, in the fields and forests, and who spend little, if any, time chasing the forms or sketching the tissues that may be seen on the narrow stage of a compound microscope.

The book then is for the beginner, and for all beginners; the college student will find that this is the guide to use when he is ready to begin studying the mushrooms; the teachers in the schools should all begin to study mushrooms now, and for the purpose they will find this book advantageous; the people who see mushrooms often but do not know them may find here a book that really is a help.

We might wish for color photography when the subject is a delicately tinted mushroom; but if with it we should lose detail in structure then the wish would be renounced. The colors can be, approximately, described, often not so the characteristic markings, shapes and forms. The halftones from the photographs will, we anticipate, prove a valuable feature of the book, especially if the plants be most carefully examined before turning to the pictures. For half an hour the pages may be turned and the illustrations enjoyed. That, however, would give one no real knowledge of mushrooms. If such use only is made of the pictures, better had they never been prepared by Mr. Hard and his friends. But if a charming little toadstool, a delicately colored mushroom, a stately agaric, be carefully removed from the bed of loam, the decaying stump, or the old tree-trunk, then turned over and over again, and upside down, every part scrutinized, the structure in every detail attentively regarded—not with repugnant feeling, rather with a sympathetic interest that should naturally find all organisms inhabiting our globe—then in due time coming to the picture, a real picture, in the book, it must surely bring both pleasure and profit. Ponder the suggestion. Then, to conclude in a word, if Mr. Hard's book will induce people to learn and enjoy the mushrooms that we have, it will be a success, and great will be his reward.

W. A. Kellerman, Ph. D.
Botanical Department

AUTHOR'S NOTE

IN MEMORIAM

It is with feelings of profound sadness that I am impelled to supplement the above Introduction by a brief tribute to the memory of that genial gentleman and lovable companion, as well as enthusiastic scientist, the late Dr. W. A. Kellerman.

Spending his life in the pursuit of science, the Angel of Death overtook him while still in search for wider knowledge of Nature and her works, and with icy fingers sealed the lids over eyes ever on the alert for the discovery of hidden truths.

Quiet, reticent, and unassuming, it was given to but few to know the great-hearted, unselfish sweetness of nature underlying his whole life. Yet the scientific world in general and Nature students especially, recognize in Dr. Kellerman's death a loss long to be regretted and not soon to be repaired.

The foregoing "Introduction" from his pen was one of the latest, if not the last of his public writings, done but a few weeks before being stricken with the fatal fever which fell upon him in the forests of Guatemala, and so quickly ended his earthly hopes and aspirations.

It seems doubly sad that one so well and widely known in his life should be called upon to lay its burdens and its pleasures down while so far away from all who knew and loved him well; and to rest at last among strangers in a strange land.

To this beloved friend and companion of so many pleasant days in woods and fields the author of this book desires to pay the tribute of a loving remembrance and heartfelt appreciation.

The Author.

PREFACE

"Various as beauteous, Nature, is thy face;

* * * all that grows, has grace.

All are appropriate. Bog and moss and fen

Are only poor to undiscerning men.

Here may the nice and curious eye explore

How Nature's hand adorns the ruby moor;

Beauties are these that from the view retire,

But will repay th' attention they require."

Botany and geology have been favorite studies of the author since leaving college, thanks to Dr. Nelson, who lives in the hearts of all his students. He, by his teachings, made these subjects so attractive and interesting that by one, at least, every spare moment has been given to following up the studies of botany and paleontology. But the mycological part of botany was brought practically to the author's attention by the Bohemian children at Salem, Ohio, at the same time arousing a desire to know the scientific side of the subject and thus to be able to help the many who were seeking a personal knowledge of these interesting plants.

Every teacher should be able to open the doors of Nature to his pupils that they may see her varied handiwork, and, as far as possible, assist in removing the mist from their eyes that they may see clearly the beauties of meadow, wood or hillside.

In beginning the fuller study of the subject the writer labored at great disadvantage because, for a number of years, there was but little available literature. Every book written upon this subject, in this country, was purchased as soon as it came out and all have been very helpful.

The study has been a very great pleasure, and some very delightful friendships have been made while in search for as great a variety of species as possible.

For a number of years the object was simply to become familiar with the different genera and species, and no photographs of specimens were made. This was a great mistake; for, after it was determined to bring out this work, it seemed impossible to find many of the plants which the author had previously found in other parts of the state.

However, this failure has been very largely overcome through the generous courtesy of his esteemed friends,—Mr. C. G. Lloyd, of Cincinnati; Dr. Fisher, of Detroit; Prof. Beardslee, of Ashville, N. C.; Prof. B. O. Longyear, of Ft. Collins, Col., and Dr. Kellerman, of Ohio State University,—who have most kindly furnished photographs representing those species found earlier in other parts of the state. The species represented here have all been found in this state within the past few years.

The writer is under great obligation to Prof. Atkinson, of Cornell University, for his very great assistance and encouragement in the study of mycology. His patience in examining and determining plants sent him is more fully appreciated than can be expressed here. Dr. William Herbst, Trexlertown, Pa., has helped to solve many difficult problems; so also have Mr. Lloyd, Prof. Morgan, Capt. McIlvaine and Dr. Charles H. Peck, State Botanist of New York.

The aim of the book has been to describe the species, as far as possible, in terms that will be readily understood by the general reader; and it is hoped that the larger number of illustrations will make the book helpful to those who are anxious to become acquainted with a part of botany so little studied in our schools and colleges.

No pains have been spared to get as representative specimens as it was possible to find. A careful study of the illustrations of the plants will, in most cases, very greatly assist the student in determining the classification of the plant when found; but the illustration should not be wholly relied upon, especially in the study of Boleti. The description should be carefully studied to see if it tallies with the characteristics of the plant in hand.

In many plants where notes had not been taken or had been lost, the descriptions given by the parties naming the plants were used. This is notably so of many of the Boleti. The author felt that Dr. Peck's descriptions would be more accurate and complete, hence they were used, giving him credit.

Care has been taken to give the translation of names and to show why the plant was so called. It is always a wonder to the uninitiated how the Latin name is remembered, but when students see that the name includes some prominent characteristic of the plant and thus discover its applicability, its recollection becomes comparatively easy.

The habitat and time of growth of each plant is given, also its edibility. The author was urged by his many friends throughout the state, while in institute work and frequently talking upon this subject, to give them a book that would assist them in becoming familiar with the common mushrooms of their vicinity. The request has been complied with.

It is hoped that the work will be as helpful as it has been pleasant to perform.

M. E. H.

CONTENTS

CHAPTER I

WHY STUDY MUSHROOMS

Some years ago, while in charge of the schools of Salem, Ohio, we had worked up quite a general interest in the study of botany. It was my practice to go out every day after flowers, especially the rarer ones, of which there were many in this county, and bring in specimens for the classes. There was in the city a wire nail mill, running day and night, whose proprietors brought over, from time to time, large numbers of Bohemians as workers in the mill. Very frequently, when driving to the country early in the morning, I found the boys and girls of these Bohemian families searching the woods, fields and pastures at some distance from town, although they had not been in this country more than a week or two and could not speak a word of English. I soon found that they were gathering mushrooms of various kinds and taking them home for food material. They could not tell me how they knew them, but I quickly learned that they knew them from their general characteristics,—in fact, they knew them as we know people and flowers.

I resolved to know something of the subject myself. I had no literature on mycology, and, at that time, there seemed to be little obtainable. About that time there appeared in Harper's Monthly an article by W. Hamilton Gibson upon Edible Toadstools and Mushrooms—an article which I thoroughly devoured, soon after purchasing his book upon the subject.

Salem, Ohio, was a very fertile locality for mushrooms and it was not long till I was surprised at the number that I really knew. I remembered that where there is a will there is a way.

In 1897 I moved to Bowling Green, Ohio; there I found many species which I had found about Salem, Ohio, but the extremely rich soil, heavy timber and numerous old lake beaches seemed to furnish a larger variety, so that I added many more to my list. After remaining three years in Bowling Green, making delightful acquaintance with the good people of that city as well as with the flowers and mushrooms of Wood county, Providence placed me in Sidney, Ohio, where I found many new species of fungi and renewed my acquaintance with many of those formerly met.

Since coming to Chillicothe I have tried to have the plants photographed as I have found them, but having to depend upon a photographer I could not always do this. I have not found in this vicinity many that I have found elsewhere in the state, although I have found many new things here, a fact which I attribute to the hilly nature of the county. For prints of many varieties of fungi obtained before coming here, I am indebted to my friends. I should advise any one intending to make a study of this subject to have all specimens photographed as soon as they are identified, thus fixing the species for future reference.

It seems to me that every school teacher should know something of mycology.Some of my teachers have during the past year made quite a study of this interesting subject, and I have found that their pupils kept them busy in identifying their finds. Their lists of genera and species, as exhibited on the blackboards at the close of the season were quite long. I found from my Bohemian boys and girls that their teachers in their native country had opened for them the door to this very useful knowledge. Observation has proven to me conclusively that there is a large and increasing interest in this subject throughout the greater part of Ohio.

Every professional man needs a hobby which he may mount in his hours of relaxation, and I am quite sure there is no field that offers better inducement for a canter than the subject of botany, and especially this particular department of botanical work.

I have a friend, a professional man who has an eye and a heart for all the beauties of nature. After hours of confinement in his office at close and critical work he is always anxious for a ramble over the hillsides and through the woods, and when we find anything new he seems to enjoy it beyond measure.

Many ministers of the gospel have become famous in the mycological world. The names of Rev. Lewis Schweiwitz, of Bethlehem, Pa.; Rev. M. J. Berkeley and Rev. John Stevenson, of England, will live as long as botany is known to mankind. Their influence for good and helpfulness to their fellowmen will be everlasting.

With such an inspiration, how quickly one is lost to all business cares, and how free and life-giving are the fields, the meadows and the woods, so that one must exclaim with Prof. Henry Willey in his "Introduction to the Study of the Lichen":

"If I could put my woods in song,

And tell what's there enjoyed,

All men would to my garden throng,

And leave the cities void.

In my lot no tulips blow;

Snow-loving pines and oaks instead;

And rank the savage maples grow,

From Spring's first flush to Autumn red;

My garden is a forest ledge,

Which older forests bound."

MUSHROOMS AND TOADSTOOLS

How To Tell Mushrooms From Toadstools

In all probability no student of mycology has any one query more frequently or persistently pressed upon his attention than the question, "How do you tell a toadstool from a mushroom?"—or if in the woods or fields, in search for new species, with an uninitiated comrade, he has frequently to decide whether a certain specimen "is a mushroom or a toadstool," so firmly fixed is the idea that one class of fungi—the toadstools—are poisonous, and the other—the mushrooms—are edible and altogether desirable; and these inquiring minds frequently seem really disappointed at being told that they are one and the same thing; that there are edible toadstools and mushrooms, and poisonous mushrooms and toadstools; that in short a toadstool is really a mushroom and a mushroom is only a toadstool after all.

Hence the questions with the beginner is, how he may tell a poisonous fungus from an edible one. There is but one answer to this question, and that is that he must thoroughly learn both genera and species, studying each till he knows its special features as he does those of his most familiar friends.

Certain species have been tested by a number of people and found to be perfectly safe and savory; on the other hand, there are species under various genera which, if not actually poisonous, are at least deleterious.

It is the province of all books on fungi to assist the student in separating the plants into genera and species; in this work special attention has been given to distinguishing between the edible and the poisonous species. There are a few species such as Gyromitra esculenta, Lepiota Morgani, Clitocybe illudens, etc., which when eaten by certain persons will cause sickness soon after eating, while others will escape any disagreeable effects. Chemically speaking, they are not poisonous, but simply refuse to be assimilated in some stomachs. It is best to avoid all such.

HOW MUSHROOMS GROW

There is a strong notion that mushrooms grow very quickly, springing up in a single night. This is erroneous. It is true that after they have reached the button stage they develop very quickly; or in the case of those that spring from a mature egg, develop so rapidly that you can plainly see the motion of the upward growth, but the development of the button from the mycelium or spawn takes time—weeks, months, and even years. It would be very difficult to tell the age of many of our tree fungi.

HOW TO LEARN MUSHROOMS

If the beginner will avoid all Amanitas and perhaps some of the Boleti he need not be much worried in regard to the safety of other species.

There are three ways by which he can become familiar with the edible kinds. The first is the physiological test suggested by Mr. Gibson in his book. It consists in chewing a small morsel and then spitting it out without swallowing the juice; if no important symptoms arise within twenty-four hours, another bit may be chewed, this time swallowing a small portion of the juice. Should no irritation be experienced after another period of waiting, a still larger piece may be tried. I always sample a new plant carefully, and thus am often able to establish the fact of its edibility before being able to locate it in its proper species. This fall I found for the first time Tricholoma columbetta; it was some time after I had proven it an edible mushroom before I had settled upon its name. A better way, perhaps, is to cook them and feed them to your cat and watch the result.

Another way is to have a friend who knows the plants go with you, and thus you learn under a teacher as a pupil learns in school. This is the quickest way to gain a knowledge of plants of any kind, but it is difficult to find a competent teacher.

Still another way, and one that is open to all, is to gain a knowledge of a few species and through their description become familiar with the terms used in describing a mushroom; this done, the way is open, if you have a book containing illustrations and descriptions of the most common plants. Do not be in a hurry to get the names of all the plants, and do not make use of any about which you are not absolutely sure. In gathering mushrooms to eat, do not put into your basket with those you intend to eat a single mushroom of whose edible qualities you have any doubt. If you have the least doubt about it, discard it, or put it in another basket.

There are no fixed rules by which you can tell a poisonous from an edible mushroom. I found a friend of mine eating Lepiota naucina, not even knowing to what genus it belonged, simply because she could peel it. I told her that the most deadly mushroom can be peeled just as readily. Nor is there anything more valuable in the silver spoon test in which Mr. Gibson's old lady put so much confidence. Some say, do not eat any that have an acrid taste; many are edible whose taste is quite acrid. Others say, do not eat any whose juice or milk is white, but this would discard a number of Lactarii that are quite good. There is nothing in the white gills and hollow stem theory. It is true that the Amanita has both, but it must be known by other characteristics. Again we are told to avoid such as have a viscid cap, or those that change color quickly; this is too sweeping a condemnation for it would cut out several very good species. I think I may safely say there is no known rule by which the good can be distinguished from the bad. The only safe way is to know each species by its own individual peculiarities—to know them as we know our friends.

The student of mycology has before him a description of each species, which must tally with the plant in hand and which will soon render him familiar with the different features of the various genera and species, so he can recognize them as readily as the features of his best friends.

WHAT ANYONE MAY EAT

In the spring of the year there comes with the earliest flowers a mushroom so strongly characteristic in all its forms that no one will fail to recognize it. It is the common morel or sponge mushroom. None of them are known to be harmful, hence here the beginner can safely trust his judgment. While he is gathering morels to eat he will soon begin to distinguish the different species of the genera. From May till frost the different kinds of puff-balls will appear. All puff-balls are good while their interior remains white. They are never poisonous, but when the flesh has begun to turn yellow it is very bitter. The oyster mushroom is found from March to December and is always a very acceptable mushroom. The Fairy Rings are easily recognized and can be found in any old pasture during wet weather from June to October. In seasonable weather they are usually very plentiful. The common meadow mushroom is found from September to frost. It is known by its pink gills and meaty cap. There is a mushroom with pink gills found in streets, along the pavements and among the cobble stones. The stems are short and the caps are very meaty. It is A. rodmani. These are found in May and June. The horse mushroom has pink gills and may be found from June to September. The Russulas, found from July to October, are generally good. A few should be avoided because of their acrid taste or their strong odor. There is no time from early spring till freezing weather when you can not find mushrooms, if the weather is at all favorable. I have given the habitat and the time when each species can be found. I should recommend a careful study of these two points. Read the descriptions of plants which grow in certain places and at certain times, and you will generally be rewarded, if you follow out the description and the season is favorable.

HOW TO PRESERVE MUSHROOMS

Many can be dried for winter use, such as the Morels, Marasmius oreades, Boletus edulis, Boletus edulis, va. clavipes, and a number of others. My wife has very successfully canned a number of species, notably Lycoperdon pyriforme, Pleurotus ostreatus and Tricholoma personatum. The mushrooms were carefully picked over and washed, let stand in salt water for about five minutes, in order to free them of any insect-life which may be in the gills, then drained, cut into pieces small enough to go into the jars easily. Each jar was packed as full as possible with mushrooms and filled up with water salt enough to flavor the mushroom properly. Then put into a kettle of cold water on the stove, the lids being loosely placed on the top, and allowed to cook for an hour or more after the water in the kettle begins to boil. The tops were then fastened on securely and after trying the jars to see if there was any leak, they were set away in a cool, dark place.

In canning puff-balls they should be carefully washed and sliced, being sure that they are perfectly white all through. They do not need to stand in salt water before packing in the jar as do those mushrooms which have gills. Otherwise they were canned as the Tricholoma and oyster mushroom. Any edible mushroom can easily be kept for winter use by canning. Use glass jars with glass tops.

TERMS USED

Some Of The Most Common Terms Used

In describing mushrooms it is necessary to use certain terms, and it will be incumbent upon anyone who wishes to become familiar with this part of botanical work to understand thoroughly the terms used in describing the plants.

The substance of all mushrooms is either fleshy, membranaceous, or corky. The pileus or cap is the expanded part, which may be either sessile or supported by a stem. The pileus is not made up of cellular tissue as in flowering plants, but of myriads of interwoven threads or hyphae. This structure of the pileus will become evident at once if a thin portion of the cap is placed under the microscope.

The gills or lamellæ are thin plates or membranes radiating from the stem to the margin of the cap. When they are attached squarely and firmly to the stem they are said to be adnate. If they are attached only by a part of the width of the gills, they are adnexed. Should they extend down on the stem, they are decurrent. They are free when they are not attached to the stem. Frequently the lower edge is notched at, or near, the stem and in this case they are said to be emarginate or sinuate.

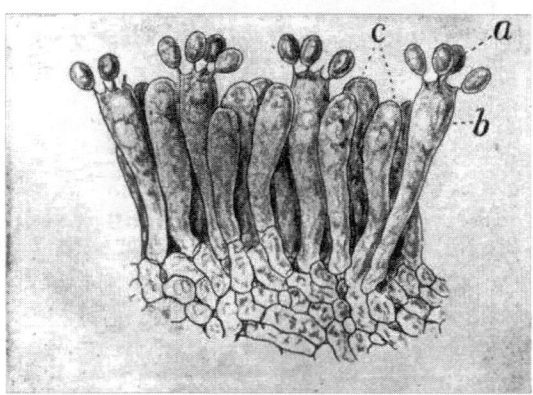

Figure 2   Small portion of a section through the spore-bearing layer of a mushroom which produces its spores on the ends of cells called basidium. (a) Spores, (b) basidium, (c) sterile cells.

In some genera the lower surface of the cap is full of pores instead of gills; in other genera the lower surface is crowded with teeth; in still others the surface is smooth, as in the Stereums. The gills, pores and teeth afford a foundation for the hymenium or fruit-bearing surface. It will be readily seen that the gills, pores and teeth simply expose in a very economical way the greatest possible spore-bearing surface.

If a section of the gills be examined by a microscope, it will be observed that upon both sides of the surface are extended hymenial layers. The hymenium consists of elongated cells or basidia (singular, basidium) more or less club-shaped. Figure 2 will show how these basidia appear on the hymenial layer when strongly magnified. It will be seen that they are placed side by side and are perpendicular to the surface of the gills. Upon each of these basidia are in some species two, usually four, slender projections upon which the spores are produced. In Figure 2 a number of sterile cells will be seen which resemble the basidia except that the latter bear four sterigmata upon which the spores rest. Among these basidia and sterile cells will frequently be seen an overgrown bladder-like sterile basidium which projects beyond the rest of the hymenium, and whose use is not as yet fully known. They are called cystidia (singular, cystidium). They are never numerous, but they are scattered over the entire surface, becoming more numerous along the edge of the gills. When they are colored, they change the appearance of the gills.

Figure 3   Rootlike strands of mycelium of the pear-shaped puff-ball growing in rotten wood. Young puff-balls in the form of small white knots are forming on the strands. Natural size.—Longyear.

The spores are the seeds of the mushroom. They are of various sizes and shapes, with a variety of surface markings. They are very small, as fine as dust, and invisible to the naked eye, except as they are seen in masses on the grass, on the ground, or on logs, or in a spore print. It is the object of every fungus to produce spores. Some fall on the parent host or upon the ground. Others are wafted away by every rise of the wind and carried for days and finally settle down, it may be, in other states and continents from those in which they started. Millions perish because of not finding a suitable resting place. Those spores that do find a favorable resting-place, under right conditions, will begin to germinate by sending out a slender thread-like filament, or hyphæ, which at once branches out in search of food material, and which always forms a more or less felted mass, called mycelium. When first formed the hyphæ are continuous and ramify through the nourishing substratum from which there arises afterward a spore-bearing growth known as the sporocarp or young mushroom. This vegetative part of the fungus is usually hidden in the soil, or in decayed wood, or vegetable matter. In Figure 3 is a representation of the mycelium of the small pear-shaped puff-ball with a number of small white knobs marking the beginning of the puff-ball. The mycelium exposed here is very similar to the mycelium of all mushrooms.

In the pore-bearing genera the hymenium lines the vertical pores; in teeth-bearing fungi it lines the surface of each tooth, or is spread out over the smooth surface of the Stereum.

The development of the spores is quite interesting. The young basidia as seen in Figure 2 are filled with a granular protoplasm. Soon small projections, called sterigma (plural, sterigmata), make their appearance on the ends of the basidia and the protoplasm passes into them. Each projection or sterigma soon swells at its extremity into a bladder-like body, the young spore, and, as they enlarge, the protoplasm of the basidium is passed into them. When the four spores are full grown they have consumed all the protoplasm in the basidium. The spores soon separate by a transverse partition and fall off. All spores of the Hymenomycetous fungi are arranged and produced in a similar manner, with their spore-bearing surface exposed early in life by the rupture of the universal veil.

In the puff-balls the spores are arranged in the same way, but the hymenium is inclosed within an outer sack. When the spores are ripe the case is ruptured and the spores escape into the air as a dusty powder. The puff-balls, therefore, belong to the Gastromycetous fungi because its spores are inclosed in a pouch until they are matured.

Another very large group of fungi is the Ascomycetes, or sac fungi. It is very easily determined because all of its members develop their spores inside of small membranous sacs or asci. These asci are generally intermixed with slender, empty asci, or sterile cells, called paraphyses. These asci are variously shaped bodies and are known in different orders by different names, such as ascoma, apothecium, perithecium, and receptacle. The Ascomycetes often include among their numbers fungi ranging in size from microscopic one-celled plants to quite large and very beautiful specimens. To this group belong the great number of small fungi producing the various plant diseases.

In a work of this kind especial attention is naturally given to the order of Discomycetes or cup fungi. This order is very large and is so called because so many of the plants are cup shaped. These cups vary greatly in size and form; some are so small that it requires a lens to examine them; some are saucer-shaped; some are like goblets, and some resemble beakers of various shapes. The saddle fungi and morels belong to this order. Here the sac surface is often convoluted, lobed, and ridged, in order to afford a greater sac-bearing surface.

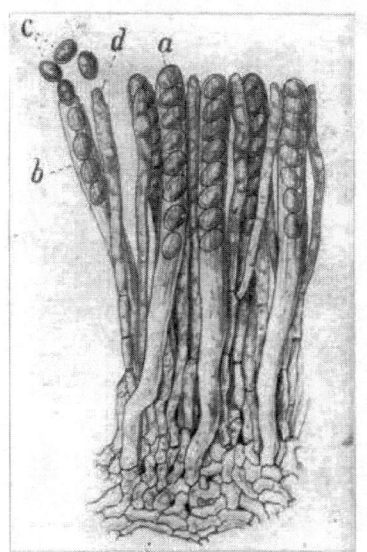

Figure 4   Small portion of a section through the spore-bearing part of a morel in which the spores are produced in little sacs or asci. (a) An ascus, (b) an ascus discharging its spores, (c) the spores, (d) sterile cells. Highly magnified.—Longyear.

In the mushrooms, puff-balls, etc., we find the spores were borne on the ends of basidia, usually four spores on each. In this group the spores are formed in minute club-shaped sacs, known as asci (singular, ascus). These asci are long, cylindrical sacs, standing side by side, perpendicular to the fruiting surface. Figure 4 will illustrate their position together with the sterile cells on the fruiting surface of one of the morels. They usually have eight spores in each sac or ascus.

The stem of the mushroom is usually in the center of the cap, yet it may be eccentric or lateral; when it is wanting, the pileus is said to be sessile. The stem is solid when it is fleshy throughout, or hollow when it has a central cavity, or stuffed when the interior is filled with pithy substance. The stems are either fleshy or cartilaginous. When the former, it is of the same consistency as the pileus. If the latter, its consistency is always different from the pileus, resembling cartilage. The stem of the Tricholoma affords a good example of the fleshy stemmed mushroom, and that of the Marasmius illustrates the cartilaginous.

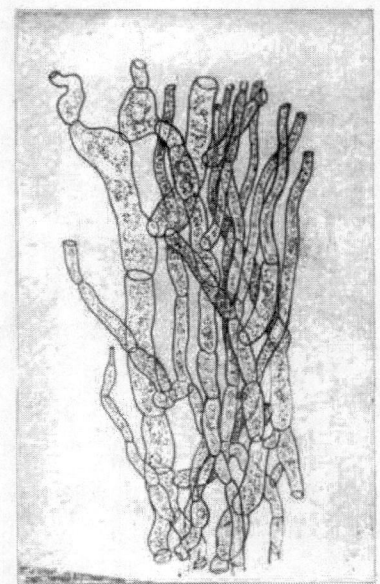

Figure 5   Small portion of a stem of a morel showing cell filaments. Highly magnified.—Longyear.

If the cap or stem of a mushroom is examined with a microscope of high magnifying power it will be found to be made up of a continuation of the mycelial filaments, interlaced and interwoven, branching, and the tubular filaments often delicately divided, giving the appearance of cells. Figure 5 represents a small portion of a Morel stem highly magnified showing the cell filaments. In soft fungi the mycelial threads are more loosely woven and have thin walls with fewer partitions.

The veil is a thin sheet of mycelial threads covering the gills, sometimes remaining on the stem, forming a ring or annulus. This sometimes remains for a time on the margin of the cap when it is said to be appendiculate. Sometimes it resembles a spider's web when it is called arachnoid.

The volva is a universal wrapper, surrounding the entire plant when young, but which is soon ruptured, leaving a trace in the form of scales on the cap and a sheath around the base of the stem, or breaking up into scales or a scaly ring at the base of the stem. All plants having this universal volva should be avoided, further than for the purpose of study. Care should be taken that, in their young state, they are not mistaken for puff-balls. Frequently when found in the egg state they resemble a small puff-ball. Figure 6 represents a section of an Amanita in the egg-state and also the Gemmed puff-ball. As soon as a section is made and carefully examined the structure of the inside will reveal the plant at once. There is but little danger of confusing the egg stage of an Amanita with the puff-ball, for they resemble each other only in their oval shape, and not in the least in their marking on the surface.

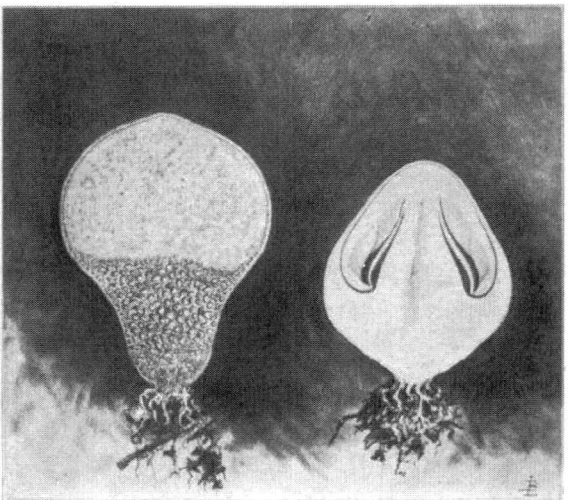

Figure 6   The lefthand figure represents a vertical section through a young plant of the gemmed puff-ball showing the cellular structure of the stem-like lower half, called the subgleba. The righthand figure shows a vertical section of the egg stage of an Amanita, a very poisonous fungus which grows in woods and which might be mistaken for a young puff-ball if not cut open. The fungus forms just below the surface of the soil, finally bursting the volva, sending up a parasol mushroom. Natural size.—Longyear.

WHAT IS A FUNGUS OR A MUSHROOM?

It is a cellular, flowerless plant, nourished by the mycelium which permeates the soil or other substances on which the fungus or mushrocm grows. All fungi are either parasites or saprophytes which have lost their chlorophyll, and are incapable of supporting an independent existence.

There is a vast number of genera and species, and many have the parasitic habit which causes them to enter the bodies of other plants and of animals. For this reason all fungi are of economic importance, especially the microscopic forms classed under the head of Bacteria. Some recent writers are inclined to separate the Bacteria and slime-molds from the fungus group, and call them fungus animals. However this may be, they are true plants and have many of the characteristics of the fungi. They may differ from the fungi in their vegetative functions, yet they have so many things in common that I am inclined to place them under this group.

Many, such as the yeast fungus, the various fermentative fungi, and the Bacteria concerned in the process of decomposition, are indeed very useful. The enrichment and preparation of soils for the uses of higher plants, effected by Bacteria, are very important services.

Parasites derive their nourishment from living plants and animals. They are so constituted that when their nourishing threads come within range of the living plant they answer a certain impulse by sending out special threads, enveloping the host and absorbing nutrition. Saprophitic plants do not experience this reaction from the living plants. They are compelled to get their nourishment from decaying products of plants or animals, consequently they live in rich ground or leaf mold, on decayed wood, or on dur.g. Parasites are usually small, being limited by their host. Saprophytes are not thus limited for food supply and it is possible to build up large plants such as the common mushroom group, puff-balls, etc.

The spores are the seeds or reproductive bodies of the mushroom. They are very fine, and invisible to the naked eye except when collected together in great masses. Underneath mushrooms, frequently, the grass or wood will be white or plainly discolored from the spores. The hymenium is the surface or part of the plant which bears the spores. The hymenophore is the part which supports the hymenium.

In the common mushroom, and in fact many others, the spores develop on a certain club-like cell, called basidium (plural, basidia), on each of which four spores usually develop. In morels these cells are elongated into cylindrical membranous sacs called asci, in each of which eight spores are usually developed. The spores will be found of various colors, shapes, and sizes, a fact which will be of great assistance to the student in locating strange species and genera. In germination the spores send out slender threads which Botanists call mycelium, but which common readers know as spawn.

The method and place of spore development furnish a basis for the classification of fungi. The best way to acquire a thorough knowledge of both our edible and poisonous mushrooms is to study them in the light of the primary characters employed in their classification and their natural relation to each other.

There is a wide difference of opinion as to the classification of mushrooms. Perhaps the most simple and satisfactory is that of Underwood and Cook. They arrange them under six groups:

- Basidiomycetes—those in which the spores or reproductive bodies are naked or external as shown in illustration 2 on page 15.

- Ascomycetes—those in which the spores are inclosed in sacs or asci. These sacs are very clearly represented in illustration Figure 4 on This will include the Morels, Pezizæ, Pyrenomycetes, Tuberaceæ, Sphairiacei, etc.

- Physcomycetes—including the Mucorini, Saprolegniaceæ, and Peronosporeæ. Potato rot and downy mildew on grape vines belong to this family.

- Myxomycetes—Slime moulds.

- Saccharomycetes—Yeast fungi.

- Schizomycetes—are minute, unicellular Protophytes which reproduce mainly by transverse fission.

- Class, Fungi—Sub-Class, Basidiomycetes.

This class will include all gill-bearing fungi, Polyporus, Boletus, Hydnum, etc.

Fungi of this class are divided into four natural groups

- Hymenomycetes.
- Gasteromycetes.
- Uredinæ.
- Ustilagineæ.

GROUP 1—HYMENOMYCETES

Under this group will be placed all fungi composed of membranes, fleshy, woody, or gelatinous, whether growing on the ground or on wood. The hymenium, or spore-bearing surface, is external at an early stage in the life of the plant. The spores are borne on basidia as explained in Figure 2, page 6. When the spores ripen they fall to the ground or are carried by the wind to a host that presents all the conditions necessary for germination; there they produce the mycelia or white thread-like vines that one may have noticed in plowing sod, in old chip piles, or decayed wood. If one will examine these threads there will be found small knots which will in time develop into the full grown mushroom. Hymenomycetes are divided into six families:

- Agaricaceæ. Hymenium with gills.
- Polyporaceæ. Hymenium with pores.
- Hydnaceæ. Hymenium with spines.
- Thelephoraceæ. Hymenium horizontal and mostly on the under surface.
- Clavariaceæ. Hymenium on a smooth club-shaped surface.
- Tremellaceæ. Hymenium even and superior. Gelatinous fungi.

FAMILY 1—AGARICACEAE

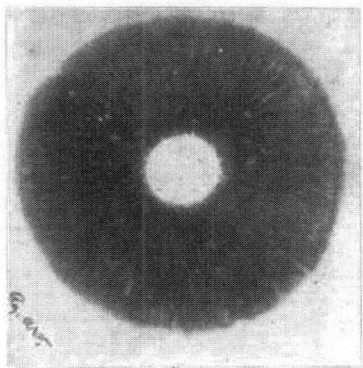

Figure 7 Spore-print of Agaricus arvensis.

In the Agaricaceæ or common mushrooms, and in all other of similar structure, the spore-producing membranes are found on the under surface of the cap. They consist of thin lamellæ, or gills, attached by the upper edge to the cap and extending from the stem to the margin of the cap. Very frequently that space may be entirely utilized by shorter lamellæ, or gills, intervening between the longer, especially toward the margin of the cap. In a few species where the stem seems to be wanting, or where it is attached to the side of the cap, the lamellæ, or gills, radiate from the point of attachment or from the lateral stem to other parts of the circumference of the cap. Berkeley gives the following characteristics: Hymenium, inferior, spread over easily divisible gills or plates, radiating from a center or stem, which may be either simple or branched.

This family includes the following genera

1. Agaricus—Gills, not melting, edge acute; including all the sub-genera which have been elevated to the rank of genera.
2. Coprinus—Gills deliquescent, spores black.
3. Cortinarius—Gills persistent, veil spider-web-like, terrestrial.
4. Paxillus—Gills separating from the hymenophorum and decurrent.
5. Gomphidius—Gills branched and decurrent, pileus top-shaped.
6. Bolbitius—Gills becoming moist, spores colored.
7. Lactarius—Gills milky, terrestrial.
8. Russula—Gills equal, rigid, and brittle, terrestrial.
9. Marasmius—Gills thick, tough, hymenium dry.
10. Hygrophorus—Stem confluent with the hymenophorum; gills sharp edged.
11. Cantharellus—Gills thick, branched, rounded edge.
12. Lentinus—Pileus hairy, hard, tough; gills, tough, unequal, toothed; on logs and stumps.
13. Lenzites—Whole plant corky; gills simple or branched.
14. Trogia—Gills venose, fold-like, channelled.

15. Panus—Gills corky, with acute edge.

16. Nyctalis—Veil universal; gills broad, often parasitic.

17. Schizophyllum—Gills corky, split longitudinally.

18. Xerotus—Gills tough, fold-like.

Therefore the gill-bearing fungi are known under the family name, Agaricaceæ, or more generally known as Agarics.

Figure 8   Spore-print of Hypholoma sublatertium.

This family is divided into five series, according to the color of their spores. The spores when seen in masses possess certain colors, white, rosy, rusty, purple-brown and black. Therefore the first and most important part to be determined in locating a mushroom is to ascertain the color of the spores. To do this, take a fresh, perfect, and fully developed specimen, remove the stem from the cap. Place the cap with the gills downward on the surface of dark velvety paper, if you suspect the spores to be white. Invert a finger bowl or a bell glass over the cap to keep the air from blowing the spores away. If the spores should be colored, white paper should be used. If the specimen is left too long the spore deposit will continue upward between the gills and it may reach an eighth of an inch in height, in which case if great care is taken in removing the cap there will be a perfect likeness of the gills and also the color of the spores.

Figure 9   Spore-print of a Flammula.

There are two ways of making these spore prints quite permanent. First take a piece of thin rice paper, muscilage it and allow it to dry, then proceed as above. In this way the print will stand handling quite a little. Another way, and that used to prepare the spore-prints in these photographs, is to obtain the spore-print upon Japanese paper as in the preceding method, then by an atomizer spray the print gently and carefully with a fixative such as is used in fixing charcoal drawings. Success in making spore-prints requires both time and care, but the satisfaction they give is ample recompense for the trouble. It is more difficult to obtain good prints from the white-spored mushrooms than from those bearing colored spores, because it is hard to obtain a black paper having a dull velvety surface, and the spores will not adhere well to a smooth-finished, glossy paper. For the prints illustrated I am indebted to Mrs. Blackford.

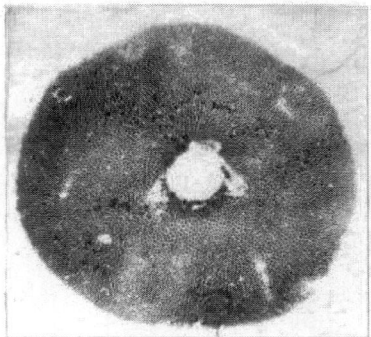

Figure 10   Spore-print of a Boletus.

If the plant is dry it is well to moisten the fingerbowl or bell-glass on the inside before placing it over the mushroom. The spores of Boleti, and, indeed, all fungi can be caught and fixed in the same way.

From the study of these spore-prints we shall find five different colors of spores. This family is, therefore, divided into five series, determined by the color of the spores, which are always constant in color, size and shape.

The five series will be treated in the following order:

1.   The white-spored Agarics.

2.   The rosy-spored Agarics.

3.   The rusty-spored Agarics.

4.   The purple-brown-spored Agarics.

5.   The black-spored Agarics.

ANALYTICAL KEY

This key is largely based upon Cooke's analytical key. Its use will help to locate the plant in hand in the genus to which it belongs.

The first thing the student should do is to determine the color of the spore if it is not evident. This is best done according to the plan described on page 15.

The plant should be fresh and mature. Careful attention should be given to different stages of development. The habit of the plant should be considered; then, as soon as the color of the spores is determined, it will be an easy matter to locate the genus by means of the key.

GROUP I—HYMENOMYCETES

Mycelium floccose, giving rise to a distinct hymenium, fungus fleshy, membranaceous, woody or gelatinous. Spores naked.

Hymenium, normally inferior—

Hymenium with gills	Agaricaceæ.
Hymenium with pores	Polyporaceæ.
Hymenium with teeth	Hydnaceæ.
Hymenium even	Thelophoraceæ.

Hymenium, superior—

Hymenium on smooth surface, club-shaped,	Clavariaceæ.
Hymenium lobed, convolute, gelatinous,	Tremellaceæ.

FAMILY 1—AGARICACEAE

Hymenium inferior, pileus more or less expanded, convex, bell-shaped. Gills radiating from the point of attachment of the pileus with the stem, or from a lateral stem to other parts of the cap, simple or branched.

I. Spores white or slightly tinted.

    A. Plants fleshy, more or less firm, decaying soon.

    a. Stem fleshy, pileus easily separating from the stem.

        Volva present and ring on the stem.

Pileus bearing warts or patches free from the cuticle	Amanita.
Volva present, ring wanting	Amanitopsis.
Pileus scaly, scales concrete with cuticle,	
Volva wanting, ring present	Lepiota.

        Hymenophore confluent,

        Without cartilaginous bark,

    b. Stem central, ring present (sometimes vague),

Volva wanting, gills attached	Armillaria.

        Without a ring,

Gills sinuate	Tricholoma.

        Gills decurrent,

Edges acute	Clitocybe.

Edges swollen   Cantharellus.

Gills adnate,
    Parasitic on other mushrooms   Nyctalis.

    Not parasitic,

    Milky   Lactarius.

    Not exuding juice when bruised,

    Rigid and brittle   Russula.

    Quite viscid, waxy consistency   Hygrophorus.

  c.   Stem lateral or none, rarely central Pleurotus.
  d.   Stem with cartilaginous bark,
    Gills adnate   Collybia.
    Gills sinuate   Mycena.
    Gills decurrent   Omphalia.
Plants tough, fleshy, membranaceous, leathery,
Stem central,
    Gills simple   Marasmius.
    Gills branched   Xerotus.
  B.   Plants gelatinous and leathery   Heliomyces.
    Stem lateral or wanting,
      Edge of gills serrate   Lentinus.
      Edge of gills entire   Panus.
      Gills fold-like, irregular   Trogia.
      Edge of gills split longitudinally   Schizophyllum.
  C.   Plants corky or woody,
    Gills anastomosing.   Lenzites.
II. Spores rosy or salmon color.
  A.   Stem central.
    Gills free, stem easily separating from pileus.
    Without cartilaginous stem,
    Volva present and distinct, no ring   Volvaria.
    Without a volva, with a ring   Annularia.
    Without a volva and without a ring   Pluteus.

B.   Stem fleshy to fibrous, margin of pileus at first incurved,

      Gills sinuate or adnate   Entoloma.

      Gills decurrent   Clitopilus.

C.   Stem eccentric or none, pileus lateral   Claudopus.

      Gills decurrent, pileus umbilicate   Eccilia.

      Gills not decurrent, pileus torn into scales, and slightly
convex, margin at first involute   Leptonia.

      Pileus bell-shaped, margin at first straight   Nolanea.

III. Spores rusty-brown or yellow-brown.

   A.   Stem not cartilaginous,

   a.   Stem central,

      With a ring,

      Ring continuous   Pholiota.

      Veil arachnoid,

      Gills adnate, powdery from spores   Cortinarius.

      Gills decurrent or adnate, mostly epiphytal   Flammula.

      Gills somewhat sinuate, cuticle of the pileus silky, or bearing fibrils   Inocybe.

      Cuticle smooth, viscid   Hebeloma.

      Gills separating from the hymenophore and decurrent   Paxillus.

   b.   Stem lateral or absent   Crepidotus.

   B.   Stem cartilaginous,

      Gills decurrent   Tubaria.

      Gills not decurrent,

      Margin of the pileus at first incurved   Naucoria.

      Margin of pileus always straight,

      Hymenophore free   Pluteolus.

      Hymenophore confluent   Galera.

      Gills dissolving into a gelatinous condition   Bolbitius.

IV. Spores purple-brown.

   A.   Stem not cartilaginous,

Pileus easily separating from the stem,

      Volva present, ring wanting   Chitonia.

      Volva and ring wanting   Pilosace.

      Volva wanting, ring present   Agaricus.

      Gills confluent, ring present on stem   Stropharia.

      Ring wanting, veil remaining attached to margin of pileus   Hypholoma.

B. Stem cartilaginous,

Gills decurrent .... Deconia.

Gills not decurrent, margin of pileus at first incurved .... Psilocybe.

Margin of pileus at first straight .... Psathyra.

V. Black spored mushrooms.

Gills deliquescent .... Coprinus.

Gills not deliquescent,

Gills decurrent .... Gomphidius.

Gills not decurrent, pileus striate .... Psathyrella.

Pileus not striate, ring wanting, veil often present on margin ... Panæolus.

Ring wanting, veil appendiculate .... Chalymotta.

Ring present Anellaria.

CHAPTER II

THE WHITE-SPORED AGARICS

The species bearing the white spores seem to be higher in type than those producing colored spores. Most of the former are firmer, while the black spored specimens soon deliquesce. The white spores are usually oval, sometimes round, and in many cases quite spiny. All white-spored specimens will be found in clean places.

AMANITA. PERS

Amanita is supposed to be derived from Mount Amanus, an ancient name of a range separating Cilicia from Syria. It is supposed that Galen first brought specimens of this fungus from that region.

The genus Amanita has both a volva and veil. The spores are white and the stem is readily separable from the cap. The volva is universal at first, enveloping the young plant, yet distinct and free from the cuticle of the pileus.

This genus contains some of the most deadly poisonous mushrooms, although a few are known to be very good. There is a large number of species—about 75 being known, 42 of which have been found in this country—a few being quite common in this state. All the Amanita are terrestrial plants, mostly solitary in their habits, and chiefly found in the woods, or in well wooded grounds.

In the button stage it resembles a small egg or puff-ball, as will be seen in Figure 6, page 11, and great care should be taken to distinguish it from the latter, if one is hunting puff-balls to eat; yet the danger is not great, since the volva usually breaks before the plant comes through the ground.

AMANITA PHALLOIDES. FR.

The Deadly Amanita.

Figure 11   Amanita phalloides. Fr. Showing volva at the base, cap dark.

Figure 12   Amanita phalloides. Fr. White form showing volva, scaly stem, ring.

Phalloides means phallus-like. This plant and its related species are deadly poisonous. For this reason the plant should be carefully studied and thoroughly known by every mushroom hunter. In different localities, and sometimes in the same locality, the plant will appear in very different shades of color. There are also variations in the way in which the volva is ruptured, as well as in the character of the stem.

The beginner will imagine he has a new species often, till he becomes thoroughly acquainted with all the idiosyncrasies of this plant.

The pileus is smooth, even, viscid when young and moist, frequently adorned with a few fragments of the volva, white, grayish white, sometimes smoky-brown; whether the pileus be white, oyster-color or smoky-brown, the center of the cap will be several shades darker than the margin. The plant changes from a knob or egg-shape when young, to almost flat when fully expanded. Many plants have a marked umbo on the top of the cap and the rim of the cap may be slightly turned up.

The gills are always white, wide, ventricose, rounded next to the stem, and free from it.

The stem is smooth, white unless in cases where the cap is dark, then the stem of those plants are apt to be of the same color, tapering upward as in the specimen (Fig. 11); stuffed, then hollow, inclined to discolor when handled.

The volva of this species is quite variable and more or less buried in the ground, where careful observation will reveal it.

One need never confound this species with the meadow mushroom, for the spores of that are always purple-brown, while a spore-print of this will always reveal white spores. I have seen a slight tint of pink in the gills of the A. phalloides but the spores were always white. Until one knows thoroughly both Lepiota naucina and A. phalloides before eating the former he should always hunt carefully for the remains of a volva and a bulbous base in the soil.

This plant is quite conspicuous and inviting in all of its various shades of color. It is found in woods, and along the margin of woods, and sometimes on lawns. It is from four to eight inches high and the pileus from three to five inches broad. There is a personality about the plant that renders it readily recognizable after it has once been learned. Found from August to October.

AMANITA RECUTITA. FR.

THE FRESH-SKINNED AMANITA. POISONOUS

Recutita, having a fresh or new skin. Pileus convex, then expanded, dry, smooth, often covered with small scales, fragments of the volva; margin almost even, gray or brownish.

The gills forming lines down the stem.

The stem stuffed, then hollow, attenuated upward, silky, white, ring distant, edge of volva not free, frequently obliterated.

Rather common where there is much pine woods. August to October.

This species differs from A. porphyria in ring not being brown or brownish.

AMANITA VIROSA. FR.

THE POISONOUS AMANITA.

Virosa, full of poison. The pileus is from four to five inches broad; the entire plant white, conical, then expanded; viscid when moist; margin often somewhat lobed, even.

The gills are free, crowded.

The stem is frequently six inches long, stuffed, round, with a bulbous base, attenuated upward, squamulose, ring near apex, volva large, lax.

The spores are subglobose, 8–10μ. This is probably simply a form of A. phalloides. It is found in damp woods. August to October.

AMANITA MUSCARIA. LINN.

THE FLY AMANITA. POISONOUS.

Figure13   Amanita muscaria.—Linn. Cap reddish or orange, showing scales on the cap and at base of stem.

Muscaria, from musca, a fly. The fly Amanita is a very conspicuous and handsome plant. It is so called because infusions of it are used to kill flies. I have frequently seen dead flies on the fully developed caps, where they had sipped of the dew upon the cap, and, like the Lotos-eaters of old, had forgotten to move away. It is a very abundant plant in the woods of Columbiana county, this state. It is also found frequently in many localities about Chillicothe. It is often a very handsome and attractive plant, because of the bright colors of the cap in contrast with the white stem and gills, as well as the white scales on the surface of the cap. These scales seem to behave somewhat differently from those of other species of Amanita. Instead of shrivelling, curling, and falling off they are inclined to adhere firmly to the smooth skin of the pileus, turning brownish, and in the maturely expanded plant appear like scattered drops of mud which have dried upon the pileus, as you will observe in Figure 13.

The pileus is three to five inches broad, globose at first, then dumb-bell in shape, convex, then expanded, nearly flat in age; margin in matured plants slightly striate; the surface of the cap is covered with white floccose scales, fragments of the volva, these scales being easily removed so that old plants are frequently comparatively smooth. The color of the young plant is normally red, then orange to pale yellow; late in the season, or in old plants, it fades to almost white. The flesh is white, sometimes stained yellow close to the cuticle.

The gills are pure white, very symmetrical, various in length, the shorter ones terminating under the cap very abruptly, crowded, free, but reaching the stem, decurrent in the form of lines somewhat broader in front, sometimes a slight tinge of yellow will be observed in the gills.

The stem is white, often yellowish with age, pithy and often hollow, becoming rough and shaggy, finally scaly, the scales below appearing to merge into the form of an obscure cup, the stem four to six inches long.

The veil covers the gills of the young plant and later is seen as a collar-like ring on the stem, soft, lax, deflexed, in old specimens it is often destroyed. The spores are white and broadly elliptical.

The history of this plant is as interesting as a novel. Its deadly properties were known to the Greeks and Romans. The pages of history record its undoing and its accessory to crime. Pliny says, alluding to this species, "very conveniently adapted for poisoning." This was undoubtedly the species that Agrippina, the mother of Nero, used to poison her husband, the Emperor Claudius; and the same that Nero used in that famous banquet when all his guests, his tribunes and centurions, and Agrippina herself, fell victims to its poisonous properties.

However, it is said this mushroom is habitually eaten by certain people as an intoxicant; indeed, it is used in Kamchatka and Asiatic Russia, generally, where the Amanita drunkard takes the place of the opium fiend and the alcohol bibber in other countries. By reading Colonel George Kennan in his "Tent-life in Siberia," and Cooke's "Seven Sisters of Sleep," you will find a full description of the toxic employment of this fungus which will far surpass any possible imagination.

It caused the death of the Czar Alexis of Russia; also Count de Vecchi, with a number of his friends, in Washington in 1896. He was in search of the Orange Amanita and found this, and the consequences were serious.

In size, shape, and color of the cap there is similarity, but in other respects the two are very different. They may be contrasted as follows:

Orange Amanita, edible.—Cap smooth, gills yellow, stem yellow, wrapper persistent, membranaceous, white.

Fly Amanita, poisonous.—Cap warty, gills white, stem white, or slightly yellowish, wrapper soon breaking into fragments or scales, white or sometimes yellowish brown.

Found along roadsides, wood margins, and in thin woods. It prefers poor soil, and is more abundant where poplar and hemlock grow. From June to frost.

Figure 14   Amanita muscaria.—Linn. One-half natural size, showing development of the plant.

AMANITA FROSTIANA. PK.

FROST'S AMANITA. POISONOUS.

Figure 15   Amanita Frostiana. Photo by C. G. Lloyd.

Frostiana, named in honor of Charles C. Frost.

The pileus is convex, expanded, bright orange or yellow, warty, sometimes smooth, striate on the margin, pileus one to three inches broad.

The gills are free, white, or slightly tinged with yellow.

The stem is white or yellow, stuffed, bearing a slight, sometimes evanescent, ring, bulbous, at the base, the bulb slightly margined by the volva. The spores globose, 8–10μ in diameter. Peck.

Great care should be taken to distinguish this species from A. cæsarea because of its often yellow stem and gills. I found some beautiful specimens on Cemetery Hill and on Ralston's Run. It is very poisonous and should be carefully avoided, or rather, it should be thoroughly known that it may be avoided. The striations on the margin of its yellow tinge might lead one to mistake it for the Orange Amanita. It is found in shady woods and sometimes in open places where there is underbrush. June to October.

AMANITA VERNA. BULL.

THE SPRING AMANITA. POISONOUS.

Figure 16   Amanita verna. Two-thirds natural size, showing the volva cup and the ring.

Verna, pertaining to spring. This species is considered by some only a white variety of Amanita phalloides. The plant is always a pure white. It can only be distinguished from the white form of the A. phalloides by its closer sheathing volva and perhaps a more ovate pileus when young.

The pileus is at first ovate, then expanded, somewhat depressed, viscid when moist, even, margin naked, smooth. The gills are free.

The stem is stuffed, with advancing age hollow, equal, floccose, white, ringed, base bulbous, volva closely embracing the stem with its free margin, ring forming a broad collar, reflexed. The spores are globose, 8μ broad.

This species is very abundant on the wooded hills in this section of the state. Its pure white color makes it an attractive plant, and it should be carefully learned. I have found it before the middle of June.

AMANITA MAGNIVELARIS. PK.

THE LARGE VEILED AMANITA. POISONOUS.

Magnivelaris is from magnus, large; velum, a veil.

The pileus is convex, often nearly plane, with even margin, smooth, slightly viscid when moist, white or yellowish-white.

The gills are free, close, white.

The stem is long, nearly equal, white, smooth, furnished with a large mebranaceous volva, the bulbous base tapering downward and rooting. The spores are broadly elliptical.

This species very closely resembles Amanita verna, from which it can be distinguished by its large, persistent annulus, the elongated downward-tapering bulb of its stem, and, especially, by its elliptical spores.

It is found solitary and in the woods. I found several on Ralston's Run under beech trees. Found from July to October.

AMANITA PELLUCIDULA. BAN.

Pileus at first campanulate, then expanded, slightly viscid, fleshy in center, attenuated at the margin; color a smooth bright red, deeper at the top, shaded into clear transparent yellow at the margin; glossy, flesh white, unchanging.

The gills are ventricose, free, numerous, yellow.

The stem is stuffed, ring descending, fugacious. Peck's 44th Report.

This species differs from Amanita cæsarea in having an even margin and a white stem. It is only a form of the cæsarea. The white stem will attract the attention of the collector.

AMANITA SOLITARIA. BULL.

THE SOLITARY AMANITA.

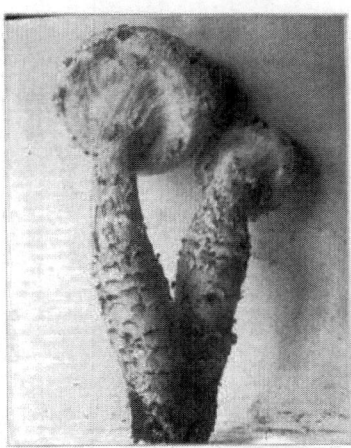

Figure 17   Amanita solitaria. Two-thirds natural size, showing the peculiar veil.

Figure 18   Amanita solitaria. Two-thirds natural size, showing scaly cap and stem.

Plate II. Figure 19   Amanita solitaria.
Natural size, showing scaly cap and stem, plant white.

Solitary, growing alone. I have found this plant in various parts of the state and have always found them growing alone. In Poke Hollow, where I found the specimens in the illustrations, I found several on the hillside on different occasions, but I have never found them growing in groups. It is quite large in size, white or whitish, very woolly or floccose. Usually the cap, stem, and the gills are covered with a floccose substance which will serve to identify the species. This fluffy exterior adheres readily to your hands or clothing. The cap is sometimes tinged with brown, but the flesh is white and smells quite strong, not unlike chloride of lime. The annulus is frequently torn from the stem and is found adhering to the margin of the cap.

The pileus is from three to five inches broad, or more, when fully expanded, at first globose to hemispherical, as will be seen in Figures 17 and 18, convex, or plane, warty, white or whitish, the pointed scales being easily rubbed off, or washed off by heavy rains, these scales varying in size from small granules to quite large conical flakes, and differing in condition and color in different plants.

The gills are free, or are not attached by the upper part, the edges are frequently floccose where they are torn from the slight connection with the upper surface of the veil; white, or slightly tinged with cream-color, broad.

The stem is four to eight inches high, solid, becoming stuffed when old, bulbous, rooting deep in the soil, very scaly, ventricose sometimes in young plants, white, very mealy. Volva friable. Ring, large, lacerated, usually hanging to the margin of the cap, but in Figure 19 it adheres to the stem.

This is a large and beautiful plant in the woods, and easily identified because of its floccose nature and the large bulb at the base of the stem. It is not so warty and the odor is not nearly so strong as the Amanita strobiliformis. It is edible but very great caution should be used to be sure of your species. Found from July to October in woods and roadsides.

AMANITA RADICATA. PK.

Figure 20   Amanita radicata. Two-thirds natural size, showing scaly cap, bulbous stem and root broken off and peculiar veil.

Radicata means furnished with a root. The root of the specimen in Figure 20 was broken off in getting it out of the ground.

The pileus is subglobose, becoming convex, dry, verrucose, white, margin even, flesh firm, white, odor resembling that of chloride of lime.

The gills are close, free, white.

The stem is solid, deeply radicating, swollen at the base or bulbous, floccose or mealy at the top, white; veil thin, floccose, or mealy, white, soon lacerated and attached in fragments to the margin of the pileus or evanescent. The spores are broadly elliptic, $7.5–10\mu$ long, $6–7\mu$ broad. Peck.

This is quite a large and beautiful plant, very closely related to Amanita strobiliformis, but readily distinguished from it because of its white color, its clearly radiating stem, and small spores.

The stem shows to be bulbous and the cap covered with warts. I found the plant frequently in Poke Hollow and on Ralston's Run. July and August.

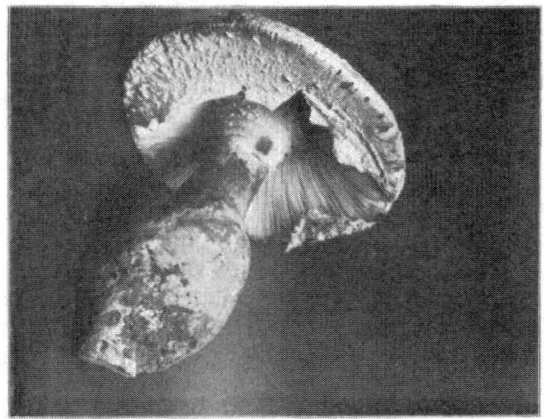

Figure 21 Amanita radicata.

AMANITA STROBILIFORMIS. FR.

THE FIR-CONE AMANITA.

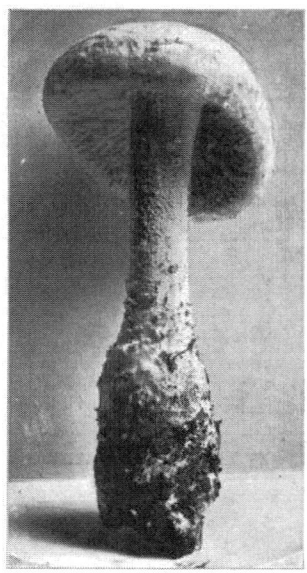

Plate III. Figure 22 Amanita strobiliformis.
Young plant showing veil covering the entire gill-surface of the plant. Cap covered with persistent warts, stem rough and rooting, odor strong of chloride of lime.

Plate IV. Figure 23   Amanita strobiliformis.
Showing long root.

Strobiliformis means fir-cone form; so called from the similarity of its undeveloped form to that of the strobile of the pine.

The pileus is six to eight inches broad, when young, subglobose, then convex, expanded, nearly plane, with persistent warts, white, ash-color, sometimes yellow on the cap, the margin even and extending beyond the gills; warts hard, angular, pointed, white; flesh white, compact.

The gills are free, crowded, rounded, white, becoming yellow.

The stem is five to eight inches long, frequently longer, tapering upward, floccosely scaly, bulbous, rooting beyond the bulb; ring large, torn; volva forming concentric rings. The spores are $13-14 \times 8-9\mu$.

This is one of the most stately plants in the woods. It is said to be edible, but the strong pungent odor, like chloride of lime, has deterred me from eating it. This, however, is said to disappear in cooking. It grows to be very large. Dr. Kellerman and I found a specimen in Haynes's Hollow whose stem measured over eleven inches, and cap nine inches. It is found in open woods and wood margins. Great caution should be used before the plant is eaten to know it beyond doubt. Found July to October.

AMANITA MAPPA. FR.

THE DELICATE AMANITA. POISONOUS.

Figure 24   Amanita mappa. Natural size, showing long smooth stem, cap yellowish-white and ring.

Mappa means a napkin, so called from the volva. The pileus is two to three inches broad, convex, then expanded, plane, obtuse or depressed, without separable cuticle; margin nearly even; white or yellowish, usually with patches of the volva dry.

The gills are adnexed, close, narrow, shining, white.

The stem is two to three inches long, stuffed, then hollow, cylindrical, nearly smooth, bulbous, nearly globose at the base, white, almost equal above the bulb.

The volva with its free margin is acute and narrow. The ring is membranaceous, superior, soft, lax, ragged.

Its color is quite as variable and its habits are much like A. phalloides, from which it can only be distinguished by its less developed volva, which, instead of being cup-shaped, is little more than a mere rim fringing the bulb. The odor at times is very strong. It is found in open woods and under brush. Label it poisonous.

AMANITA CRENULATA. PK.

Figure 25   Amanita crenulata.

Crenulata means bearing notches, referring to the crenulate form of the gills, which are very distinct.

The pileus is thin, two to two and a half inches broad, broadly ovate, becoming convex, or nearly plane, somewhat striate on the margin, adorned with a few thin whitish floccose warts or with whitish flocculent patches, whitish or grayish, sometimes tinged with yellow.

The gills are close, reaching the stem, and sometimes forming decurrent lines upon it, floccose crenulate on the edge, the short ones truncate at the inner extremity, white.

The stem is equal, bulbous, floccose mealy above, stuffed or hollow, white, the annulus slight, evanescent. Spores broadly elliptic or subglobose, 7.5–10 long, nearly as broad, usually containing a single large nucleus. Peck, Bull. Tor. Bot. Club.

The stem is bulbous at the base but the volva is rarely seen upon it although slight patches are frequently seen on the pileus. The ring is very evanescent and soon disappears. The specimens I have received from Mrs. Blackford look good enough to eat and she speaks highly of the edible qualities of this species. So far as I know this plant is confined to the New England states. Found from September to November. It grows in low damp ground under trees.

AMANITA COTHURNATA. ATKINSON.

THE BOOTED AMANITA.

Figure 26   Amanita cothurnata. Slightly reduced from natural size, showing different stages of development.

Cothurnata means buskined; from corthunus, a high shoe or buskin worn by actors. This species is easily separated from the other Amanitas. I shall give Prof. Atkinson's description of it in full: "The pileus is fleshy and passes from nearly globose to hemispherical, convex, expanded, and when specimens are very old sometimes the margin is elevated. It is usually white, though specimens are found with a tinge of citron yellow in the center or of tawny yellow in the center of other specimens. The pileus is viscid, strongly so when moist. It is finely striate on the margin, and covered with numerous, white, floccose scales from the upper half of the volva, forming more or less dense patches, which may wash off in heavy rains.

The gills are rounded next the stem, and quite remote from it. The edge of the gills is often eroded or frazzly from the torn-out threads with which they were loosely connected to the upper side of the veil in the young or button stage. The spores are globose or nearly so, with a large "nucleus" nearly filling the spore.

The stem is cylindrical, even, and expanded below into quite a large oval bulb, the stem just above the bulb being margined by a close-fitting roll of the volva, and the upper edge of this presenting the appearance of having been sewed at the top like the rolled edge of a garment or buskin. The surface of the stem is minutely floccose, scaly or strongly so, and decidedly hollow even from a very young stage or sometimes when young with loose threads in the cavity.

A. cothurnata resembles in many points A. frostiana and it will afford the collector a very interesting study to note the points of difference. I found the two species growing on Cemetery Hill. Figure 26 is from plants collected in Michigan and photographed by Dr. Fisher. Found in September and October.

AMANITA RUBESCENS. FR.

THE REDDISH AMANITA. EDIBLE.

Figure 27   Amanita rubescens. One-third natural size, caps a dingy reddish-brown, stains reddish when bruised.

Rubescens is from rubesco, to become red. It is so called because of the dingy reddish color of the entire plant, and also because when the plant is handled or bruised it quickly changes to a reddish color. It is often a large bulky plant and rather uninviting.

The pileus is four to six inches broad, dingy reddish, often becoming pale flesh color, fleshy, oval to convex, then expanded; sprinkled with small pale warts, unequal, mealy, scattered, white, easily separating; margin even, faintly striate, especially in wet weather; flesh soft, white, becoming red when broken.

The gills are white or whitish, free from the stem but reaching it and forming at times decurrent lines upon it, thin, crowded.

The stem is four or five inches long, nearly cylindrical, solid, though inclined to be soft within, tapering from the base up, with a bulbous base which often tapers abruptly below, containing reddish scales, color dull red. It has seldom any distinct evidence of a volva at the base but abundant evidence on the cap. Ring large, superior, white, and fragile.

The plant is quite variable in color, sometimes becoming almost white with a slight reddish or brownish tint. The strong distinguishing character of the species is the almost entire absence of any remains of the volva at the base of the stem. By this, and by the dull red hues and the bruised portions quickly changing to a reddish color, it is easily distinguished from any of the poisonous Amanitas.

According to Cordier it is largely used as an article of food in France. Stevenson and Cooke speak well of it. I noticed the small Bohemian boys gathered it about Salem, Ohio, not having been in this country more than a week and not being able to speak a word of English. It convinced me that it was an article of diet in Bohemia and that our species is similar to theirs. I have found the plants in woods about Bowling Green and Sidney, Ohio. The plants in Figure 27 were collected on Johnson's Island, Sandusky, Ohio, and photographed by Dr. Kellerman. It is found from June to September.

AMANITA ASPERA. FR.

Rough Amanita.

Aspera means rough. The pileus is convex, then plane; warts minute, somewhat crowded, nearly persistent; margin even, rather thin, increasing in thickness toward the stem; scarcely umbonate, reddish with various tints of livid and gray; flesh rather solid, white, with tints of reddish-brown immediately next to the epidermis.

The gills are free, with sometimes a little tooth behind, running down the stem, white, broad in front.

The stem is white, squamulose, bulb rugulose, ring superior and entire. The spores are $8 \times 6\mu$.

When the flesh is bruised or eaten by insects it assumes a reddish-brown color, and in this respect it resembles A. rubescens. The odor is strong but the taste is not unpleasant. In woods from June till October. The collector should be sure he knows the plant before he eats it.

AMANITA CÆSAREA. SCOP.

THE ORANGE AMANITA. EDIBLE.

Figure 28 Amanita cæsarea. From a drawing showing the different stages of the plant. Caps, gills, stem and collar yellow, volva white.

Figure 29   Amanita cæsarea.

The Orange Amanita is a large, attractive, and beautiful plant. I have marked it edible, but no one should eat it unless he is thoroughly acquainted with all the species of the genus Amanita, and then with great caution. It is said to have been Cæsar's favorite mushroom. The pileus is smooth, hemispherical, bell-shaped, convex, and when fully expanded nearly flat, the center somewhat elevated and the margin slightly curved downward; red or orange, fading to yellow on the margin; usually the larger and well-developed specimens have the deeper and richer color, the color being always more marked in the center of the pileus; margin distinctly striate; gills rounded at the stem end and not attached to the stem, yellow, free and straight. The color of the gills of matured plants usually is an index to the color of the spores but it is an exception in this case as the spores are white.

The stem and the flabby membranaceous collar that surrounds it toward the top are yellow like the gills, the depth of the color varying more with the size of the plant than is the case with color of the cap. Sometimes in small and inferior plants the color of both stem and gills is nearly white, and if the volva is not distinct it is difficult to distinguish it from the fly mushroom, which is very poisonous. The stem is hollow, with a soft cottony pith in the young plants.

In very young plants the edge of the collar is attached to the margin of the cap and conceals the gills, but with the upward growth of the stem and the expansion of the cap the collar separates from the margin and remains attached to the stem, where it hangs down upon it like a ruffle.

The expanded cap is usually from three to six inches broad, the stem from four to six inches long and tapering upward.

When in the button stage, the plant is ovate; and the white color of the volva, which now entirely surrounds the plant, presents an appearance much like a hen's egg in size, color, and shape. As the parts within develop, the volva ruptures in its upper part, the stem elongates and carries upward the cap, while the remains of the volva surrounds the base of the stem in the form of a cup.

When the volva first breaks at the apex, it reveals the point of the cap with its beautiful red color and in contrast with the white volva makes quite a pretty plant, but with advancing age the red or orange red fades to a yellow. In drying the specimens the red often entirely disappears. In young, as well as in old plants, the margin is often prominently marked with striations, as will be seen in Figures 28 and 29. The flesh of the plant is white but more or less stained with yellow next to the epidermis and the gills, which are of that color.

The plant grows in wet weather from July to October. It grows in thin woods and seems to prefer pine woods and sandy soil. I have found it from the south tier of counties to the north of our state. It is not, however, a common plant in Ohio.

From its several names—Cæsar's Agaric, Imperial Mushroom, Cibus Deorum, Kaiserling—one would infer that for ages it had been held in high esteem as an esculent.

Too great caution cannot be used in distinguishing it from the very poisonous fly mushroom.

AMANITA SPRETA. PK.

HATED AMANITA. POISONOUS.

Spreta, hated. The pileus at first is nearly ovate, slightly umbonate, then convex, smooth, sometimes fragments of the volva adhering, the margin striate, whitish or pale-brown toward and on the umbo, soft, dry, more or less furrowed on the margin.

The flesh is white, thin on the edges, and increasing in thickness toward the center. Gills close, white, reaching the stem.

The stem is equal, smooth, annulate, stuffed or hollow, whitish, finely striate at the top from the decurrent lines of the gills, not bulbous at the base, the volva rather large and inclined to yellowish color. The spores are elliptical.

The plant resembles the dark forms of the Amanitopsis in having the marked striations and the entire and closely fitting volva at the base, but can be easily distinguished by its ring. I found it on Cemetery Hill in company with the Amanitopsis. It does not seem to root as deep in the ground as the Amanitopsis. It is very poisonous and should be carefully studied so that it may be readily recognized and avoided.

It is found in open woods from July to September.

AMANITOPSIS. ROZE.

Amanitopsis is from Aminita and opsis, resembling; so called because it resembles the Amanita. The principal feature wherein the genus differs from the Amanita is the absence of a collar on the stem. Its species are included among the Amanita by many authors. The spores are white. The gills are free from the stem, and it has a universal veil at first completely enveloping the young plant, which soon breaks it, carrying remnants of it on the pileus, where they appear as scattered warts. It differs from the Lepiota in having a volva.

AMANITOPSIS VAGINATA. BULL.

THE SHEATHED AMANITOPSIS. EDIBLE.

Figure 30   Amanita vaginata. One-third natural size. Notice a portion of the volva adhering to the cap.

Vaginata—from vagina, a sheath. The plant is edible but should be used with very great caution. It is quite variable in color, ranging from white to mouse color, brownish or yellowish.

The pileus is ovate at first, bell-shaped, then convex and expanded, thin, quite fragile, smooth, when young with a few fragments of the volva adhering to its surface, deeply and distinctly striate.

The gills are free, white, then pallid, ventricose, broadest in front, irregular. The flesh is white, but in the darker forms stained under the easily separating skin. The spores are white and nearly round, 7–10μ.

The stem is cylindrical, even or slightly tapering upward, hollow or stuffed, smooth or sprinkled with downy scales, not bulbous at the base.

The volva is long, thin, fragile, forming a permanent sheath which is quite soft and readily adheres to the base of the stem.

The striations on the margin are deep and distinct, as in the Orange Amanita. The cup is quite regular but it is fragile, easily broken and usually deep in the ground. In some plants a slight umbo is developed at the center.

The mushroom-eater wants to distinguish very carefully between this species and Amanita spreta, which is very poisonous.

It is found in woods, in open places where there is much vegetable mould, sometimes found in stubble and pastures, especially in meadows under trees. Found from June to November.

The plant varies considerably in color, and there are several varieties, separable by means of their color:

A. vaginata, var. alba. The whole plant is white.

A. vaginata var. fulva. The cap tawny yellow or pale ochraceous.

A. vaginata var. livida. The cap leaden brown; gills and stem tinged with smoky brown.

Plate V. Figure 31   Amanita vaginata

AMANITOPSIS STRANGULATA. FR.

THE GRAY AMANITOPSIS. EDIBLE.

Strangulata means choked, from the stuffed stem. The pileus is two to four inches broad, soon plane, livid-bay or gray, with patches of the volva, margin striate or grooved.

The gills are free, white, close.

The stem is stuffed, silky above, scaly below, slightly tapering upwards. The volva soon breaking up, forming several ring-like ridges on the stem. The spores are globose, 10–13μ.

This is a synonym for A. ceciliae. B. and Br. and perhaps nothing more than a vigorous growth of Amanitopsis vaginata. It has almost no odor and a sweet taste and cooks deliciously.

Found in the woods and in open places from August to October.

LEPIOTA. FR.

Lepiota means a scale. In the Lepiota the gills are typically free from the stem, as in Amanita and Amanitopsis, but they differ in having no superficial or removable warts on the cap, and no sheathing or scaly remains of a volva at the base of the stem. In some species the epidermis of the cap breaks into scales which persistently adhere to the cap, and this feature, indeed, suggests the name of the genus, which is derived from the Latin word lepis, a scale.

The stem is hollow or stuffed, its flesh being distinct from the pileus and easily separable from it. There are a number of edible species.

LEPIOTA PROCERA. SCOP.

THE PARASOL MUSHROOM. EDIBLE.

Plate VI. Figure 32   Lepiota procera.

Procera means tall.

The pileus is thin, strongly umbonate, adorned with brown spot-like scales.

The gills are white, sometimes yellowish-white, free, remote from the stem, broad and crowded, ventricose, edge sometimes brownish.

The stem is very long, cylindrical, hollow or stuffed, even, very long in proportion to its thickness and is, therefore, suggestive of the specific name, procera. The ring is rather thick and firm, though in mature plants it becomes loosened and movable on the stem. This and the form of the plant suggest the name, parasol. The cap is from three to five inches broad and the stem from five to nine inches high. I found one specimen among fallen timber that was eleven inches tall and whose cap was six inches broad.

It has a wide distribution. It is found in all parts of Ohio but is not abundant anywhere. It is a favorite with those who have eaten it, and, indeed, it is a delicious morsel when quickly broiled over coals, seasoned to taste with salt and pepper, butter melted in the gills and served on toast. This mushroom is especially free from grubs and it can be dried for winter use.

There is no poisonous species with which one is likely to confound it. The very tall, slender stem with a bulbous base, the very peculiar spotted cap with the prominent dark colored umbo and the movable ring on the stem, are ear-marks sufficient to identify this species.

Spores white and elliptical, $14 \times 10\mu$. Lloyd. It is found in pastures, stubble, and among fallen timber. July to October.

I am indebted to C. G. Lloyd for the photograph given here.

LEPIOTA NAUCINA. FR.

SMOOTH LEPIOTA. EDIBLE.

Figure 33   Lepiota naucina. The entire plant white.

Pileus soft, smooth, white or smoky-white; gills free, white, slowly changing with age to a dirty pinkish-brown color; stem annulate, slightly thickened at the base, attenuated upward, clothed with fibres pure white. The Smooth Lepiota is generally very regular in shape and of a pure white color. The central part of the cap is sometimes tinged with yellow or a smoky white hue. Its surface is nearly always very smooth and even. The gills are somewhat narrower toward the stem than they are in the middle. They are rounded and not attached to the stem.

Cap two to four inches broad; stem two to three inches long. It grows in clean grassy places in lawn, pastures, and along roadsides. I have seen the roadside white with this species around Sidney, Ohio. The specimens represented in figure were found in Chillicothe, August to November.

This is one of the best mushrooms, not inferior to the meadow mushroom. It has this advantage over the former that the gills retain their white color and do not pass from a pink to a repulsive black. The halftone and the description ought to make the plant known to the most casual reader.

LEPIOTA AMERICANA. PK.

THE AMERICAN LEPIOTA. EDIBLE.

Figure 34   Lepiota americana. Center of disk red or reddish-brown, stem frequently swollen. Plant turning red when drying.

This plant is quite common about Chillicothe, especially upon sawdust piles. It grows both singly and in clusters. The umbonate cap is adorned with reddish or reddish-brown scales except on the center where the color is uniformly reddish or reddish-brown because the surface is not broken up into scales; gills close, free, white, ventricose; stem smooth, enlarged at the base. In some plants the base of the stem is abnormally large; ring white, inclined to be delicate.

Wounds and bruises are apt to assume brownish-red hues. Dr. Herbst says: "This is truly an American plant, not being found in any other country. This is the pride of the family. There is nothing more beautiful than a cluster of this fungi. To look over the beautiful scaly pileus is a sight equally as fascinating as a covey of quail."

Found in grassy lawns and on old sawdust piles, in common with Pluteus cervinus. It is found almost all over the state. It is quite equal to the Parasol mushroom in flavor. It has a tendency to turn the milk or cream in which it is cooked to a reddish color. It is found from June to October. Mr. Lloyd suggests the name Lepiota Bodhami. It is the same as the European plant L. hæmatosperma. Bull.

LEPIOTA MORGANI. PK.

IN HONOR OF PROF. MORGAN.

Plate VII. Figure 35 Lepiota morgani.
Entire plant white or brownish-white. Gills white at first then greenish.

Pileus fleshy, soft, at first subglobose, then expanded or even depressed, white, the brownish or yellowish cuticule breaking up into scales on the disk; gills close, lanceolate, remote, white, then green; stem firm, equal or tapering upward, subbulbous, smooth, webby-stuffed, whitish tinged with brown; ring rather large, movable as you will observe in Figure 35. Flesh of both pileus and stem white, changing to a reddish, then to yellowish hue when cut or bruised. Spores ovate or subelliptical, mostly uninucleate, sordid green. 10–13×7–8. Peck.

This plant is very abundant about Chillicothe and I found it equally so at Sidney. I have known several families to eat of it, making about half of the children in each family sick. I regard it as a dangerous plant to eat. It grows very large and I have seen it growing in well marked rings a rod in diameter. If you are in doubt whether the plant you have is Morgani or not, let it remain in the basket over night and you will plainly see that the gills are turning green. The gills are white until the spores begin to fall. The plant is found in pastures and sometimes in pasture woods. June to October.

LEPIOTA GRANULOSA. BATSCH.

GRAINY LEPIOTA. EDIBLE.

Granulosa—from granosus, full of grains. Pileus thin, convex or nearly plain, sometimes almost umbonate, rough, with numerous granular scales, often radiately wrinkled, rusty-yellow or reddish-yellow, often growing paler with age. Flesh white or reddish tinged. Gills close, rounded behind and usually slightly adnexed, white. Stem equal or slightly thickened at the base, stuffed or hollow, white above the ring, colored and adorned like the pileus below it. Ring slight and evanescent. Spores elliptical, .00016 to .0002 inch long, .00012 to .00014 inch broad.

Plant one to two and one-fifth inches high; pileus one to two and one-fifth inches broad; stem one to three lines thick. Common in woods, copses, and waste places. August to October.

"This is a small species with a short stem and granular reddish-yellow pileus, and gills slightly attached to the stem. The annulus is very small and fugacious, being little more than the abrupt termination to the coating of the stem. The species was formerly made to include several varieties which are now regarded as distinct."—Peck's Report.

Found in the open woods about Salem, Ohio. The plant is small but quite meaty and of a pleasing quality.

LEPIOTA CRISTATELLA. PK.

Pileus thin, convex, subumbonate, minutely mealy, especially on the margin, white disk slightly tinged with pink.

Gills close, rounded behind, free, white; stem slender, whitish, hollow; spores subelliptical, .0002 inch long.

Mossy places in woods. October.—Peck's Report. No one will fail to recognize the crested Lepiota the moment he sees it. It has many of the ear marks of the Lepiota family.

LEPIOTA GRANOSA. MORG.

Plate VIII. Figure 36   Lepiota granosa.

Granosa means covered with granules.

The pileus is convex, obtuse or umbonate, even, radiately rugose-wrinkled, generally even and regular on the margin, reddish-yellow or light bay.

The gills are attached to the stem, slightly decurrent, somewhat crowded, whitish, then reddish-yellow.

The stem is thickened at the base, tapering toward the cap, flesh of the stem is yellow. The veil is membranous and forms a persistent ring on the stem.

It grows on decayed wood. I found it in large quantities, and tried to make it L. granulosa, but I found it fit better L. amianthinus, which it resembles very closely, but it is much larger and its habit is not the same. I was not satisfied with this description and sent the specimens to Prof. Atkinson, who set me right. It is a beautiful plant found on decayed wood in September and October.

LEPIOTA CEPÆSTIPES. SOW.

THE ONION STEMMED LEPIOTA. EDIBLE.

Figure 37   Lepiota cepæstipes. Pileus thin, white or yellowish.

Cepæstipes is from cepa, an onion and stipes, a stem, Pileus is thin at first ovate, then bell-shaped or expanded, umbonate, soon adorned with numerous minute brownish scales, which are often granular or mealy, folded into lines on the margin, white or yellow, the umbo darker.

The gills are thin, close, free, white, becoming dingy with age or drying.

The stem is rather long, tapering toward the apex, generally enlarged in the middle or near the base, hollow. The ring is thin and subpersistent. The spores are subelliptical, with a single nucleus, $8-10 \times 5-8\mu$.

The plants often cespitose, two to four inches high. Pileus is one to two inches broad. It is found in rich ground and decomposing vegetable matter. It is also found in graperies and conservatories. Peck.

This plant derives its specific name from the resemblance of its stem to that of the seed-stalk of an onion. One form has a yellow or yellowish cap, while the other has a white or fair cap. It seems to delight to grow in well rotted sawdust piles and hot houses. The specimens represented in Figure 37 were collected in Cleveland and photographed by Prof. H. C. Beardslee.

LEPIOTA ACUTESQUAMOSA. WEIN.

THE SQUARROSE LEPIOTA. EDIBLE.

Figure 38   Lepiota acutesquamosa. Two-thirds natural size, showing small pointed scales.

Acutesquamosa is from acutus, sharp, and squama, a scale; so called from the many bristling, erect scales on the pileus. The pileus is two to three inches broad, fleshy, convex, obtuse, or broadly umbonate; pale rusty with numerous small pointed scales, which are usually larger and more numerous at the disk.

The gills are free, crowded, simple, white or yellowish.

The stem is two to three inches or more long; stuffed or hollow, tapering upward slightly from a swollen base; below the ring rough or silky, pruinose above, ring large. The spores are $7-8 \times 4\mu$.

They are found in the woods, in gardens, and frequently in greenhouses. There is a slight difference between the specimens growing in the woods and those in the greenhouse. In the latter the pubescent covering is less dense and the erect scales are more numerous than in the former. In older specimens these scales fall off and leave small scars on the cap where they were attached. The specimens in Figure 38 were gathered in Michigan and were photographed by Dr. Fisher of Detroit.

ARMILLARIA. FR.

Armillaria, from armilla, a bracelet—referring to the ring upon the stem. This genus differs from all the foregoing white-spored species in having the gills attached to the stem by their inner extremity. The spores are white and the stem has a collar, though a somewhat evanescent one, but no wrapper at the base of the stem as in the Amanita and Amanitopsis. By the collar the genus differs from the other genera which are to follow.

The Amanita and Lepiota have the flesh of the stem and the pileus not continuous, and their stems are, therefore, easily separated from the cap, but in the Armillaria the gills and the pileus are attached to the stem.

ARMILLARIA MELLEA. VAHL.

THE HONEY-COLORED ARMILLARIA. EDIBLE.

Figure39   Armillaria mellea. Two-thirds natural size. Honey colored. Tufted with dark-brown fugitive hairs. Flesh white.

Mellea, from melleus, of the color of honey. Cap fleshy, honey colored, or ochraceous, striate on the margin, shaded with darker brown toward the center, having a central boss-like elevation and sometimes a central depression in full grown specimens, tufted with dark-brown fugitive hairs. Color of the cap varies, depending upon climatic conditions and the character of the habitat. Gills distant, ending in a decurrent tooth, pallid or dirty white, very often showing brown or rust colored spots when old. Spores white and abundant. Frequently the ground under a clump of this species will be white from the fallen spores. Stem elastic and scaly, four inches or more in length. Ring downy. Diameter of cap from two to five inches. Manner of growth is frequently in tufts, and, as with most of the Armillarias, generally parasitic on old stumps.

The veil varies greatly. It may be membranaceous and thin, or quite thick, or may be wanting entirely, as will be seen in Figure 39; in Figure 40 only a slight trace of the ring can be seen. The two plants grew under very different environment; the last grew in the woods and Figure 39 on a lawn in the city. The species is very common and grows either in thin woods or in cleared lands, on the ground or on decaying wood. Its favorite habit is about stumps. It is either solitary, gregarious, or in dense clusters. It is very abundant about Chillicothe, where I have seen stumps literally surrounded with it. It has a slight acridity while raw, which it seems to lose in cooking. Those who like it may eat it without fear, all varieties being edible.

Prof. Peck gives the following varieties:

- A. mellea var. obscura—has the cap covered with numerous small black scales.
- A. mellea var. flava—has a cap yellow or reddish-yellow, otherwise normal.

- A. mellea var. glabra—has a smooth cap, otherwise normal.
- A. mellea var. radicata—has a tapering root penetrating the soil.
- A. mellea var. bulbosa—has a bulbous base.
- A. mellea var. exannulata—has the cap smooth and even on the margin, and the stem tapering at the base.

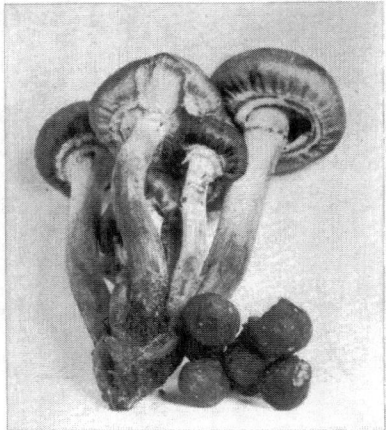

Figure 40   Armillaria mellea. Two-thirds natural size, showing double ring present.

ARMILLARIA BULBIGERA. A. & S.

MARGINATE-BULBED ARMILLARIA.

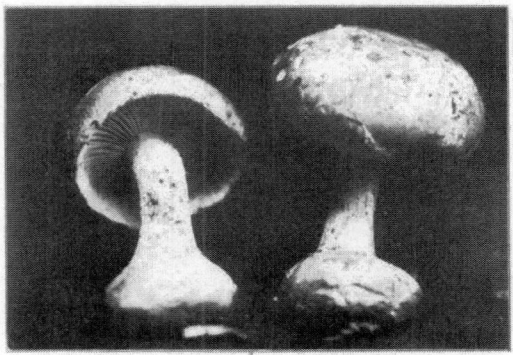

Figure 41   Armillaria bulbigera. Reddish-gray caps and short bulbous stems.

Bulbigera is from bulbus, a bulb, and gero, to bear.

The pileus is fleshy, three to four inches across, convex, then expanded, obtuse, even, brownish, gray, sometimes reddish, dry, fibrillose near the margin.

The gills are notched at the stem, pallid, crowded at first, at length rather distant, becoming slightly colored.

The stem is distinctly bulbous, two to three inches long, stuffed, pallid, fibrillose, ring oblique, fugacious. The spores are $7-10 \times 5\mu$.

I have found some very fine specimens in Poke Hollow, near Chillicothe. The stems were short and very bulbous, having hardly any trace of the ring on the older specimens. The caps were obtusely convex and of a grayish rufescent color. This species can readily be distinguished by the distinctly marginate bulb at the base of the stem. The specimens in Figure 41 were found in Poke Hollow, near Chillicothe, October 2d. I have no doubt of their edibility but I have not eaten them.

ARMILLARIA NARDOSMIA. ELLIS.

SPIKENARD-SMELLING ARMILLARIA. ELLIS.

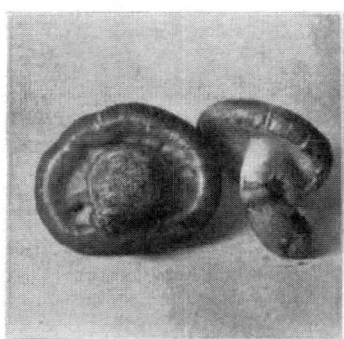

Figure 42   Armillaria nardosmia. One-half natural size, showing the veil and incurved margin.

Nardosmia is from nardosmius, the odor of nardus or spikenard.

The pileus is quite thick, firm and compact, thinner toward the margin, strongly involute when young, grayish white and beautifully variegated with brown spots, like the breast of a pheasant, rather tough, with a separable epidermis, flesh white.

The gills are crowded, slightly notched or emarginate, somewhat ventricose, white.

The stem is solid, short, fibrous, sheathed by a veil forming a ring more or less evanescent. The spores are nearly round, 6μ in diameter.

This is the most beautiful species of the genus, and from its pheasant-like spotted cap, as well as its strong odor and taste of spikenard or almonds, it is easily determined. The almond taste and odor disappears in cooking. I found some very fine specimens around a pond in Mr. Shriver's woods, east of Chillicothe. In older specimens the cuticule of the caps frequently breaks into scales. Found in woods in September and October.

ARMILLARIA APPENDICULATA. PK.

Appendiculata, bearing small appendages. Pileus is broadly convex, glabrous, whitish, often tinged with rust-color or brownish rust-color on the disk. Flesh white or whitish. Gills close, rounded behind, whitish. Stem equal or slightly tapering upward, solid, bulbous, whitish, the veil either membraneous or webby, white, commonly adhering in fragments to the margin of the pileus. Spores subelliptical, 8×5.

Pileus two to four inches broad. Stem 1.5–3.5 inches long; 5–10 lines thick.

The general appearance of this species is suggestive of Tricholoma album, but the appearance of a veil separates it from that fungus and places it in the genus Armillaria. The veil, however, is often slightly lacerated, or webby, and adherent to the margin of the pileus. Peck's Report.

I have found this at Salem and Chillicothe.

TRICHOLOMA. FR.

Tricholoma is from two Greek words meaning hair and fringe. This genus is known by its stout, fleshy stem, without any evidence of a ring, and by the gills being attached to the stem and having a notch in their edges near or at the extremity. The veil is absent, or, if present, it is downy and adherent to the margin of the cap. The cap is generally quite fleshy; the stem is homogeneous and confluent with the pileus, central and nearly fleshy, without either ring or volva, and with no distinct bark-like coat. The spores are white or grayish-white.

The distinguishing features are the fleshy stem, continuous with the flesh of the pileus, and the sinuate or notched gills. This is quite a universal genus. All the species grow on the ground, so far as I know them.

There are many edible species under this genus, there being only two, so far as I know, not edible; and no one is likely to touch those on account of their strong odor. They are T. sulphureum and T. saponaceum.

TRICHOLOMA TRANSMUTANS. PK.

THE CHANGING TRICHOLOMA. EDIBLE.

Transmutans means changing, from changes of color in both stem and gills in different stages of the plant. This species has a cap two to four inches broad, viscid or sticky when moist. It is at first tawny-brown, especially with advancing age. The flesh is white and has a decided farinaceous odor and taste.

The gills are crowded, rather narrow, sometimes branched, becoming reddish-spotted with age.

The stem is equal or slightly tapering upward; bare, or slightly silky-fibrillose; stuffed or hollow; whitish, often marked with reddish stains or becoming reddish-brown toward the base, white within. Spores subglobose, 5μ.

The species grows in woods and open places, also in clover pastures, either singly or in tufts. I have seen large tufts of them, and in that case the caps are more or less irregular on account of their crowded condition. I found it frequently about Salem, and this fall, 1905, I found it quite plentiful in a clover pasture near Chillicothe. Found in wet weather from August to September.

TRICHOLOMA EQUESTRE. LINN.

THE KNIGHTLY TRICHOLOMA. EDIBLE.

Figure 43   Tricholoma equestre.

Equestre means belonging to a horseman; so called from its distinguished appearance in the woods.

The pileus is three to five inches broad, fleshy, compact, convex, expanded, obtuse, viscid, scaly, margin incurved at first, pale yellowish, with sometimes a slight tinge of green in both cap and gills. Flesh white or tinged with yellow.

The gills are free, crowded, rounded behind, yellow.

The stem is stout, solid, pale yellow or white, white within. The spores are $7–8 \times 5\mu$.

It differs from T. coryphæum in having gills entirely yellow, while the edges only of the latter are yellow. It differs from T. sejunctum in the latter having pure white gills and a more slender stem.

It is found but occasionally here, and then only a specimen or two. It is an attractive plant and no one would pass it in the woods without admiring it. Found from August to October.

TRICHOLOMA SORDIDUM. FR.

Figure 44   Tricholoma sordidum.

Sordidum means dingy, dirty.

The pileus is two to three inches broad, rather tough, fleshy, convex, bell-shaped, then depressed, subumbonate, smooth, hygrophanous, margin slightly striate, brownish lilac, then dusky.

The gills are rounded, rather crowded, dingy violet then dusky, notched with a decurrent tooth.

The stem is colored like the pileus, fibrillose striate, usually slightly curved, stuffed, short, often thickened at the base.

The spores are 7–8 × 3–4, minutely rugulose.

This species differs from T. nudum in being smaller, tougher, and often hygrophanous.

It is found in richly manured gardens, about manure piles, and in hot-houses. The specimens in Figure 44 were found in a hot-house near Boston, Mass., and sent to me by Mrs. E. Blackford. They are found in September and October.

TRICHOLOMA GRAMMOPODIUM. BULL.

THE GROOVED STEM TRICHOLOMA. EDIBLE.

Figure 45   Tricholoma grammopodium. Natural size.

Grammopodium is from two Greek words meaning line and foot.

The pileus is three to six inches broad, flesh thick at the center, thin at the margin, solid yet tender; brownish, blackish-umber, almost a dingy-lavender when moist, whitish when dry; at first bell-shaped, then convex, sometimes slightly wavy, obtusely umbonate; margin at first inclined to be involute, and extending beyond the gills.

The gills are attached to the stem, broadly notched as will be seen in the specimen, closely crowded, quite entire, shorter ones numerous, a few branched, white or whitish.

The stem is three to four inches long, thickened at the base, smooth, firm, longitudinally grooved from which it gets its specific name, whitish.

The spores are nearly round, $5-6\mu$.

It closely resembles T. fuligineum but can be distinguished by the grooved stem and crowded gills. The specimens in Figure 45 were found near Boston, and were sent to me by Mrs. Blackford. The plants keep well and are easily dried. They were found the first of June. They have an excellent flavor.

TRICHOLOMA PÆDIDUM. FR.

Paedidum means nasty, stinking.

The pileus is small, about one and a half inches broad, rather fleshy, tough; convex, then flattened, soon depressed around the conical umbo; fibrillose, becoming smooth; smoky gray, somewhat streaked; moist; margin involute, naked.

The gills are adnexed, crowded, narrow, white, then grayish, somewhat sinuate with a slight decurrent tooth.

The stem is short, slightly striate, dingy gray, thickened at the base. The spores are elliptical or fusiform, $10–11 \times 5–6\mu$.

The specific name, "nasty" or "stinking," has really no application to the plant. It is said to be very good when cooked. It is found in well manured gardens and fields, or about manure piles.

It differs from T. sordidum in having no trace of violet color. T. lixivium differs in the free truncate gills.

TRICHOLOMA LIXIVIUM. FR.

Lixivium means made into lye; hence, of the color of ashes and water.

The pileus is two to three inches broad; flesh thin; convex then plane; umbonate, never depressed; even; smooth; grayish-brown when moist, then umber; margin membranaceous, at length slightly striate, sometimes wavy.

The gills are rounded behind and adnexed, free, soft, distant, often crisped, gray.

The stem is about two inches long, fibrous, hollow, or stuffed, equal, at first covered with a white down, fragile, gray.

The spores are elliptical, $7 \times 4–5\mu$.

The umbonate pileus and the nearly free, broad, gray gills will distinguish it. They are a late grower and are found under pine trees in November.

TRICHOLOMA SULPHUREUM. BULL.

SULPHURY TRICHOLOMA. POISONOUS.

Figure 46   Tricholoma sulphureum.

Sulphureum, sulphur; so called from the general color of the plant.

The pileus is one to three inches broad, fleshy, convex, then expanded, plane, slightly umbonate, sometimes depressed, or flexuous and irregular, mar gin at first involute, dingy or reddish-yellow, at first silky, becoming smooth and even.

The gills are rather thick, narrowed behind, emarginate or acutely adnate, sulphur-colored.

The stem is two to four inches long, somewhat bulbous, sometimes curved, frequently slightly striate; stuffed, often hollow; sulphur-yellow, yellow within; furnished at the base occasionally with many rather strong, yellow, fibrous roots. Odor strong and disagreeable. Flesh thick and yellow. Spores are $9–10 \times 5\mu$.

It grows in mixed woods. I find it frequently where logs have decayed. The specimen in Figure 46 was found in Haynes' Hollow and photographed by Dr. Kellerman. Found in October and November.

TRICHOLOMA QUINQUEPARTITUM. FR.

Quinquepartitum means divided into five parts. There is no apparent reason for the name. Fries could not identify Linnæus' Agaricus quinquepartitus and he attached the name of this species.

The pileus is three or four inches broad, slightly fleshy; convex, rather involute, then flattened, somewhat repand; viscid, smooth, even, pale yellowish.

The gills are notched at the point of attachment to the stem, broad, white.

The stem is three to four inches long, solid, striate or grooved, smooth. The spores are $5–6 \times 3–4$.

This species differs from T. portentosum in the pileus not being virgate, and from T. fucatum in the smooth, striate or grooved stem. This plant is found in thin woods where logs have decayed. I have not eaten this species but I have no doubt of its edibility. The taste is pleasant. Found in October and November.

TRICHOLOMA LATERARIUM. PK.

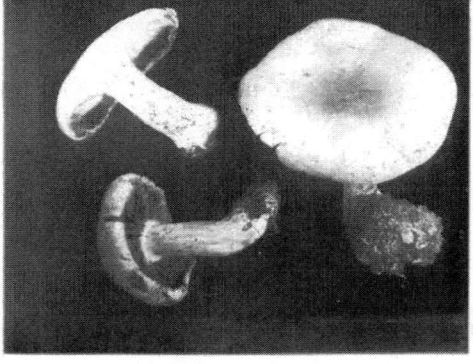

Figure 47 Tricholoma laterarium.

Laterarium is from later, a brick; so called because there is nearly always a slight tinge of brick red on the disk.

The pileus is two to four inches broad, convex, then expanded, sometimes slightly depressed in the center; pruinose, whitish, the disk often tinged with red or brown, the thin margin marked with slight subdistant, short, radiating ridges.

The gills are narrow, crowded, white, prolonged in little decurrent lines on the stem. The stem is nearly equal, solid, white. The spores are globose, .00018 inch in diameter. Peck's 26th Rep.

This plant is quite widely distributed in the United States. It is found quite frequently in Ohio and is rather abundant on the hillsides about Chillicothe, where it is frequently somewhat bulbous. The tinge of brownish-red on the disk, and the short radiating ridges on the margin of the pileus will serve to identify the plant. It is edible and fairly good. Found on leaf-mold in rather damp woods from July to November.

TRICHOLOMA PANÆOLUM. FR.

Figure 48   Tricholoma panæolum.

Panæolum, all variegated. The pileus is from three to four inches broad, deeply depressed, dusky with a gray bloom, hygrophanous; margin at first inrolled, sometimes wavy or irregular when fully expanded.

The gills are quite crowded, adnate, arcuate, white at first, turning to a light gray tinged with an intimation of red, notched with a decurrent tooth.

The stem is short, slightly bulbous, tapering upward, solid, smooth, about the same color as the cap. The spores are subglobose, 5–6.

I found the specimens in Figure 48 under pine trees, growing on a bed of pine needles, on Cemetery Hill. They were found on the 9th of November.

Var. calceolum, Sterb., has the pileus spongy, deformed, thin, soft, expanded, edge incurved, sooty-gray; gills smoky; stem excentric, fusiform, very short.

TRICHOLOMA COLUMBETTA. FR.

THE DOVE-COLORED TRICHOLOMA. EDIBLE.

Figure 49   Tricholoma columbetta. One-third natural size. Caps white. Stems bulbous.

Columbetta is the diminutive of columba, a dove; so called from the color of the plant. The pileus is from one to four inches broad, fleshy, convex, then expanded; at first smooth, then silky; white, center sometimes a dilute mouse color shading to a white, frequently a tinge of pink will be seen on the margin, which is at first inrolled, tomentose in young plants, sometimes cracked.

The gills are notched at the junction of the stem, crowded, thin, white, brittle.

The stem is two inches or more long, solid, white, cylindrical, unequal, often compressed, smooth, crooked, silky especially in young plants, bulbous. Spores .00023 by .00018 inch. Flesh white, taste mild.

This is a beautiful plant, seeming to be quite free from insects, and will remain sound for several days on your study table. I had no end of trouble with it till Dr. Herbst suggested the species. It is quite plentiful here. Dr. Peck gives quite a number of varieties. Curtis, McIlvaine, Stevenson, and Cooke all speak of its esculent qualities. Found in the woods in September and October.

TRICHOLOMA MELALEUCUM. PERS.

THE CHANGEABLE TRICHOLOMA.

Figure 50   Tricholoma melaleucum. Two-thirds natural size.

Melaleucum, black and white; from contrasted colors of the cap and gills.

This Tricholoma grows in abundance in northern Ohio. I have found it in the woods near Bowling Green, Ohio. The specimens in the halftone were found near Sandusky, Ohio, and were photographed by Dr. Kellerman. It is usually found in sandy soil, growing singly in shady woods.

The pileus fleshy, thin, from one to three inches broad, convex, rather broadly umbonate, smooth, moist, with variable color, usually pale, nearly white at first, later much darker, sometimes slightly wavy.

The gills are notched, adnexed, ventricose, crowded, white.

The stem is stuffed, then hollow, elastic, from two to four inches long, somewhat smooth, whitish, sprinkled with a few fibrils, usually thickened at the base. The flesh is soft and white. There is no report, so far as I know, regarding its edibility, and I have no doubt as to this, but would advise caution.

TRICHOLOMA LASCIVUM. FR.

THE TARRY TRICHOLOMA.

Lascivum, playful, wanton; so called because of its many affinities, none of which are very close. The pileus is fleshy, convex, then expanded, slightly obtuse, somewhat depressed, silky at first, then smooth, even. The gills are notched, adnexed, crowded, white; the stem is solid, equal, rigid, rooting, white, tomentose at the base. Found in the woods, Haynes' Hollow near Chillicothe. September and October.

TRICHOLOMA RUSSULA. SCHÆFF.

THE REDDISH TRICHOLOMA. EDIBLE.

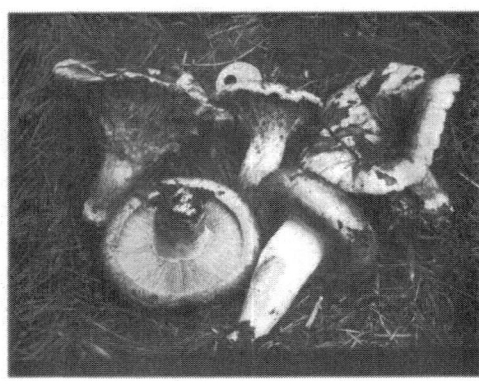

Figure 51   Tricholoma Russula. Natural size. Caps reddish or flesh color.

Russula is so named because of its likeness in color to some species of the genus Russula.

The pileus is three to four inches broad, fleshy, convex, then depressed, viscid, even or dotted with granular scales, red or flesh color, the margin somewhat paler, involute and minutely downy in the young plant.

The gills are rounded or slightly decurrent, rather distant, white, often becoming red-spotted with age.

The stem is two to three inches long, solid, firm, whitish rosy-red, nearly equal, scaly at the apex. The spores are elliptical, $10 \times 5\mu$.

This plant is quite variable in many of its peculiar characteristics, yet it usually has enough to readily distinguish it. The cap may be flesh-color and the stem rosy-red, the cap may be red and the stem white or whitish with stains of red. During wet weather the caps of all are viscid; when dry, all may be cracked more or less. The stems may not be scaly at the apex, often rosy when young. They are found in the woods solitary, in groups, or frequently in dense clusters. The specimens in Figure 51 were found in Michigan and photographed by Dr. Fischer.

I found this plant in Poke Hollow. The gills were quite decurrent.

TRICHOLOMA ACERBUM. BULL.

THE BITTER TRICHOLOMA.

Acerbum means bitter to the taste.

The pileus is three to four inches broad, convex to expanded, obtuse, smooth, more or less spotted, margin thin, at first involute, rugose, sulcate, viscid, whitish, often tinged rufous, or yellow, quite bitter to the taste.

The gills are notched, crowded, pallid or rufescent, narrow.

The stem is solid, rather short, blunt, yellowish, squamulose above or about the apex. The spores are subglobose, 5–6μ.

These plants were found growing in a thick bed of moss along with Armillaria nardosmia. They were not perfect plants but I judged them to be T. acerbum from their taste and involute margin. I sent some to Prof. Atkinson, who confirmed my classification. They grow in open woods in October and November.

TRICHOLOMA CINERASCENS. BULL.

Cinerascens means becoming the color of ashes.

The pileus is two to three inches broad, fleshy, convex to expanded, even, obtuse, smooth, white, then grayish, margin thin.

The gills are emarginate, crowded, rather undulate, dingy, reddish often yellowish, easily separating from the pileus.

The stem is stuffed, equal, smooth, elastic.

They grow in clusters in mixed wood. They are mild to the taste.

TRICHOLOMA ALBUM. SCHÆFF.

THE WHITE TRICHOLOMA. EDIBLE.

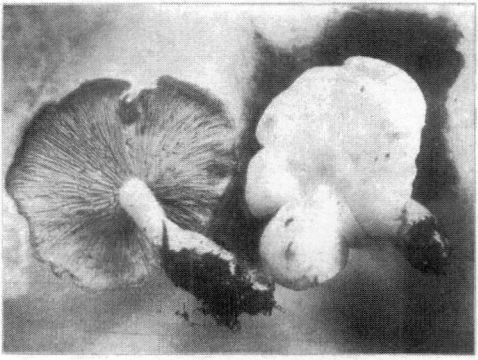

Figure 52   Tricholoma album. Entirely white.

Album means white.

The pileus is two to four inches broad, fleshy, entirely white, convex, then depressed, obtuse, smooth, dry, disc frequently tinged with yellow, margin at first involute, at length repand.

The gills are rounded behind, rather crowded, thin, white, broad.

The stem is two to four inches long, solid, firm, narrowed upwards, smooth.

This plant is quite plentiful in our woods, growing usually in groups. It grows upon the leaf mould and is frequently quite large. It is quite acrid to the taste when raw, but this is overcome in cooking. It is found from August to October.

These plants are quite plentiful on the wooded hillsides about Chillicothe. Those in Figure 52 were found on Ralston's Run and photographed by Dr. Kellerman.

TRICHOLOMA IMBRICATUM. FR.

THE IMBRICATED TRICHOLOMA. EDIBLE.

Figure 53   Tricholoma imbricatum.

Imbricatum means covered with tiles, imbreces, referring to the lacerated condition of the cap. This species is very closely related to T. transmutans in size, color and taste. It is, however, easily separated by its dry cap and solid stem. Its cap is reddish-brown or cinnamon-brown, and its surface often presents a somewhat scaly appearance because the epidermis becomes lacerated or torn into small irregular fragments which adhere and seem to overlap like shingles on a roof. The flesh is firm, white, and has a farinaceous taste as well as odor. The gills are white, becoming red or rusty spotted, rather close, and notched. The stem is solid, firm, nearly equal, except slightly swollen at the base, colored much like the cap but usually paler. When old it is sometimes hollow on account of the insects mining it. The spores are white and elliptical, .00025 inch long.

I found this mushroom near Salem, Ohio, Bowling Green, Ohio, and on Ralston's Run near Chillicothe. Found in mixed woods from September to November.

TRICHOLOMA TERRIFERUM. PK.

THE EARTH-BEARING TRICHOLOMA. EDIBLE.

Terriferum, earth-bearing, alluding to the viscid cap's holding particles of loam and pine needles to it as it breaks through the soil. This is a meaty mushroom, and when properly cleaned makes an appetizing dish.

The pileus is convex, irregular, wavy on the margin and rolled inward, smooth, viscid, pale yellow, sometimes whitish, generally covered with loam on account of the sticky surface of the cap, flesh white.

The gills are white, thin, close, slightly adnexed.

The stem is short, fleshy, solid, equal, mealy, very slightly bulbous at the base.

Found near Salem, Ohio, on Hon. J. Thwing Brooks' farm September to October.

TRICHOLOMA FUMIDELLUM. PK.

THE SMOKY TRICHOLOMA. EDIBLE.

Fumidellum—smoky, because of the clay-colored caps clouded with brown.

The pileus is one to two inches broad, convex, then expanded, subumbonate, bare, moist, dingy-white or clay-color clouded with brown, the disk or umbo generally smoky brown.

The gills are crowded, subventricose, whitish.

The stem is one and a half to two and a half inches long, equal, bare, solid whitish. The spores minute, subglobose, $4–5 \times 4\mu$. Peck, 44 Rep.

The specimens I found grew in a mixed woods in the leaf-mold. They are found only occasionally in our woods in September and October.

TRICHOLOMA LEUCOCEPHALUM. FR.

THE WHITE-CAPPED TRICHOLOMA. EDIBLE.

Leucocephalum is from two Greek words meaning white and head, referring to the white caps.

The pileus is one and a half to two inches across, convex, then plane; even, moist, smooth when the silky veil is gone, water-soaked after a rain; flesh thin, tough, smell mealy, taste mild and pleasant.

The gills are rounded behind and almost free, crowded, white.

The stem is about two inches long, hollow, solid at the base, smooth, cartilaginous, tough, rooting. The spores are $9-10 \times 7-8\mu$.

It differs from T. album in having the odor of new meal strongly marked. It is found in open woods during September and October.

TRICHOLOMA FUMESCENS. PK.

SMOKY TRICHOLOMA. EDIBLE.

Figure 54   Tricholoma fumescens.

Fumescens means growing smoky.

Pileus convex or expanded, dry, clothed with a very minute appressed tomentum, whitish.

The gills are narrow, crowded, rounded behind, whitish or pale cream color, changing to smoky blue or blackish where bruised.

The stem is short, cylindrical, whitish. Spores are oblong-elliptical, $5-6 \times 5\mu$. Pileus is one inch broad. Stem one to one and a half inches high. Peck, 44th Rep. N. Y. State Bot.

The caps are quite a bit larger in the specimens found in Ohio than those described by Dr. Peck. So much so that I was in doubt as to the correct identification. I sent some specimens to Dr. Peck for his determination. The species will be readily identified by the fine crowded gills and the smoky blue or blackish hue they assume when bruised. The caps are frequently wavy, as will be seen in Figure 54.

I found the plants in Poke Hollow near Chillicothe, September to November.

TRICHOLOMA TERREUM. SCHAEFF.

THE GRAY TRICHOLOMA. EDIBLE.

Figure 55 Tricholoma terreum. Cap grayish-brown or mouse color.

Terreum is from terra, the earth; so called from the color. This is quite a variable species in color and size, as well as manner of growth.

The pileus is one to three inches broad, dry, fleshy, thin, convex, expanded, nearly plane, often having a central umbo; floccose-scaly, ashy-brown, grayish-brown or mouse-color.

The gills are adnexed, subdistant, white, becoming grayish, edges more or less eroded. Spores, 5–6μ.

The stem is whitish, fibrillose, equal, paler than the cap, varying from solid to stuffed or hollow, one to three inches high.

I find this plant on north hillsides, in beech woods. It is not plentiful. There are several varieties:

Var. orirubens. Q. Edge of gills reddish.

Var. atrosquamosum. Chev. Pileus gray with small black scales; g. whitish.

Var. argyraceum. Bull. Entirely pure white, or pileus grayish.

Var. chrysites. Jungh. Pileus tinged yellowish or greenish.

The plants in Figure 55 were found in Poke Hollow near Chillicothe. Their time is September to November.

TRICHOLOMA SAPONACEUM. FR.

Figure 56   Tricholoma saponaceum.

Saponaceum is from sapo, soap, so called from its peculiar odor.

The pileus is two to three inches broad, convex, then plane, involute at first as will be seen in Figure 56, smooth, moist in wet weather but not viscid, often cracked into scales or punctate, grayish or livid-brown, often with a tinge of olive, flesh firm, becoming more or less red when cut or wounded.

The gills are uncinately emarginate, thin, quite entire, not crowded, white, sometimes tinged with green. Spores subglobose, $5 \times 4\mu$.

The stem is solid, unequal, rooting, smooth, sometimes reticulated with black fibrils or scaly.

This species is found quite frequently about Chillicothe. It is quite variable in size and color, but can be readily recognized from its peculiar odor and the flesh's becoming reddish when wounded. It is not poisonous but its odor will prevent any one from eating it. Found in mixed woods from August to November.

TRICHOLOMA CARTILAGINEUM. BULL.

THE CARTILAGINOUS TRICHOLOMA. EDIBLE.

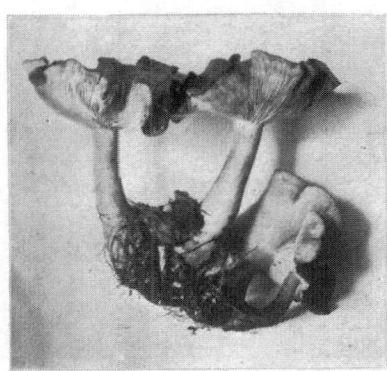

Figure 57   Tricholoma cartilagineum. Two-thirds natural size.

Cartilagineum means gristly or cartilaginous.

The pileus is two to three inches broad, cartilaginous, elastic, fleshy, convex, soon expanded, wavy, as seen in Figure 57, margin incurved, smooth, inclined to be blackish at first, then broken up into small black spots.

The gills are slightly notched, adnexed, somewhat crowded, grayish.

The stem is one to two inches long, rather firm, stuffed, equal, smooth, white, often striate and mealy. Taste and odor pleasant.

A number of my friends ate it because of its inviting taste and odor. It grew in quantities among the clover in our city park during the wet weather of the last of May and the first of June.

TRICHOLOMA SQUARRULOSUM. BRES.

Figure 58   Tricholoma squarrulosum. Caps showing black squamules.

Squarrulosum means full of scales.

The pileus is two to three inches broad, convex, then expanded, umbonate, dry; fuscous then lurid tan, center black, with black squamules; edge fibrillose, exceeding gills.

The gills are broad, crowded, whitish-gray, reddish when bruised.

The stem is of the same color as the pileus, punctato-squamulose. The spores are elliptical, $7–9\times4–5\mu$.

This is a beautiful plant, growing in mixed woods among the leaves. The stem is short and apparently the same color as the pileus. The latter is covered with black squamules which give rise to the name of the species. I have succeeded in finding the plants only in October. The specimens in Figure 58 were found in Poke Hollow, near Chillicothe.

TRICHOLOMA MACULATESCENS. PK.

SPOTTED TRICHOLOMA.

Figure 59 Tricholoma maculatescens. One-third natural size.

Maculatescens means growing spotted; so called because when the specimen is dried the cap becomes more or less spotted.

The pileus is one and a half to three inches broad, compact, spongy, reddish-brown, convex, then expanded, obtuse, even, slightly viscid when wet, becoming rivulose and brown spotted in drying, flesh whitish, margin inflexed, exceeding the gills.

The gills are slightly emarginate, rather narrow, cinereous.

The stem is spongy-fleshy, equal, sometimes abruptly narrowed at the base, solid, stout, fibrillose, pallid or whitish. The spores are oblong or subfusiform, pointed at the ends, uninucleate, .0003 inch long, .00016 broad. Peck.

I found the plant on several occasions in the month of November, but was unable to fix it satisfactorily until Prof. Morgan helped me out. The specimens in Figure 59 were found on Thanksgiving day in the Morton woods, in Gallia County, Ohio. I had found several specimens about Chillicothe, previous to this.

This species seems to be very near T. flavobrunneum, T. graveolens, and T. Schumacheri, but may be distinguished from them by the spotting of the pileus when drying and the peculiar shape of the spores.

It is found among the leaves in mixed woods even during freezing weather. It is no doubt edible, but I should try it cautiously for the first time.

TRICHOLOMA FLAVOBRUNNEUM. FR.

THE YELLOW-BROWN TRICHOLOMA. EDIBLE.

Flavobrunneum is from flavus, yellow; brunneus, brown; so called from the brown caps and yellow flesh.

The pileus is three to four or more inches broad, fleshy, conical, then convex, expanded, subumbonate, viscid, brownish-bay, scaly-streaked, flesh yellow, then tinged with red.

The gills are pale yellow, emarginate, slightly decurrent, somewhat crowded, and often tinged with red.

The stem is three to four inches long, hollow, slightly ventricose, brownish, flesh yellow, at first viscid, sometimes reddish-brown. The spores are 6–7×4–5. Found in mixed woods among leaves.

TRICHOLOMA SCHUMACHERI. FR.

Schumacheri in honor of C. F. Schumacher, author of "Plantarum Sællandiæ." The pileus is from two to three inches broad, spongy, convex, then plane, obtuse, even, livid gray, moist, edge beyond gills incurved.

The gills are narrow, close, pure white, slightly emarginate.

The stem is three to four inches long, solid, fibrillosely-striate, white and fleshy.

This seems to be a domestic plant, found in greenhouses.

TRICHOLOMA GRANDE. PK.

THE LARGE TRICHOLOMA. EDIBLE.

Grande, large, showy. This was quite abundant in Haines' Hollow and on Ralston's Run during the wet weather of the fall of 1905. It seems to be very like T. columbetta and is found in the same localities.

The pileus is thick, firm, hemispherical, becoming convex, often irregular, dry, scaly, somewhat silky-fibrillose toward the margin, white, the margin at first involute. Flesh grayish-white, taste farinaceous.

The gills are close, rounded behind, adnexed, white.

The stem is stout, solid, fibrillose, at first tapering upward, then equal or but slightly thickened at the base, pure white. The spores are elliptical, $9-11 \times 6\mu$.

The pileus is four to five inches broad, the stem two to four inches long, and an inch to an inch and a half thick. Peck, 44th Rep.

This is a very large and showy plant, growing among leaves after heavy rains. Both this and T. columbetta, as well as a white variety of T. personatum, were very plentiful in the same woods. They grow in groups so closely crowded that the caps are often quite irregular. The darker and scaly disk and larger sized spore will help you to distinguish it from T. columbetta. The very large specimens are too coarse to be good. Found in damp woods, among leaves, from August to November.

TRICHOLOMA SEJUNCTUM. SOW.

THE SEPARATING TRICHOLOMA. EDIBLE.

Figure 60   Tricholoma sejunctum. One-half natural size.

Sejunctum means having separated. It refers to the separation of the gills from the stem. Pileus fleshy, convex, then expanded, umbonate, slightly viscid, streaked with innate brown or blackish fibrils, whitish or yellow, sometimes greenish-yellow, flesh white and fragile.

The gills are broad, subdistant, rounded behind or notched, white.

The stem is solid, stout, often irregular, white. The spores are subglobose, .00025 inch broad. The pileus is one to three inches broad; stem one to four inches long and from four to eight lines thick. Peck's Report.

This is quite common about Salem, Ohio; on the old Lake Shore line in Wood County near Bowling Green, Ohio; and I have found it frequently near Chillicothe. When cooked it has a pleasant flavor. It is always an attractive specimen. I find it under beech trees in the woods, September to November.

TRICHOLOMA UNIFACTUM. PK.

UNITED TRICHOLOMA. EDIBLE.

Unifactum means united or made into one, referring to the stems united in one base root or stem.

The pileus is fleshy but thin, convex; often irregular, sometimes eccentric from its mode of growth; whitish, flesh whitish, taste mild.

The gills are thin, narrow, close, rounded behind, slightly adnexed, sometimes forked near the base, white.

The stems are equal or thicker at the base, solid, fibrous, white, united at the base in a large fleshy mass.

Spores are white, subglobose, .00016 to .0002 of an inch broad. Peck.

I found a beautiful specimen in Poke Hollow, in a beech woods with some oak and chestnut. There was but one cluster growing from a large whitish fleshy mass. There were fifteen caps growing from this fleshy mass. I could not identify species until too late to photograph.

TRICHOLOMA ALBELLUM. FR.

THE WHITISH TRICHOLOMA. EDIBLE.

The pileus is two to three inches broad, becoming pale-white, passing into gray when dry, fleshy, thick at the disk, thinner at the sides, conical then convex, gibbous when expanded, when in vigor moist on the surface, spotted as with scales, the thin margin naked, flesh soft, floccose, white, unchangeable.

The gills are very much attenuated behind, not emarginate, becoming broad in front; very crowded, quite entire, white.

The stem is one to two inches long, solid, fleshy-compact, ovate-bulbous (conical to the middle, cylindrical above), fibrillose-striate, white. Spores elliptical, $6-7 \times 4\mu$.

TRICHOLOMA PERSONATUM. FR.

THE MASKED TRICHOLOMA. EDIBLE.

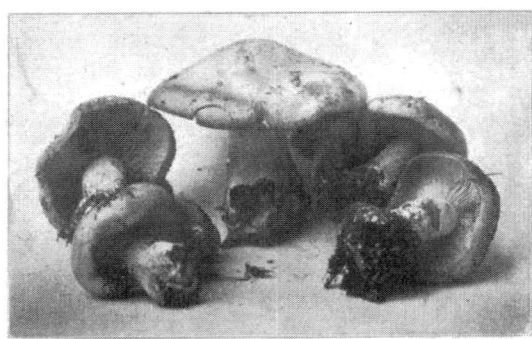

Figure 61 Tricholoma personatum. One-third natural size. Caps usually tinged with lilac or violet. Stems bulbous.

Figure 62 Tricholoma personatum. Two-thirds natural size. The entire plant white.

Personatum means wearing a mask; so called because of the variety of colors it undergoes. This is a beautiful mushroom, and is excellently flavored; it has a wide range and is frequently found, in great abundance. I have often seen it growing in almost a straight line for over twenty feet, the caps so thoroughly crowded that they had lost their form. When young the cap is convex and quite firm, with the margin minutely downy or adorned with mealy particles, and incurved. In the mature plant it is softer, broadly convex, or nearly plane, with the thin margin spreading and more or less turned upward and wavy. When young it is pale lilac in color, but with advancing age it changes to a tawny or rusty hue, especially in the center. Sometimes the cap is white, whitish or gray, or of a pale violaceous color.

The gills are crowded, rounded next to the stem, and nearly free but approaching close to the stem, more narrow toward the margin, with a faint tinge of lilac or violet tint when young, but often white.

The stem is short, solid, adorned with very minute fibers, downy or mealy particles when young and fresh, but becoming smooth with advancing age. The color of the stem is much like the cap but perhaps a shade lighter.

The cap is from one to five inches broad, and the stem from one to three inches high. It grows singly or in groups. It is found in thin woods and thickets. It delights to grow where an old saw mill has stood.

The finest specimens of this species that I ever saw grew on a pile of compost of what had been green cobs from the canning factory. They had lain in the pile for about three years and late in November the compost was literally covered with this species, many of whose caps exceeded five inches while the color and figuration of the plants were quite typical.

In English books this plant is spoken of as Blewits and in France as Blue-stems, but the stems in this country are inclined to be lilac or violet, and then only in the younger plants.

The spores are nearly elliptical and dingy white, but in masses on white paper they have a salmon tint. Its smooth, almost shining, unbroken epidermis and its peculiar peach-blossom tint distinguish it from all other species of the Tricholoma. There is a white variety, very plentiful in our woods, which is illustrated in Figure 62. They are found only in leaf-mould in the woods. September to freezing weather.

TRICHOLOMA NUDUM. BULL.

THE NAKED TRICHOLOMA. EDIBLE.

Nudum, naked, bare; from the character of the margin. The pileus is two to three inches broad, fleshy, rather thin, convex, then expanded, slightly depressed; smooth, moist, the whole plant violet at first, changing color, margin involute, thin, naked, often wavy.

The gills are narrow, rounded behind, slightly decurrent when the plant becomes depressed, crowded, violet at first, changing to a reddish-brown without any tinge of violet.

The stem is two to three inches long, stuffed, elastic, equal, at first violaceous, then becoming pale, more or less mealy. Spores $7 \times 3.5 \mu$

I found some very fine specimens among the leaves in the woods in Haynes' Hollow, near Chillicothe. October and November.

TRICHOLOMA GAMBOSUM. FR.

ST. GEORGE'S MUSHROOM. EDIBLE.

Gambosum, with a swelling of the hoof, gamba. The pileus is three to six inches broad, sometimes even larger; very thick, convex, expanded, depressed, commonly cracked here and there; smooth, suggesting soft kid leather; margin involute at first, pale ochre or yellowish white.

The gills are notched, with an adnexed tooth, densely crowded, ventricose, moist, various lengths, yellowish white.

The stem is short, solid, flocculose at apex, substance creamy white; swollen slightly at the base. The spores are white.

It is called St. George's mushroom in England because it appears about the time of St. George's day, April 23d. It frequently grows in rings or crescents. It has a very strong odor. Its season is May and June.

TRICHOLOMA PORTENTOSUM. FR.

THE STRANGE TRICHOLOMA. EDIBLE.

Portentosum means strange or monstrous.

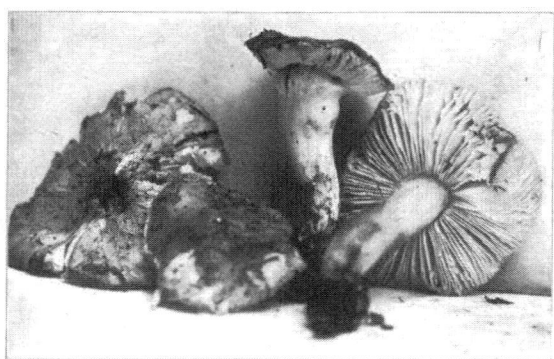

Figure 63   Tricholoma portentosum.

The pileus is three to five inches broad, fleshy, convex, then expanded, subumbonate, viscid, sooty, often with purple tinge, frequently unequal and turned up, streaked with dark lines, the thin margin naked, flesh not compact, white, fragile, and mild.

The gills are white, very broad, rounded, almost free, distant, often becoming pale-gray or yellowish.

The stem is three to six inches long, solid, quite fibrous, sometimes equal, often tapering toward the base, white, stout, striate, villous at base. The spores are subglobose, $4–5 \times 4\mu$.

The plants grow in pine woods and along the margins of mixed woods, frequently by roadsides. It is usually found in October and November. The plants in Figure 63 were found near Waltham, Mass., and were sent to me by Mrs. E. B. Blackford. This is said to even excel T. personatum in edible qualities.

CLITOCYBE. FR.

Clitocybe is from two Greek words, a hillside, or declivity, and a head; so called from the central depression of the pileus.

The genus Clitocybe differs from Tricholoma in the character of the gills. They are attached to the stem by the whole width and usually are prolonged down the stem or decurrent. This is the first genus with decurrent gills. The genus has neither a volva nor a ring and the spores are white. The stem is elastic, spongy within, frequently hollow and extremely fibrous, continuous with the pileus.

The pileus is generally fleshy, growing thin toward the margin, plane or depressed or funnel-shaped, and with margin incurved. The universal veil, if present at all, is seen only on the margin of the pileus like frost or silky dew.

These plants usually grow on the ground and frequently in groups, though a few may be found on decayed wood.

The Collybia, Mycena, and Omphalia have cartilaginous stems, while the stem of the Clitocybe is extremely fibrous, and the Tricholoma is distinguished by its notched gills.

This genus, because of the variations in its species, will always be puzzling to the beginner, as it is to experts. We may easily decide it is a Clitocybe because of the gills squarely meeting the stem, or decurrent upon it, and its external fibrous stem, but to locate the species is quite a different matter.

CLITOCYBE MEDIA. PK.

THE INTERMEDIATE CLITOCYBE. EDIBLE.

Figure 64   Clitocybe media. One-half natural size.

Media is from medius, middle; it is so called because it is intermediate between C. nebularis and C. clavipes. It is not as plentiful as either of the others in our woods.

The pileus is grayish-brown or blackish-brown, always darker than C. nebularis. The flesh is white and farinaceous in taste.

The gills are rather broad, not crowded, adnate and decurrent, white, with few transverse ridges or veins in the spaces between the gills.

The stem is one to two inches long, usually tapering upward, paler than the pileus, rather elastic, smooth. The spores are plainly elliptical, $8 \times 5\mu$.

This resembles very closely the two species mentioned above and is hard to separate. I found the specimens in Figure 64 along Ralston's Run where the ground is mossy and damp. Found in September and October.

CLITOCYBE INFUNDIBULIFORMIS. SCHAEFF.

THE FUNNEL-FORMED CLITOCYBE. EDIBLE.

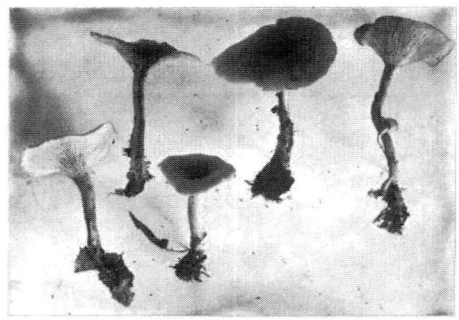

Plate IX. Figure 65   Clitocybe infundibuliformis.

Infundibuliformis means funnel-shaped. This is a beautiful plant and very abundant in woods after a heavy rain. It grows upon the leaves and especially among pine needles.

The pileus is at first convex and umbonate and as the plant advances in age the margin becomes elevated until the plant becomes funnel-shaped. The margin is frequently incurved and finally wavy. The flesh is soft and white. The color of the cap is a pale tan. If the cap is examined carefully it will be seen to be covered with a slight down or silky substance, especially on the margin. The color of the cap is apt to fade so that specimens will be found almost white.

The gills are thin, close, white or whitish, and very decurrent.

The stem is quite smooth, and generally tapers upward from the base. It is sometimes white or whitish, but more frequently like the cap. Mycelium will usually be found at the base on the leaves, forming a soft white down. I have found this species in several parts of the state. It is frequently found in clusters, when the caps will be irregular on account of the crowded condition. They are very tender and of excellent flavor. Found from August to October.

CLITOCYBE ODORA. BULL.

SWEET-SMELLING CLITOCYBE. EDIBLE.

Figure 66   Clitocybe odora. One-third natural size. Cap pale green.

Odora means fragrant. This is one of the easiest of the Clitocybes to identify. The collector will very readily recognize it by its olive-green color and its odor. The color in the old plant is quite variable but in young plants is well marked. The pileus is one to two and a half inches broad, flesh quite thick; at first convex, then expanded, plane, often depressed, sometimes inclined to be wavy; even, smooth, olive-green.

The gills are adnate, rather close, sometimes slightly decurrent, broad, pallid.

The stem is one to one and a half inches long, often slightly bulbous at the base.

These plants are found from August to October, in the woods, on leaves. They are quite common about Chillicothe after a rain. When cooked by themselves the flavor is a bit strong, but when mixed with other plants not so strong in flavor, they are fine.

CLITOCYBE ILLUDENS. SCHW.

THE DECEIVING CLITOCYBE. NOT EDIBLE.

Plate X. Figure 67 Clitocybe illudens.

Caps reddish-yellow to deep yellow. Gills yellow and decurrent.

Illudens means deceiving. Pileus of a beautiful yellow, very showy and inviting. Many a basketful has been brought to me to be identified with the hope of their edibility. The cap is convex, umbonate, spreading, depressed, smooth, often irregular from its crowded condition of growth; in older and larger plants the margin of the pileus is wavy. The flesh is thick at the center but thinner toward the margin. In old plants the color is brownish.

The gills are decurrent, some much further than others; yellow; not crowded; broad.

The stem is solid, long, firm, smooth tapering towards the base, as will be seen by Figure 67, sometimes the stems are very large.

The pileus is from four to six inches broad. The stem is six to eight inches high. It occurs in large clusters and the rich saffron color of the entire plant compels our admiration and we are reminded that "not all is gold that glitters." It will be interesting to gather a large cluster to show its phosphorescence and the heat which the plant will generate. You can show the phosphorescence by putting it in a dark room and by placing a thermometer in the cluster you can show the heat. It is frequently called "Jack-o'-lantern."

I have known people to eat it without harm, but the chances are that it will make most persons sick. It ought to be good, since it is so abundant and looks so rich. Found from July to October.

CLITOCYBE MULTICEPS. PK.

THE MANY-HEADED CLITOCYBE. EDIBLE.

Figure 68   Clitocybe multiceps. One-half natural size. Caps grayish-white.

Multiceps means many heads; so called because many caps are found in one cluster. It is a very common plant around Chillicothe. It has been found within the city limits. It is quite a typical species, too, having all the characteristics of the genus. I have often seen over fifty caps in one cluster.

The pileus is white or gray, brownish-gray or buff; smooth, thin at the margin, convex, slightly moist in rainy weather.

The gills are white, crowded, narrow at each end, decurrent.

The stem is tough, elastic, fleshy, solid, tinged with the same color as the cap.

The pileus is one to three inches broad; grows in dense tufts. Spores are white, smooth and globose.

When found in June the plants are a shade whiter than in the fall. The fall plants are very much the oyster color. The early plant is a more tender one and better for table use, however, I do not regard it as excellent. They are found in woods, in old pastures by logs and stumps, and in lawns. June to October.

CLITOCYBE CLAVIPES. PERS.

Figure 69   Clitocybe clavipes.

Clavipes is from clava, a club, and pes, a foot.

The pileus is one to two and a half inches broad, fleshy, rather spongy, convex to expanded, obtuse, even, smooth, gray or brownish, sometimes whitish toward the margin.

The gills are decurrent, descending, rather distant, nearly entire, rather broad, white.

The stem is two inches long, swollen at the base, attenuated upward, stuffed, spongy, fibrillose, livid sooty. Spores are elliptical, $6-7 \times 4\mu$.

I found specimens on Cemetery Hill underneath pine trees. I sent some to Dr. Herbst and Prof. Atkinson; both pronounced them C. clavipes. They resemble quite closely C. nebularis. I have also found this plant in mixed woods. Edible and fairly good.

CLITOCYBE TORNATA. FR.

Tornata means turned in a lathe; so called because of its neat and regular form.

The pileus is orbicular, plane, somewhat depressed, thin, smooth, shining, white, darker on the disk, very regular.

The gills are decurrent adnate, rather crowded, white.

The stem is stuffed, firm, slender, smooth, pubescent at the base.

The spores are elliptical, $4-6 \times 3-4\mu$.

These are small, very regular, and inodorous plants. They are found in open fields in the grass about elm stumps. July to September. They are edible and cook readily.

CLITOCYBE METACHROA. FR.

THE OBCONIC CLITOCYBE. EDIBLE.

Figure 70 Clitocybe metachroa. Caps dark gray. Gills pale gray.

Metachroa means changing color.

The pileus is one to two and a half inches broad, somewhat fleshy, convex, then plane, depressed, smooth, hygrophanous, brownish-gray, then livid, growing pale.

The gills are attached to the stem, crowded, pale gray, slightly decurrent.

The stem is one to two inches long, stuffed, then hollow, apex mealy, equal, gray.

It differs from C. ditopa in being inodorous and having a thicker and depressed pileus.

The caps are quite smooth and are frequently concentrically cracked or wrinkled, much as in Clitopilus noveboracensis.

It is found growing on leaves in mixed woods, after a rain, in August and September. When young the margin is incurved but wavy in age. It is quite a hardy plant.

CLITOCYBE ADIRONDACKENSIS. PK.

Figure 71   Clitocybe adirondackensis. Three-fourths natural size. Caps white.

Adirondackensis, so called because the plant was first found in the Adirondack Mountains of New York.

The pileus is thin, submembranaceous, funnel-form, with the margin decurved, nearly smooth, hygrophanous, white, the disk often darker.

The gills are white, very narrow, scarcely broader than the thickness of the flesh of the pileus, crowded, long, decurrent, subarcuate, some of them forked.

The stem is slender, subequal, not hollow, whitish, mycelio-thickened at the base. Peck.

The pileus is one to two inches broad and the stem is one to two and a half long. This is quite a pretty mushroom and has the Clitocybe appearance in a marked degree. The long, narrow, decurrent gills, sometimes tinged with yellow, some of them forked, margin of the pileus sometimes wavy, will assist in distinguishing it. I have no doubt of its edibility. Found among leaves in woods after heavy rains. With us it is confined to the wooded hillsides. The specimens in Figure 71 were found in Michigan and photographed by Dr. Fischer. Found in July and August.

CLITOCYBE OCHROPURPUREA. BERK.

THE CLAY-PURPLE CLITOCYBE. EDIBLE.

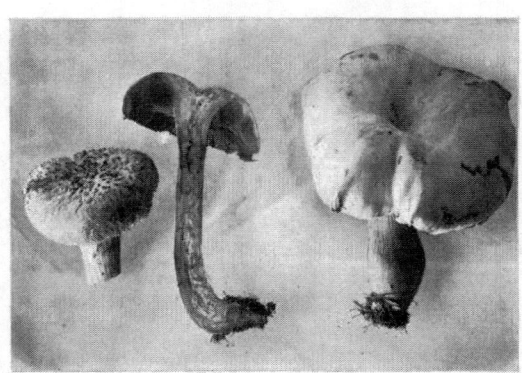

Plate XI. Figure 72   Clitocybe ochropurpurea.

Ochropurpurea is from ochra, ocher or clay color; purpureus, purple; it is so called because the caps are clay-color and the gills are purple. The caps are convex, fleshy, quite compact, clay-colored, sometimes tinged with purple around the margin, cuticle easily separating, margin involute, often at first tomentose, old forms often repand or wavy.

The gills are purple, sometimes whitish in old specimens from the white spores, broad behind, decurrent, distant.

The stem is paler than the cap, often tinted with purple, solid, frequently long and swollen in the middle, fibrous. The spores white or pale yellow.

The first time I found this species I never dreamed that it was a Clitocybe. It was especially abundant on our wooded clay banks or hillsides, near Chillicothe, during the wet weather in July and August of 1905. It is a hardy plant and will keep for days. Insects do not seem to work in it readily. When cooked carefully it is rather tender and fairly good.

CLITOCYBE SUBDITOPODA. PK.

Subditopoda is so called because it is nearly (sub) like Fries' C. ditopus, which means living in two places, perhaps referring to the stem being sometimes central and sometimes eccentric.

The pileus is thin, convex or nearly plane, umbilicate, hygrophanous, grayish-brown, striate on the margin when moist, paler when dry, flesh concolorous, odor and taste farinaceous.

The gills are broad, close, adnate, whitish or pale cinereous.

The stem is equal, smooth, hollow, colored like the pileus. The spores are elliptical, .0002 to .00025 inch long, .00012 to .00016 broad. Peck.

It is found on mossy ground in woods. I have found them under pine trees on Cemetery Hill. Dr. Peck says he separated this species from C. ditopoda because of the "striate margin of the pileus, paler gills, longer stem, and elliptical spores." The plant is edible. September and October.

CLITOCYBE DITOPODA. FR.

Ditopoda is from two Greek words, di-totos, living in two places, and pus or poda, foot, having reference to the stem being central at times and again eccentric.

The pileus is rather fleshy, convex, then plane, depressed, even, smooth, hygrophanous.

The gills are adnate, crowded, thin, dark, cinereous.

The stem is hollow, equal, almost naked.

This species resembles in appearance C. metachroa but can be separated by the mild taste and farinaceous odor. Its favorite habit is on pine needles. August and September. I found this species in various places about Chillicothe and on Thanksgiving day I found it in a mixed wood in Gallia County, Ohio, along with Hygrophorus lauræ and Tricholoma maculatescens. I sent some specimens to Dr. Herbst, who pronounced it C. ditopoda.

CLITOCYBE PITHYOPHILA. FR.

THE PINE-LOVING CLITOCYBE.

Figure 73 Clitocybe pithyophila. Two-thirds natural size. Cap white and showing the pine needles upon which they grow.

Pithyophila means pine-loving. This plant is very abundant under pine trees on Cemetery Hill. They grow on the bed of pine needles. The pileus is very variable in size, white, one to two inches broad; fleshy, thin, becoming plane, umbonate, smooth, growing pale, at length irregularly shaped, repand, wavy, sometimes slightly striate.

The stem is hollow, terete, then compressed, smooth, equal, even, downy at the base.

The gills are adnate, somewhat decurrent, crowded, plane, always white. The spores are $6-7 \times 4\mu$. The plants in Figure 73 are small, having been found during the cold weather in November. They are said to be good, but I have not eaten them.

CLITOCYBE CANDICANS. FR.

Candicans, whitish or shining white. Pileus is one inch broad, entirely white, somewhat fleshy, convex, then plane, or depressed, even, shining, with regularly deflexed margin.

The gills are adnate, crowded, thin, at length decurrent, narrow.

The stem is nearly hollow, even, waxy, shining, nearly equal, cartilaginous, smooth, incurved at the base. The spores are broadly elliptical, or subglobose, $5-6 \times 4\mu$. Found in damp woods on leaves.

CLITOCYBE OBBATA. FR.

THE BEAKER-SHAPED CLITOCYBE. EDIBLE.

Obbata means shaped like an obba or beaker.

The pileus is somewhat membranaceous, umbilicate, then rather deeply depressed, smooth, inclined to be hygrophanous, sooty-brown, margin at length striate.

The gills are decurrent, distant, grayish-white, pruinose.

The stem is hollow, grayish-brown, smooth, equal, rather tough.

I found plants growing on Cemetery Hill under pine trees. I had some trouble to identify the species until Prof. Atkinson helped me out. August to September.

CLITOCYBE GILVA. PERS.

THE YELLOW CLITOCYBE. EDIBLE.

Gilva means pale yellow or reddish yellow.

The pileus is two to four inches broad, fleshy, compact, soon depressed and wavy, smooth, moist, dingy ocher, flesh same color, sometimes spotted, margin involute.

The gills are decurrent, closely crowded, thin, sometimes branched, narrow but broader in the middle, ochraceous yellow.

The stem is two to three inches long, solid, smooth, nearly equal, somewhat paler than the cap, and inclined to be villous at the base.

The spores are nearly globose, 4–5μ.

This plant is sometimes found in mixed woods, but it seems to prefer pine trees. It has a wide distribution, found in the east and south as well as the west. I have found it in several localities in Ohio. Found from July to September.

CLITOCYBE FLACCIDA. SOW.

THE LIMP CLITOCYBE. EDIBLE.

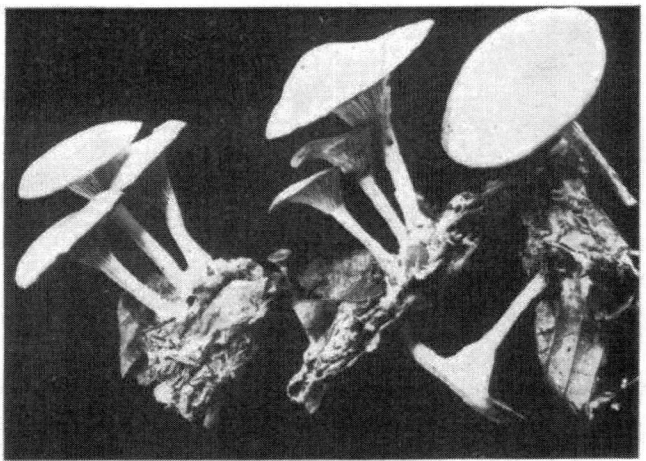

Figure 74   Clitocybe flaccida. One-half natural size.

Flaccida means flabby, limp.

The pileus is two to three inches broad, rather fleshy, thin, limp, umbilicate, then funnel-shaped, even, smooth, sometimes cracking into minute scales, tawny or rust-colored, margin broadly reflexed.

The gills are strongly decurrent, yellowish, to whitish, close, arcuate.

The stem is tufted, unequal, rusty, somewhat wavy, tough, naked, villous at the base. The spores are globose or nearly so, 4–5×3–4μ.

This resembles the C. infundibuliformis very closely, both in its appearance and its habit. It grows among leaves in mixed woods during wet weather. It is gregarious, often many stems growing from one mass of mycelium. The plants in Figure 74 were collected in Ackerman's woods near Columbus, Ohio, and were photographed by Dr. Kellerman. They are found on all the hillsides about Chillicothe. Found from July to late in October.

CLITOCYBE MONADELPHA. MORG.

THE ONE-BROTHERHOOD CLITOCYBE. EDIBLE.

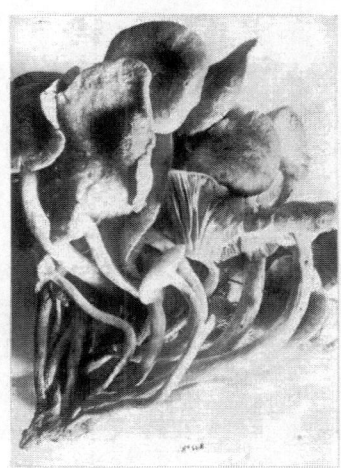

Plate XII. Figure 75   Clitocybe monadelpha.

Monadelpha is from monos, one and adelphos, brother.

Prof. Morgan of Preston, Ohio, gives the following description of the One-Brotherhood Clitocybe in the Mycological Flora of the Miama Valley: "Densely cespitose. Pileus fleshy, convex, then depressed, at first glabrous, then scaly, honey-colored, varying to pallid-brown or reddish. The stem elongated, solid, crooked, twisted, fibrous, tapering at the base, pallid-brownish or flesh color. Spores white, a little irregular, .0055MM."

It might be easily taken for the ringless Armillaria mellea, but the decidedly decurrent gills and the solid stem ought to set any one right. In very wet weather it soon becomes water-soaked, and is then not good. It is found in woods about stumps, and in newly cleared fields about roots or stumps. From spring to October. See Plate XII, Figure 75, for an illustration. Bresadola of Europe has determined this to be the same as that described by Scoparius in 1772 as Agaricus (Clitocybe) tabescens. I have preferred to retain the name given by Prof. Morgan.

CLITOCYBE DEALBATA. SOW.

THE WHITE CLYTOCYBE. EDIBLE.

Dealbata means whitewashed; so called from its white color.

The pileus is about one inch broad, rather fleshy, convex, then plane, upturned and wavy, smooth, shining, even.

The gills are crowded, white, attached to the stem.

The stem is fibrous, thin, equal, stuffed. Spores are $4-5 \times 2.5\mu$.

This is a beautiful plant and widely distributed. Found among leaves and sometimes in the grass. It makes a delicious dish.

CLITOCYBE PHYLLOPHILA. FR.

THE LEAF-LOVING CLITOCYBE. EDIBLE.

Phyllophila means leaf and fond of. It is so called because it is found on leaves in the woods during wet weather.

The pileus is one and a half to three inches in diameter, whitish-tan, rather fleshy, convex, then plane, at length depressed, even, dry, noticeably white around the margin.

The gills are attached to the stem, decurrent especially after the cap is depressed, somewhat distant, rather broad, white, becoming yellowish or ocher tinged, thin.

The stem is two to three inches long, stuffed, becoming hollow, silky, rather tough, whitish. The spores are elliptical., $6 \times 4\mu$.

The whitish-tan cap with its white, silvery zone near the margin will serve to identify the species. August to October.

CLITOCYBE CYATHIFORMIS. BULL.

THE CUP-SHAPED CLITOCYBE. EDIBLE.

Cyathiformis is from cyathus, a drinking cup; formis, form or shape.

The pileus is two to three inches broad, fleshy, rather thin; at first depressed, then funnel-shaped; even, smooth, moist, hygrophanous; the margin involute, sooty or dark brown when moist, becoming pale when dry, often dingy ochraceous or tan-color, inclined to be wavy.

The gills are attached to the stem, decurrent from the depressed form of the pileus, united behind, somewhat dingy, sparingly branched.

The stem is stuffed, elastic, tapering upward, fibrillose, base villous. The spores are elliptical, $9 \times 6\mu$.

This plant has a wide distribution and is found in woods or wood margins. I found some very fine specimens on Ralston's Run, near Chillicothe. September to October.

CLITOCYBE LACCATA. SCOP.

WAXY CLITOCYBE. EDIBLE.

Figure 76   Clitocybe laccata. Two-thirds natural size. Caps violet or reddish-brown. Gills broad and distant.

Laccata means made of shellac or sealing-wax. This is a very common, variable plant. Sometimes of a bright amethyst but usually of a reddish brown. The pileus is from one to two inches broad, almost membranaceous, convex, then plane, depressed in the center, downy with short hairs, violet or reddish-brown.

The gills are broad, distant, attached to the stem by the entire width; pale fleshy-red in hue which is more constant than the color of the cap and which forms an ear-mark to tell the species; adnate with a decurrent tooth, plane, the white spores being very abundant.

The stem is tough, fibrous, stuffed, crooked, white-villous at the base, rather long and slender, dull reddish yellow or reddish-flesh-colored, sometimes pallid or dull ochraceous, slightly striate; when the season is wet it is often watery.

This waxy Clitocybe has a wide range and is frequently very abundant. It is found through almost the entire season. It will grow almost anywhere, in woods, pastures, and lawns, and sometimes on naked ground. The plants in Figure 76 were found in tall grass in a grove in August. Those in Figure 77 were found the last of November on Cemetery Hill, under pine trees.

Figure 77   Clitocybe laccata. Two-thirds natural size. Specimens growing late in the fall.

Prof. Peck gives the following varieties:

- ⊙ Var. amethystina—in which the cap is much darker in color.

- ⊙ Var. pallidifolia—gills much paler than usual.

- ⊙ Var. striatula—cap smooth, thin, so that shadowy lines are seen on cap, radiating from near the center to the margin. This grows in damp places. Some authors make Clitocybe laccata a type for a new genus and call it Lacaria laccata.

COLLYBIA. FR.

Collybia is from a Greek word meaning a small coin or a small round cake. The ring and volva are both wanting in this genus. The pileus is fleshy, generally thin, and when the plant is young the margin of the pileus is incurved.

The gills are adnate or nearly free, soft, membranaceous. Many species of Collybia will revive to some extent when moistened, but they are not coriaceous.

The stem differs in substance from the pileus, cartilaginous or has a cartilaginous cuticle, while the inside is stuffed or hollow. This is quite a large genus, containing fifty-four American species.

COLLYBIA RADICATA. REHL.

THE ROOTING COLLYBIA. EDIBLE.

Plate XIII. Figure 78   Collybia radicata.

This, in its season, is one of the most common mushrooms in the woods. It grows in the ground, frequently around old stumps, sometimes on lawns.

Those in Figure 78 were found in the woods on the ground. One plant, as will be seen by the square, is a foot high.

It is easily recognized by its long root and flat cap. The root extends into the ground and will frequently break before pulling up. This root gives name to the species.

The pileus is fleshy, rather thin, convex, then plane, often with margin upturned in old plants as in Figure 78, and frequently wrinkled at and toward the umbo, smooth, viscid when moist.

The color is quite variable, from almost white to gray, grayish-brown; flesh thin, very white, elastic.

The gills are usually snow white, broad, rather distant, broad in the middle, joined to the stem by the upper angle, unequal.

The stem is frequently long, of the same color as the cap, yet sometimes paler; smooth, firm, sometimes grooved, often twisted, tapering upward, ending in a long tapering root, deeply planted in the soil.

The spores are elliptical, $15 \times 10\mu$.

They grow singly, but generally have many neighbors. They are found in open woods and around old stumps. I seldom have any trouble in getting enough for a large family and some for my neighbor, who may not know what to get but does know how to appreciate them. Found from June to October and from the New England states through the middle west. They differ from C. hariolarum in the densely tufted habit of the latter.

COLYBIA INGRATA. SCHUM.

Ingrata means unpleasant; from its somewhat unpleasant odor.

The pileus is one to two inches broad, globose, bell-shaped, then convex, umbonate, even, brownish-tan.

The gills are free, narrow, crowded, pallid.

The stem is twisted, subcompressed, sprinkled with a mealy tomentum above, umber below, hollow, rather long, unequal.

I found this plant quite abundant on Cemetery Hill, growing under pine trees, from the mass of pine needles. Found in July and August.

COLLYBIA PLATYPHYLLA. FR.

BROAD-GILLED COLLYBIA. EDIBLE.

Figure 79   Collybia platyphylla. One-third natural size.

Platyphylla is from two Greek words meaning broad and leaf, referring to the broad gills. It is a much larger and stouter plant than Collybia radicata. It is found in new ground on open pastures about stumps, also in woods, on rotten logs and about stumps.

The pileus is three to four inches broad, at first convex, then expanded, plane, margin often upturned, smoky brown to grayish, streaked with dark fibrils, watery when moist, flesh white.

The gills are adnexed, very broad, obliquely notched behind, distant, soft, white, in age more or less broken or cracked.

The stem is short, thick, often striated, whitish, soft, stuffed, sometimes slightly powdered at the apex, root blunt. The spores are white and elliptical.

It is easily distinguished from C. radicata by the blunt base of the root and the very broad gills. Like C. radicata they need to be cooked well or there is a slightly bitter taste to them. They are found from June to October.

COLLYBIA DRYOPHILA. BULL.

OAK-LOVING COLLYBIA. EDIBLE.

Figure 80   Collybia dryophila. Natural size. Caps bay-brown.

Dryophila is from two Greek words, oak and fond of. The pileus is bay-brown, bay red, or tan color, one or two inches broad, convex, plane, sometimes depressed and the margin elevated, flesh thin and white.

The gills are free with a decurrent tooth, crowded, narrow, white, or whitish, rarely yellow.

The stem is cartilaginous, smooth, hollow, yellow, or yellowish, equal, sometimes thickened at the base as will be seen in Figure 80. The color of the stem is usually the same as the cap. This is a very common plant about Chillicothe. They are found in woods, especially under oak trees, but are also found in open places. I found them on the High School lawn in Chillicothe. Some very fine specimens that were found growing in a well marked ring, in an old orchard, were brought to me about the first of May. Their season is from the first of May to October.

COLLYBIA ZONATA. PK.

THE ZONED COLLYBIA. EDIBLE.

Plate XIV. Figure 81 Collybia Zonata.

Zonata, zoned; referring to the concentric zones on the cap which show faintly in Figure 81.

The pileus is about one inch broad, sometimes more, sometimes less; rather fleshy, thin, convex, when expanded nearly plane, slightly umbilicate, covered with fibrous down; tawny or ochraceous tawny, sometimes marked with faintly darker zones; even in the very young specimens the umbilicate condition is usually present.

The gills are narrow, close, free, white or nearly white, usually with a pulverulent edge.

The stem is one to three inches long, rather firm, equal, hollow, covered like the cap with a fibrous down, tawny, or brownish tawny. The spores are broadly elliptical, .0002 inch long, .00016 broad.

This species closely resembles C. stipitaria, but is easily distinguished from it because of its habits of growth, different gills, and shorter spores. It is found on or near decaying wood in mixed woods. I have found it frequently on Ralston's Run but always only a few specimens in one place. It does not grow in a cespitose manner with us. Found in August.

COLLYBIA MACULATA. ALB. & SCHW.

THE SPOTTED COLLYBIA. EDIBLE.

Figure82   Collybia maculata. Two-thirds natural size. Reddish-brown spots on caps and stems.

Maculata, spotted; referring to the reddish spots or stains both on the cap and on the stem. The pileus is two to three inches broad, at first white, then spotted (as well as the stem) with reddish brown spots or stains, fleshy, very firm, convex, sometimes nearly plane, even, smooth, truly carnose, compact, at first hemispherical and with an involute margin, often repand.

The gills are somewhat crowded, narrow, adnexed, often free, linear, white or whitish, often brownish cream, gills not reaching to the margin of the cap.

The stem is three to four inches long, nearly solid, more or less grooved, stout, unequal, sometimes ventricose, frequently partially bulbous, lighter than the gills, usually spotted in age, white at first. The spores are subglobose, 4–6μ. The plant is a hardy one. It will keep for several days. The plants in Figure 82 grew in the woods where a log had rotted down.

Var. immaculata, Cooke, differs from the typical form in not changing color or being spotted, and in the broader and serrated gills. This variety delights in fir woods. September to November.

COLLYBIA ATRATA. FR.

CHARCOAL COLLYBIA.

Figure 83   Collybia atrata. One-half natural size. Caps dull blackish-brown. Gills grayish-white.

Atrata, clothed in black; from the pileus being very black when young. The pileus is from one to two inches broad, at first regular and convex, when expanded becoming, as a rule, irregular in shape, sometimes partially lobed or wavy; in young plants the cap is a dull blackish brown, faded in older specimens to a lighter brown, umbilicate, smooth, shining.

The gills are adnate, slightly crowded, with many short ones, rather broad, grayish-white.

The stem is smooth, equal, even, hollow, or stuffed, tough, short, brown within and without, but lighter than the cap. The plant grows in pastures where stumps have been burned out, always, so far as I have noticed, on burned ground. Spores .00023×.00016.

COLLYBIA AMBUSTA. FR.

THE SCORCHED COLLYBIA.

Ambusta, burned or scorched, from its being found on burned soil.

The pileus is nearly membranaceous, convex, then expanded, nearly plane, papillate, striatulate, smooth, livid brown, hygrophanous, umbonate.

The gills are adnate, crowded, lanceolate, white, then of a smoky tinge.

The stem is somewhat stuffed, tough, short, livid. Spores 5–6×3–4.

This species differs from C. atrata in having an umbonate pileus.

COLLYBIA CONFLUENS. PERS.

THE TUFTED COLLYBIA. EDIBLE.

Figure 84 Collybia confluens. Natural size, showing reddish stems.

Confluens means growing together; so called from the stems often being confluent or adhering to each other.

The pileus is from an inch to an inch and a quarter broad, reddish-brown, often densely cespitose, somewhat fleshy, convex, then plane, flaccid, smooth, often watery, margin thin, in old specimens slightly depressed and wavy.

The gills are free and in old plants remote from the stem, rather crowded, narrow, flesh colored, then whitish.

The stem is two to three inches long, hollow, pale red, sprinkled with a mealy pubescence. The spores are slightly ovate, inclined to be pointed at one end, $5-6 \times 3-4\mu$.

These plants grow among leaves in the woods after warm rains, growing in tufts, sometimes in rows or lines. They are not as large as C. dryophylla, the stem is quite different and the plants seem to have the ability to revive like a Marasmius. They can be dried for winter use.

COLLYBIA MYRIADOPHYLLA. PK.

MANY-LEAVED COLLYBIA.

Figure 85   Collybia myriadophylla.

Myriadophylla is from two Greek words, meaning many leaves. It has reference to its numerous gills.

The pileus is very thin, broadly convex, then plane or centrally depressed, sometimes umbillicate, hygrophanous, brown when moist, ochraceous or tan-color when dry.

The gills are very numerous, narrow, linear, crowded, rounded behind or slightly adnexed, brownish-lilac.

The stem is slender, but commonly short, equal, glabrous, stuffed or hollow, reddish-brown. The spores are minute, broadly elliptical, .00012 to .00016-inch long, .0008-inch broad. Peck, 49th Rep.

I found only a few specimens in Haynes's Hollow. The caps were about an inch broad and the stems were an inch and a half long. It will be easily identified if one has the description of it, because of its peculiarly colored gills. I found my plants on a decayed stump in August. In the dried specimens the gills assume a more brownish-red hue, as in the next following species.

Collybia colorea. Pk. They sometimes appear to have a glaucous reflection, probably from the abundance of the spores. The stem is more or less radicated and often slightly floccose-pruinose toward the base. The basidia are very short, being only .0006 to .0008-inch long.

COLLYBIA ATRATOIDES. PK.

THE BLACKISH COLLYBIA.

Figure 86   Collybia atratoides. Two-thirds natural size. Caps blackish to grayish-brown.

Atratoides means like the species atrata, which means black; so called because the caps when fresh are quite black. Atratoides has a different habitat and is not so dark.

The pileus is thin, convex, subumbilicate, glabrous, hygrophanous, blackish-brown when moist, grayish-brown and shining when dry.

The gills are rather broad, subdistant, adnate, grayish-white, often transversely veiny above and venosely connected.

The stem is equal, hollow, smooth, grayish-brown with a whitish mycelioid tomentum at the base. The spores are nearly globose, about .0002-inch broad. The pileus is six to ten lines broad and the stem is about one inch long. Peck.

The plant is gregarious, growing on decayed wood and on mossy sticks in mixed woods. The margin of the cap is often serrated, as you will see in Figure 86, yet this does not seem to be a constant characteristic of the species. It is closely related to C. atrata, but its habitat and the color of its pileus and gills differ very greatly. I have not eaten it, but have no doubt of its good qualities.

Found in August and September. Quite common in all our woods.

COLLYBIA ACERVATA. FR.

THE TUFTED COLLYBIA. EDIBLE.

Figure 87   Collybia acervata. Two-thirds natural size. Caps pale, tan or dingy pink.

Acervata, from acervus, a mass, a heap.

Pileus fleshy but thin, convex, or nearly plane, obtuse, glabrous, hygrophanous, pale, tan-color or dingy pinkish-red, and commonly striate on the margin when moist, paler or whitish when dry.

Gills narrow, close, adnexed or free, whitish or tinged with flesh-color.

The stem slender, rigid, hollow, glabrous, reddish, reddish-brown or brown, often whitish at the top, especially when young, commonly with a matted down at the base. Spores elliptical, 6×3–4μ.

The plant is cespitose. Pileus one-half inch broad. Stem two to three inches long. Peck's 49th Report.

This is a beautiful plant when growing in large tufts. The entire plant is tender and has a delicate flavor. I found the plant figured here on the Frankfort pike where an old saw mill had formerly stood. It grew abundantly there, along with Lepiota Americana and Pluteus cervinus.

Found from August to October.

COLLYBIA VELUTIPES. CURTIS.

THE VELVET-FOOT COLLYBIA. EDIBLE.

Plate XV. Figure 88   Collybia velutipes.

Natural size, showing the velvet stems, which give name to the species.

Velutipes, from vellum, velvet and pes, foot.

Pileus from one to four inches broad, tawny yellow, fleshy at the center, thick on the margin, quite sticky or viscid when moist, margin slightly striate, sometimes inclined to be excentric.

Gills rounded behind, broad, slightly adnexed, tan or pale-yellow, somewhat distant.

The stem is cartilaginous, tough, hollow, umber, then becoming blackish, with a velvety coat. Spores are elliptical, 7×3–3.5μ.

It grows on stumps, logs and roots, in the ground. It grows almost the year round. I have gathered it to eat in February. Plate XV gives a very correct notion of the plant. It is most plentiful in September, October and November, yet found throughout the winter months.

MYCENA. FR.

Mycena is from a Greek word, meaning a fungus. The plants of this genus are small and rather fragile.

Pileus more or less membranaceous, generally striate, with the margin almost straight, and at first pressed to the stem, never involute, expanded, campanulate, and generally umbonate.

The stem is externally cartilaginous, hollow, not stuffed when young, confluent with the cap. Gills never decurrent, though some species have a broad sinus near the stem.

Most species are small and inodorous, but some which have a strong alkaline odor are probably not good. Some are known to be edible.

A few species exude a colored or watery juice when bruised. The Mycena resembles the Collybia, but never has the incurved margin of the latter. The plants are usually smaller, and the caps are more or less conical.

This genus might be mistaken for Omphalia, in which the gills are but slightly decurrent, but in Omphalia the cap is umbilicate while in Mycena it is umbonate.

Their being so small makes the determination of species somewhat difficult. Some have characteristic odors which greatly assist in establishing their identity.

MYCENA GALERICULATA. SCOP.

THE SMALL PEAKED-CAP MYCENA. EDIBLE.

Plate XVI. Figure 89 Mycena galericulata.

Natural size.

Galericulata, a small peaked-cap.

The pileus is campanulate, whitish or grayish, center of the disk darker and lighter toward the margin, smooth, dry, margin striated nearly to the peak of the umbo, sometimes slightly depressed.

The gills are adnate with a tooth, connected by veins, whitish, then gray, often flesh color, rather distant, ventricose, edge sometimes entire, sometimes serrate.

The stem is rigid, cartilaginous, hollow, tough, straight, polished, smooth, hairy at the base.

It grows on logs and stumps in the woods. It is very common and sometimes found in abundance. The plants are frequently densely clustered, the numerous stems matted together by a soft hairy down at the base. There are many forms of this plant. Found from September to frost. The plants in Figure 89 were photographed by Prof. G. D. Smith, Akron, O.

MYCENA RUGOSA. FR.

THE WRINKLED MYCENA. EDIBLE.

Rugosa means wrinkled. The pileus is somewhat fleshy, darker and smaller than the galericulata, quite tough, bell-shaped, then expanded, with unequal elevated wrinkles, always dry, striate on the margin.

The gills are adnate, with a tooth, united behind, connected by veins, somewhat distant, whitish, then gray, edge sometimes entire, sometimes serrate.

The stem is short, tough, rooted with a hairy base, strongly cartilaginous, hollow, rigid, smooth. It is found on stumps or decayed logs during September and October.

MYCENA PROLIFERA. SOW.

THE PROLIFEROUS MYCENA. EDIBLE.

Prolifera is from proles, offspring, and fero, to bear. The pileus is somewhat fleshy, campanulate, then expanded, dry, with a broad, dark umbo; margin at length sulcate or furrowed and sometimes split, pale-yellowish or becoming brownish-tan.

The gills are adnexed, subdistant, white, then pallid.

The stem is firm, rigid, smooth, shining, minutely striate, rooting. Fries.

This species, as well as M. galericulata, is closely related to M. cohærens. I have found it in dense tufts or clusters, sometimes on lawns, on the bare ground, and in the woods. It is one of the plants in which the stems may be cooked with the caps.

MYCENA CAPILLARIS. SCHUM.

Capillaris means hair-like. This is a very small but beautiful white plant.

The pileus is bell-shaped, at length umbilicate, smooth.

The gills are attached to the stem, ascending, rather distant.

The stem is thread-like, smooth, short.

The spores are 7–8×4. Fries.

These plants are very small and easily overlooked. They grow on leaves in the woods after a rain. July and August. Quite common.

MYCENA SETOSA. SOW.

Setosa means full of setæ or hairs.

The pileus is very delicate, hemispherical, obtuse, smooth.

The gills are distant, white, almost free.

The stem is short, slender, and covered with spreading hairs which gives rise to its specific name.

Commonly found on dead leaves in the woods after a rain. Found in July and August.

MYCENA HÆMATOPA. PERS.

THE BLOOD-FOOT MYCENA. EDIBLE.

Figure 90 Mycena hæmatopa. Brownish-red or flesh-color. A dull red juice exudes from the stem. Margin dentate by sterile flap.

Hæmatopa is from two Greek words, meaning blood and foot.

The pileus is fleshy, one inch broad, conic, or bell-shaped, somewhat umbonate, obtuse, whitish to flesh-color, with more or less dull red, even, or slightly striate at the margin, the margin extending beyond the gills and is toothed.

The gills are attached to the stem, often with a decurrent tooth, whitish. Spores, 10×6–7.

The stem is two to four inches long, firm, hollow, sometimes smooth, sometimes powdered with whitish, soft hairy down, in color the same as the pileus, yielding a dark red juice which gives name to the species.

The color varies quite a little in these plants, owing to some having more of the red juice than others. The genus is readily identified by the dull blood-red juice, hollow stem, the crenate margin of the cap, and its dense cespitose habits. It is found on decayed logs in damp places from

August to October. The plants in Figure 90 were found in Haynes' Hollow, September 8. The plant is widely distributed over the United States. No one will have the slightest difficulty in recognizing this species after seeing the plants in the figure above.

MYCENA ALKALINA. FR.

THE STUMP MYCENA.

Figure 91   Mycena alkalina. Two-thirds natural size, often larger. Young specimens.

Solitary or cespitose; pileus one-half to two inches broad, rather membranaceous, campanulate, obtuse, naked, deeply striate, moist, shining when dry, when old expanded or depressed, but little changed in color, though occasionally with a pink or yellow hue, whitish or grayish, the center of the disk darker.

The gills are adnate, rather distant, slightly ventricose, at first pale, then glaucous, pinkish, or yellow, more or less connected by veins.

The stem is smooth, slightly sticky, shining, villous at the base with a sometimes tawny-down, sometimes firm and tenacious, hollow, attenuated upward. The plant is rigid, but brittle, and strong-scented. Found on decayed stumps and logs, you will meet it frequently. August to November.

MYCENA FILOPES. BULL.

THREADY-STEMMED MYCENA.

Pileus membranaceous, obtuse, campanulate, then expanded, striate, brown or umber, tinged with pink.

The gills are free or minutely adnexed, slightly ventricose, white or paler than the pileus, crowded.

The stem is hollow, juicy, smooth, filiform, rather brittle, whitish or brownish. Found in woods on leaves, after a rain, from July to October.

MYCENA STANNEA. FR.

THE TIN-COLORED MYCENA.

Figure 92   Mycena stannea. Natural size. Caps white, sometimes smoky.

Stannea pertaining to the color of tin. This is a delicate species that grows in the woods in tufts on rotten wood in damp places. The general character is shown in the illustration, being nearly white but many of the pilei are somewhat smoky.

The pileus is firm, membranaceous, bell-shaped, then expanded, smooth, very slightly striate, hygrophanous, quite silky, tin-color.

The gills are firmly attached to the stem, with a decurrent tooth, connected by veins, grayish-white.

The stem is smooth, even, shining, becoming pale, at length compressed. This species differs from Mycena vitrea in having a tooth to the gills. May, June, and July.

MYCENA VITREA. FR.

Vitrea, glassy. This plant is quite fragile. The pileus is membranaceous, bell-shaped, livid-brown, finely striate, no trace of umbo.

The gills are firmly attached to the stem, not connected by veins, distinct, linear, whitish.

The stem is slender, slightly striate, polished, pale, base fibrillose. This species differs from M. ætites and M. stannea in gills not having a decurrent tooth and not being connected by veins.

MYCENA CORTICOLA. FR.

Figure 93   Mycena corticola.

Corticola means dwelling on bark.

It is one of the smallest of the Mycenas, the pileus being about two to four lines across, thin, hemispherical, obtuse, becoming slightly umbilicate, deeply striate, glabrous or flocculosely pruinose, gray, tan, or brownish.

The gills are attached to the stem, with slight decurrent tooth, broad, rather ovate, pallid.

The stem, is short, slender, incurved, glabrous or minutely scurfy, somewhat paler than the pileus. The spores are elliptical, $5-6 \times 3\mu$; cystidia obtusely fusiform, $50-60 \times 8-10\mu$.

These plants are found on the bark of living trees. After rains I have seen the bark on the shade trees along the walks in Chillicothe, literally covered with these beautiful little plants. The plants in Figure 93 were taken from a maple tree the 4th of December. They are very close allied to M. hiemalis but can be distinguished by the broad, ovate gills bearing cystidia, and smaller spores.

MYCENA HIEMALIS. OSBECK.

THE WINTER MYCENA.

Hiemalis, of, or belonging to, winter. The pileus quite thin, bell-shaped, very slightly umbonate, margin striate; pinkish, rufescent, white, sometimes pruinose.

The gills are adnate, linear, white or whitish.

The stem is slender, curved, base downy, whitish, pinkish-red. The spores are $7-8 \times 3$.

This is a more delicate species than M. corticola and differs from it in its narrow gills, and striate, not sulcate, pileus, also in the color of the stem. Found on stumps and logs. October and November.

MYCENA LEAIANA. BERK.

Figure 94   Mycena leaiana. Natural size. Caps bright orange and very viscid.

Leaiana named in honor of Mr. Thomas G. Lea, who was the first man to study mycology in the Miami Valley. This is a very beautiful plant growing on decayed beech logs in rainy weather. The pileus is fleshy, very viscid, bright orange, the margin slightly striate as will be seen in the one whose cap shows.

The gills are distant, not entire, broad, notched at the stem, attached, the edge a dusky orange, or vermilion, the short gills beginning at the margin.

The stem is in most cases curved, attenuated toward the cap, smooth, hollow, rather firm, quite hirsute or strigose at the base. The spores are elliptical, apiculate, .0090×.0056 mm.

They are cæspitose, growing in dense tufts on logs somewhat decayed. It is extremely viscid, so much so that your hands will be stained yellow if you handle it much. It grows from spring to fall but is usually more abundant in August and September. Very common.

MYCENA IRIS. B.

Pileus is small, convex, expanded, obtuse, slightly viscid, striate, quite blue when young, growing brownish with blue fibrils.

The gills are free, tinged with gray.

The stem is short, bluish below, tinged with brown above, somewhat pruinose. Found in damp woods after a rain, in August.

MYCENA PURA. PERS.

Figure 95 Mycena pura.

Pura means unstained, pure.

The pileus is fleshy, thin, bell-shaped. expanded, obtusely umbonate, finely striate on the margin, sometimes having margin upturned, violet to rose.

The gills are broad, adnate to sinuate, in older plants sometimes free by breaking away from the stem, connected by veins, sometimes wavy and crenate on the edge, the edge of the gills sometimes almost or quite white, violet, rose.

The stem is even, nearly naked, somewhat villous at the base, sometimes almost white when young, later assuming the color of the cap, hollow, smooth.

The spores are white and oblong, 6–8×3–3.5. M. Pelianthina differs from this in having dark-edged gills. It differs from M. pseudopura and M. zephira in having a strong smell. M. ianthina differs in having a conical cap.

This plant is quite widely distributed. Our plants are light-violet in color, and the color seems constant. I have found it in mixed woods. It is found in September and October.

MYCENA VULGARIS. PERS.

Vulgaris means common.

The pileus is small, convex, then depressed, papillate, viscid, brownish-gray, finely striate on the margin.

The gills are subdecurrent, thin, white; the depressed cap and decurrent gills make the plant resemble an Omphalia. Spores, $5×2.5\mu$.

The stem is viscid, pale, tough, fibrillose at the base, rooting, becoming hollow. It differs from M. pelliculosa in not having a separable cuticule and the fold-like gills.

This plant will be recognized by its smoky or grayish color, umbilicate pileus, and viscid stem. It is found in woods on leaves and decayed sticks. August and September.

MYCENA EPIPTERYGIA. SCOP.

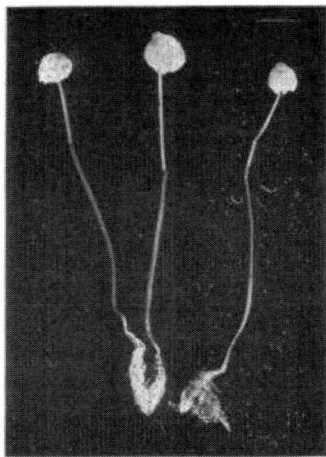

Figure 96   Mycena epipterygia.

Epipterygia is Epi, upon, and Pterygion, a small wing.

These are small, the pileus being one-half to one inch broad, membranaceous, bell-shaped, then expanded, rather obtuse, not depressed, striate, the cuticule separable in every condition and viscid in damp weather, gray, often pale yellowish-green near the margin often minutely notched when young.

The gills are attached to the stem with a decurrent tooth, thin, whitish or tinged with gray.

The stem is two to four inches long, hollow, tough, rooting, viscid, yellowish, sometimes gray or even whitish. The spores are elliptical, $8-10 \times 4-5\mu$.

These plants have a wide distribution and are found on branches, among moss and dead leaves. They are found in clusters and solitary. They resemble in many ways M. alcalina but do not have the peculiar smell.

The plants in Figure 96 were photographed by Prof. G. D. Smith of Akron.

OMPHALIA. FR.

Omphalia is from a Greek word meaning the navel; referring here to the central depression in the cap.

The pileus from the first is centrally depressed, then funnel-shaped, almost membranaceous, and watery when moist; margin incurved or straight. Stem cartilaginous and hollow, often stuffed when young, continuous with the cap but different in character. Gills decurrent and sometimes branched.

They are generally found on wood, preferring a damp woody situation and a wet season. It is easily distinguished from Collybia and Mycena by its decurrent gills. In some of the species of the Mycena where the gills are slightly decurrent, the pileus is not centrally depressed as it is in corresponding species of Omphalia. There are a few species of Omphalia whose pileus is not centrally depressed but whose gills are plainly decurrent.

OMPHALIA CAMPANELLA. BATSCH.

THE BELL OMPHALIA. EDIBLE.

Plate XVII. Figure 97 Omphalia campanella.

Campanella means a little bell.

The pileus is membranaceous, convex to extended, centrally depressed, striate, watery, rusty-yellow in color.

The gills are moderately close, decurrent, bow-shaped, connected by veins, rigid, firm, yellowish. The spores elliptical, $6-7 \times 3-4\mu$.

The stem is hollow, clothed with down, and paler above.

This plant is very common and plentiful in our woods and is widely distributed in the states. It grows on wood or on ground very heavily charged with decaying wood. It is found through the summer and fall. It is delicious if you have the patience to gather them.

OMPHALIA EPICHYSIA. PERS.

The pileus is thin, convex to expanded, depressed in the center, sooty-gray with a watery appearance, pallid to nearly white when dry.

The gills are slightly decurrent, whitish then gray, somewhat crowded.

The stem is slender, hollow, gray. The spores are elliptical, $8–10 \times 4–5\mu$.

It grows in decayed wood. Its smoky color, funnel-shaped pileus, and gray short stem will distinguish it. I have some plants sent me from Massachusetts which seem to be much smaller than our plants.

OMPHALIA UMBELLIFERA. LINN.

THE UMBEL OMPHALIA. EDIBLE.

Umbellifera—umbella, a small shade; ferro, to bear. Pileus one-half inch broad, membranaceous, whitish, convex, then plane, broadly obconic, slightly umbilicate even in the smallest plants, hygrophanous in wet weather, rayed with darker striæ.

The gills are decurrent, very distant, quite broad behind, triangular, with straight edges.

The stem is short, not more than one inch long, dilated at the apex, of same color as the pileus, at first stuffed, then hollow, firm, white, villous at the base.

It is a common plant in our woods, growing on decayed wood or ground largely made up of rotten wood. Decayed beech bark is a favorite habitat. Found from July till October.

OMPHALIA CÆSPITOSA. BOL.

Figure 98 Omphalia cæspitosa. Natural size.

Cæspitosa means growing in tufts; cæspes, turf. The pileus is submembranaceous, very small, convex, nearly hemispherical, umbilicate, thin, sulcate, light-ochre, margin crenate, smooth.

The gills are distant, rather broad, shortly decurrent, whitish.

The stem is curved, hollow, colored like the pileus, slightly bulbous at the base. The spores are 6×5.

This species is very much like Omphalia oniscus and they can only be distinguished by their habitats and color. It is found in August and September. It delights in well rotted wood. I have seen millions in one place.

OMPHALIA ONISCUS. FR.

BOLTON'S OMPHALIA. EDIBLE.

Oniscus, a name given to a species of codfish by the Greeks, so named because of their gray color. The pileus is flaccid, irregular, about one inch broad, convex, plane, or depressed, slightly fleshy, wavy, sometimes lobed, margin striate, dark cinereous, paler when dry.

The gills are adnate, decurrent, livid or whitish, arranged in groups of four, somewhat distant.

The stem is about one inch long, rather firm, straight or curved, sometimes unequal, nearly hollow. The spores are $12 \times 7-8\mu$.

This is found in damp places from August to November.

OMPHALIA PYXIDATA. BULL.

THE BOX OMPHALIA.

Pyxidata means made like a box, from pyxis, a box.

The pileus is somewhat membranaceous, clearly umbilicate, then funnel-shaped, smooth when moist, margin often striate, brick-red.

The gills are decurrent, rather distant, triangular, narrow, reddish gray, often yellowish.

The stem is stuffed, then hollow, even, tough, pale-tawny. The spores are $7-8 \times 5-6\mu$.

The plants are usually hygrophanous, but when dry, floccose or slightly silky. This is a small plant growing usually on lawns, nearly hidden in the grass. I found some very fine specimens on Dr. Sulzbacher's lawn on Second Street, Chillicothe. The plant is, however, widely distributed. I found many specimens on the 3d of November.

OMPHALIA FIBULA. BULL.

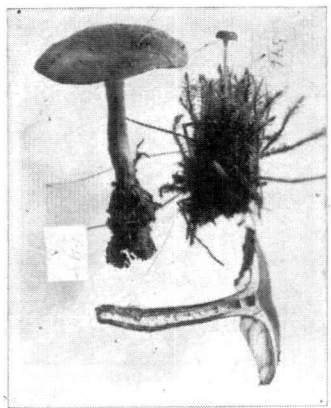

Figure 99 Omphalia fibula.

Fibula means a buckle or pin, from the pin-like stem.

The pileus is membranaceous, at first top-shaped, expanded, slightly umbilicate, striate, margin inclined to be inflexed, yellow or tawny, with a dusky center, minutely pilose.

The gills are deeply decurrent, paler, distinct.

The stem is slender, nearly orange color with a violet-brown apex, the whole minutely pilose. The spores are elliptical, $4-5 \times 2\mu$.

They are found on mossy banks where it is more or less damp. I have only found it in October.

OMPHALIA ALBOFLAVA. MOY.

THE GOLDEN-GILLED OMPHALIA.

Figure 100 Omphalia alboflava. Cap yellowish-brown, sometimes a greenish tinge. Gills golden yellow.

Alboflava is from two Greek words meaning whitish-yellow, from the yellow gills.

The pileus is one to two inches broad, thin, somewhat membranaceous, umbilicate, flaccid, covered with fine woolly material, yellow-brown, lighter when dry, margin reflexed.

The gills are distant, deep golden-yellow, occasionally forked.

The stem is hollow, equal, smooth, shining, egg-yellow.

The spores are elliptical, $8 \times 4\mu$.

This plant, is found quite frequently on decayed branches and logs about Chillicothe. I have never had the opportunity to test its edibility but I have no doubt of its being good.

The plants in Figure 100 were found in Haynes' Hollow and were photographed by Dr. Kellerman. Found from July to October.

MARASMIUS. FR.

Marasmius is a Greek participle meaning withered or shriveled; it is so called because the plant will wither and dry up, but revive with the coming of rain.

The spores are white and subelliptical. The pileus is tough and fleshy or membranaceous.

The stem is cartilaginous and continuous with the pileus, but of a different texture. The gills are thick, rather tough and distant, sometimes unequal, variously attached or free, rarely decurrent, with a sharp entire edge. It is quite a large genus and many of its species will be of great interest to the student.

MARASMIUS OREADES. FR.

THE FAIRY-RING MUSHROOM. EDIBLE.

Figure 101   Marasmius oreades. Two-thirds natural size.

Oreades, mountain nymphs. Pileus is fleshy, tough and pliable when moist, brittle when dry, convex, becoming flat, somewhat umbonate, brownish-buff at first, becoming cream-color; when old it is usually quite wrinkled.

The gills are broad and wide apart, creamy or yellowish, rounded at the stem end, unequal in length.

The stem is solid, equal, tough, fibrous, naked and smooth at base, everywhere with a downy surface. The spores are white, 8×5.

To my mind there is no more appetizing mushroom than the "Fairy Ring" mushroom. Figure 101 will give an accurate notion of the plant and Figure 102 will show how they grow in the grass. It is found in all parts of Ohio. Every old pasture field or lawn will be full of these rings. The plant is small but its plentifulness will make up for its size.

There are many conjectures why this and many other mushrooms grow in a circle. The explanation is quite obvious. The ring is started by a clump or an individual mushroom. The ground where the mushroom grew is rendered unfit for mushrooms again, the spores fall upon the ground and the mycelium spreads out from this point, consequently each year the ring is growing larger. Sometimes they appear only in a crescent form. One can tell, by looking over a lawn or pasture, where the rings are, because, from the decay of the mushroom, the grass is greener and more vigorous there.

Long ago, in England and Ireland, before the peasantry had begun to question the reality of the existence of the fairy folk and their beneficent interference in the affairs of life, these emerald-hued rings were firmly believed to be due to the fairy footsteps which nightly pressed their chosen haunts, and to mark the "little people's" favorite dancing ground. "They had always fine music among themselves, and danced in a moonshiny night around or in a ring, as one may see to this day upon every common in England where mushrooms grow," quaintly says one old writer. And the Rev. Gerard Smith still further voices the belief of the people as to the nature of these grassy rings:

"The nimble elves That do by moonshine green sour ringlets make, Whereof the ewe bites not; whose pastime 'tis To make these midnight mushrooms."

It is a very common plant, and it will pay any one to know it, as we cannot find anything in the markets that will equal it as a table delicacy.

Found in pastures and lawns during rainy weather from May till frost.

Figure 102   Marasmius oreades. Showing a fairy ring.

MARASMIUS URENS. FR.

THE STINGING MARASMIUS.

Urens means burning; so called from its acrid taste.

The pileus is pale-buff, tough, fleshy, convex or flat, becoming depressed and finally wrinkled, smooth, even, one to two inches broad.

The gills are unequal, cream-colored, becoming brownish, much closer than in the Fairy Ring, hardly reaching the stem proper, joined behind.

The stem is solid above and hollow below, fibrous, pale, its surface more or less covered with flocculent down, and densely covered with white down at the base.

It will be well for collectors to pass by this and M. peronatus, or to exercise the greatest caution in their use. They have been eaten without harm, but they also have so long been branded as poisonous that too great care cannot be taken. Its taste is acrid, and it grows in lawns and pastures from June to September.

MARASMIUS ANDROSACEUS. LINN.

Figure 103   Marasmius androsaceus. Natural size.

Androsaceus is from a Greek word which means an unidentified sea plant or zoophyte.

The pileus is three to six lines broad, membranaceous, convex, with a slight depression, pale-reddish, darker in the center, striate, smooth.

The gills are attached to the stem, frequently quite simple and few in number, about fifteen, with shorter ones between, sometimes forked, whitish.

The stem is one to two inches long, horny, filiform, hollow, quite smooth, black, often twisted when dry. The spores are 7×3–4μ.

This is a very attractive little plant found on the leaves in the woods after a rain. They are quite abundant. Found from July to October.

MARASMIUS FŒTIDUS. SOW.

Figure 104   Marasmius foetidus.

Fœtidus means stinking or fœtid.

The pileus is submembranaceous, tough, convex, then expanded, umbilicate striato-plicate, turning pale when dry, subpruinose.

The gills are annulato-adnexed, distant, rufescent with a yellow tinge.

The stem is hollow, minutely velvety, bay, base flocculose.

The caps are light brownish-red in color, fading when dry. When fresh it has a fœtid odor quite perceptible for such small plants. It is found on decayed sticks and leaves in woods. During wet weather or after heavy rains it is quite common in the woods about Chillicothe.

Found from July to October.

This is also called Heliomyces fœtens (Pat.) and is so classified by Prof. Morgan in his very excellent Monogram on North American Species of Marasmius.

MARASMIUS VELUTIPES. B. & C.

Figure 105   Marasmius velutipes.

Velutipes means velvet-footed, from the velvety stem. The pileus is thin, submembranaceous, smooth, convex, or expanded, grayish-rufous when moist, cinereous when dry, a half to one and a half inches broad.

The gills are very narrow, crowded, whitish or grayish.

The stem is slender, three to five inches long, equal, hollow, clothed with a dense grayish velvety tomentum. Peck.

They usually grow in a very crowded condition, many plants growing from one mat of mycelium. It is quite a common plant with us, found in damp woods or around a swampy place. The pileus with us is convex. Some authorities speak of an umbilicate cap. The plant is quite hardy and easily identified because of its long and slender stem, with the grayish tomentum at the base. Found from July to October.

The specimens in Figure 105 were found at Ashville, Ohio.

MARASMIUS COHÆRENS. (FR.) BRES.

THE STEMMED-MASSED MARASMIUS. EDIBLE.

Figure 106 Marasmius cohærens. Two-thirds natural size, showing how the stems are massed together.

Cohærens means holding together, referring to the stems being massed together.

The pileus is fleshy, thin, convex, campanulate, then expanded, sometimes slightly umbonate, in old specimens the margin upturned or wavy, velvety, reddish tan-color, darker in the center, indistinctly striate.

The gills are rather crowded, narrow, adnate, sometimes becoming free from the stem, connected by slight veins, pale cinnamon-color, becoming somewhat darker with age, the variation of color due to the number of cystidia scattered over the surface of the gills and on their edge. Spores, oval, white, small, $6 \times 3\mu$.

The stem is hollow, long, rigid, even, smooth, shining, reddish-brown, growing paler or whitish toward the cap, a number of the stems growing together at the base with a whitish myceloid tomentum present.

The plant grows in dense clusters among leaves and in well rotted wood. I have found it quite often about Chillicothe. It is called Mycena cohærens, Fr., Collybia lachnophylla, Berk., Collybia spinulifera, Pk. The plants in Figure 106 were found near Ashville, Ohio. September to frost.

MARASMIUS CANDIDUS. BOLT.

THE WHITE MARASMIUS.

Figure 107   Marasmius candidus. Natural size.

Candidus means shining white. This delicate species grows in moist and shady places in the woods. It grows on twigs, its habitat and structure are fully illustrated in the Figure 107.

The pileus is rather membranaceous, hemispherical, then plane or depressed, pellucid, wrinkled, naked, entirely white.

The gills are adnexed, ventricose, distant, not entire.

The stem is thin, stuffed, whitish, slightly pruinose, base tinged with brown. Spores are elliptical, $4 \times 2\mu$.

This plant has a wide distribution in this country. The specimens figured were collected by H. H. York near Sandusky, Ohio, and were photographed by Dr. Kellerman. I have found them at various points in Ohio.

MARASMIUS ROTULA. FR.

THE COLLARED MARASMIUS.

Figure 108   Marasmius rotula. Natural size. Caps white or pale-buff.

Rotula means a little wheel.

The pileus is one to three lines broad, hemispherical, umbilicate, and minutely umbonate, plaited, smooth, membranaceous, margin crenate, white, or pale buff, with a dark umbilicus.

The gills are broad, distant, few, equal, or occasionally with a few short ones, of the color of the pileus, attached to a free collar behind.

The stem is setiform, slightly flexuous, white above, then tawny, deep shining brown at the base, striate, hollow, frequently branched and sarmentose, with or without abortive pilei.—M. J. B. This plant is very common in woods on fallen twigs. The plants in Figure 108 were collected near Cincinnati. This plant has a wide distribution. It is in all our Ohio woods.

MARASMIUS SCORODONIUS. FR.

STRONG-SCENTED MARASMIUS. EDIBLE.

Figure 109   Marasmius scorodonius.

Scorodonius is from a Greek word meaning like garlic.

The pileus is one-half inch or more broad, reddish when young, but becoming pale, whitish; somewhat fleshy, tough; even, soon plane, rugulose even when young, at length rugulose and crisped.

The gills are attached to the stem, often separating, connected by veins, crisped in drying, whitish.

The stem is at least one inch long, hollow, equal, quite smooth, shining, reddish. The spores are elliptical, $6 \times 4\mu$.

It is found in woods growing on sticks and decayed wood. It is strong-smelling. It is frequently put with other plants to give a flavor of garlic to the dish. Found from July to October.

MARASMIUS CALOPUS. FR.

Calopus is from two Greek words meaning beautiful and foot, so called because of its beautiful stem.

The pileus is rather fleshy, tough, convex, plane then depressed, even, at length rugose, whitish.

The gills are emarginate, adnexed, thin, white, in groups of 2–4.

The stem is hollow, equal, smooth, not rooting, shining, reddish-bay. It is found growing on twigs and fallen leaves, in the woods. Smaller than M. Scorodonius but with longer stem.

MARASMIUS PRASIOSMUS. FR.

THE LEEK-SCENTED MARASMIUS.

Prasiosmus means smelling like a leek; from, prason, a leek. The pileus is one-half to one inch broad, somewhat membranaceous, tough, bell-shaped, pale yellow or whitish, disk often darker, wrinkled.

The gills are adnexed, somewhat close, white.

The stem is tough, hollow, pallid and smooth above, dilated at the base, tomentose and brown. It is found in woods adhering to oak leaves after heavy rains. It is very near M. porreus but differs from it in its gills being white and caps not being striated. It differs from M. terginus mainly in its habitat and leek-like scent.

MARASMIUS ANOMALUS. PK.

Anomalus, not conforming to rule, irregular. The pileus is one to two inches broad, somewhat fleshy, tough, convex, even, reddish-gray.

The stem is two to three inches long, hollow, equal, smooth, pallid above, reddish-brown below.

The gills are rotundate-free, close, narrow, whitish or pallid. Morgan.

This is quite a pretty plant, growing on sticks among leaves in the woods. It is larger than most of the small Marasmii found in similar habitats.

MARASMIUS SEMIHIRTIPES. PK.

Semihirtipes means a slightly hairy foot or stem.

The pileus is thin, tough, nearly plane or depressed, smooth, sometimes striate on the margin, hygrophanous, reddish-brown when moist, alutaceous when dry, the disk sometimes darker.

The gills are subdistant, reaching the stem, slightly venose-connected, sub-crenulate on the edge, white.

The stem is equal, even or finely striate, hollow, smooth above, velvety-tomentose toward the base, reddish-brown. Peck.

These plants are very small, often no doubt overlooked by the collector. They are gregarious in their mode of growth.

MARASMIUS LONGIPES. PK.

Longipes means long stem or foot.

The pileus is thin, convex, smooth, finely striate on the margin, tawny-red.

The gills are not crowded, attached, white.

The stem is tall, straight, hollow, equal, covered with a downy meal, rooting, brown or fawn-color, white at the top.

These plants are quite small and slender, sometimes four to five inches high. They are rather common in our woods after a rain.

MARASMIUS GRAMINUM. BERK.

Graminum is the gen. pl. of gramen, which means grass.

The pileus small, membranaceous, convex, then nearly plane, umbonate, deeply and distinctly striate or sulcate, tinged with rufous, the furrows paler, disc brown.

The gills are attached to a collar that is free around the stem, few in number, slightly ventricose, cream-color.

The stem is short, slender, equal, smooth, shining, black, whitish above.

The spores are globose, 3–4μ.

This species is very near M. rotula but it can be easily distinguished by the pale rufescent, distinctly sulcate pileus, and its growing on grass. I have frequently found it on the Chillicothe high school lawn.

MARASMIUS SICCUS. SCHW.

THE BELL-SHAPED MARASMIUS.

Plate XVII. Figure 110   Marasmius siccus.

Natural size. The cap ochraceous red, the disks somewhat darker, the stems shining and blackish-brown.

Figure 111   Marasmius siccus. Natural size. Caps deeply furrowed and pinkish.

This is a very beautiful plant found in the woods after a rain, growing from the leaves. They are found singly, but usually in groups.

The pileus is at first nearly conical, then campanulate, membranaceous, dry, smooth, furrows radiating from almost the center, growing larger as they approach the margin, ochraceous-red, the disk a little darker.

The gills are free or slightly attached, few, distant, broad, narrowed toward the stem, whitish.

The stem is hollow, tough, smooth, shining, blackish-brown, two to three inches long. The pileus is about a half inch broad.

The plant is quite common in our woods. I have not found it elsewhere. The plants in the photograph represent the pink form, which is not so common as the ochraceous-red. In the pink form the center of the cap and the apex of the stem is a delicate pink, which gives the plant a beautiful appearance.

Found from June to October. I have not tested it but have no doubt of its esculent qualities.

MARASMIUS FAGINEUS. MORGAN.

Fagineus means belonging to beech.

Pileus a little fleshy, convex then plane or depressed, at length somewhat repand, rugose-striate, reddish-pallid or alutaceous.

The gills are short-adnate, somewhat crisped, close, pale reddish.

The stem is short, hollow, pubescent, thickened upward, concolorous; the base somewhat tuberculose. Morgan, Myc. Flora M. V.

This plant is quite frequently found in our woods growing on the bark at the base of living beech trees. Its habitat, its reddish or alutaceous cap, and its paler gills will clearly identify the species.

MARASMIUS PERONATUS. FR.

THE MASKED MARASMIUS.

Figure 112 Marasmius peronatus. Natural size. Cap reddish-buff. Gills creamy or light reddish-brown.

Peronatus is from pero, a boot.

The pileus is reddish-buff, convex, slightly flattened at the top, quite wrinkled when old; diameter, at full expansion, between one and two inches, margin striate.

The gills are thin and crowded, creamy, becoming light reddish-brown, continuing down the stem by a short curve.

The stem is fibrous-stuffed, pale, densely clothed at the base with stiff yellowish hairs.

It grows in the woods, among dead leaves, from May till frost.

It is usually solitary yet is sometimes found in clusters. It has been eaten frequently without injury, but by most writers is branded poisonous. It is quite acrid, but that disappears in cooking. The dense yellow hairs at the base of the stem appear to constitute the distinguishing characteristic. Found from July to October.

MARASMIUS RAMEALIS. FR.

Figure 113 Marasmius ramealis. Natural size.

Ramealis means a branch or stick; so called because the plant is found growing on sticks, in open woods.

The pileus is very small, somewhat fleshy, plane or a trifle depressed, obtuse, not striate, slightly rugulose, opaque.

The gills are attached to the stem, somewhat distant, narrow, white.

The stem is about one inch long, stuffed, mealy, white, inclined to be rufescent at the base.

The spores are elliptical, $4 \times 2\mu$.

This is a very pretty plant, but easily overlooked. It is found on oak and beech branches, frequently in large groups. Figure 113 illustrates their mode of growth and will assist the collector in identifying the species. Not poisonous, but too small to gather. Found from July to October. The specimens in Figure 113 were found in Haynes' Hollow near Chillicothe and photographed by Dr. Kellerman.

MARASMIUS SACCHARINUS. BATSCH.

GRANULAR MARASMIUS. EDIBLE.

Saccharinus is from saccharum, sugar; it is so called because the white pileus looks very much like loaf sugar.

The pileus is entirely white, membranaceous, convex, somewhat papillate, smooth, sulcate and plicate.

The gills are broadly and firmly attached to the stem, narrow, thick, very distant, united by veins, whitish.

The stem is quite thin, thread-form, attenuated upward, at first flocculose, at length becoming smooth, inserted obliquely, reddish, pale at the apex. Spores, $5 \times 3\mu$.

Quite common in wet weather on dead oak limbs in woods. This plant differs from M. epiphyllus in its habitat, in the papillate form of its pileus and the stem's being flocculose, then smooth; also in that the gills are united in a reticulated manner. Common. July to October.

MARASMIUS EPIPHYLLUS. FR.

THE LEAF MARASMIUS. EDIBLE.

Epiphyllus means growing on leaves.

The pileus is white, membranaceous, nearly plane, at length umbilicate, smooth, wrinkled, plicate.

The gills are firmly attached to the stem, white, connected by veins, entire, distant, few.

The stem is rather horny, bay, minutely velvety, apex pale, inserted. The spores are $3 \times 2\mu$. This plant is abundant everywhere, on fallen leaves in woods during rainy weather. July to October.

MARASMIUS DELECTANS. MORGAN.

Figure 114   Marasmius delectans. Natural size. Caps white. Gills broad and distant.

Delectans means pleasing or delightful.

The pileus is subcoriaceous, convex, then expanded and depressed, glabrous, rugulose, white, changing in drying to pale alutacecus.

The gills are moderately broad, unequal, rather distant, trabeculate between, white, emarginate, adnexed; the spores are lance-oblong, hyaline, $7-9 \times 4\mu$.

The stem, arising from an abundant white-floccose mycelium, is long, slender, tapering slightly upward, smooth, brown and shining, white at the apex.

It is found growing on old leaves in woods. The plants in the figure were collected in the woods at Sugar Grove, Ohio, by R. A. Young, July 28, 1906, and photographed by Dr. Kellerman. Found from July to October.

MARASMIUS NIGRIPES. SCHW.

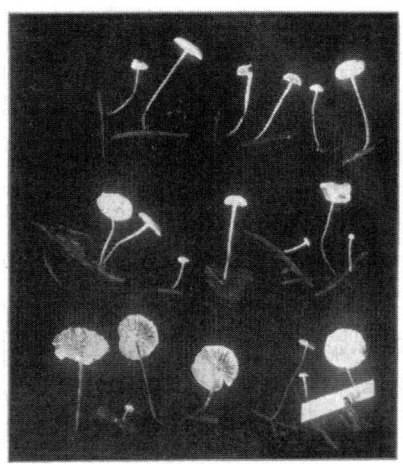

Figure 115   Marasmius nigripes. Natural size. Caps and gills white, stems black.

Nigripes means black foot, so called because the stems are black.

Tremmelloid. Pileus very thin, pure white, pruinose, rugulose-sulcate, convex then expanded.

The gills are pure white, unequal, some of them forked, adnate, the interstices venulose.

The stem is thickest at the apex, tapering downward, black, white-pruinose, the base insititious. Morgan.

It is found on old leaves, sticks, and old acorns and hickory-nuts. When dry, the stem loses its black color and the gills become flesh-color. It is quite common in thin and open woods. The spores are hyaline and stellate, 3–5-rayed. Found from July to October.

This is called Heliomyces nigripes by some authors.

PLEUROTUS. FR.

Pleurotus is from two Greek words meaning side and ear, alluding to its manner of growth on a log. This genus is very common everywhere in Ohio, and is easily determined by its eccentric, lateral, or even absent stem, but it must have white spores, and the characteristics of the Agaricini.

Pileus fleshy in the larger species and membranaceous in the smaller forms, but never becoming woody. Stem mostly lateral or wanting; when present, continuous with cap. Gills with sinus or broadly decurrent, toothed.

Grows in woods.

PLEUROTUS OSTREATUS. JACQ.

THE OYSTER MUSHROOM. EDIBLE.

Figure 116   Pleurotus ostreatus. Two-thirds natural size. Often growing very large.

Pileus two to six inches broad, soft, fleshy, convex, or slightly depressed behind, subordinate, often cespitosely imbricated, moist, smooth, margin involute; whitish, cinereous or brownish; flesh white, the whole surface shining and satiny when dry.

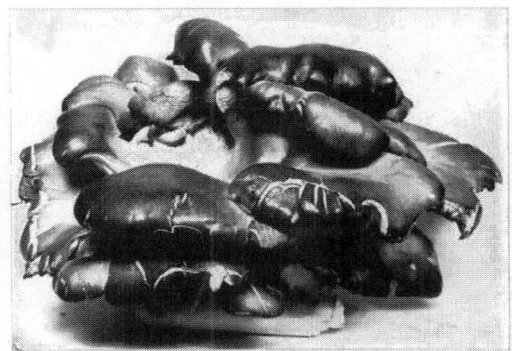

Plate XVIII. Figure 117   Pleurotus ostreatus.

Two-thirds natural size.

Gills broad, decurrent, subdistant, branching at the base, white or whitish. The stem when present is very short, firm, lateral, sometimes rough with stiff hair, hairy at the base. Spores oblong, white, .0003 to .0004 inch long, .00016 inch broad.

This is one of our most abundant mushrooms, and the easiest for the beginner to identify. In Figures 116 and 117, you will see the plant growing in imbricated form apparently without any stem. In Figure 118 is a variety that has a pronounced stem, showing how the stems grow together at the base, the slight grooving on the stems, also the decurrent gills. In most of these plants the stems are plainly lateral, but a few will appear to be central. It will be difficult to distinguish it from the Sapid mushroom and for table purposes there is little need to separate them. In Ohio the Oyster mushroom is very common everywhere. I have seen trees sixty to seventy feet high simply loaded with this mushroom. If one will locate a few logs or stumps upon which the Oyster mushroom grows, he can find there an abundant supply (when conditions are right for fungus growth) during the entire season. It is almost universally a favorite among mushroom eaters, but it must be carefully and thoroughly cooked. It grows very large and frequently in great masses. I have often found specimens whose caps were eight to ten inches broad. It is found from May to December.

Figure 118   Pleurotus ostreatus. One-half natural size, showing gills and stems.

PLEUROTUS SALIGNUS. FR.

THE WILLOW PLEUROTUS. EDIBLE.

Salignus, from salix, a willow. Pileus is compact, nearly halved, horizontal, at first cushion-shaped, even, then with the disk depressed, substrigose, white or fuliginous. The stem, eccentric or lateral, sometimes obsolete, short, white-tomentose. The gills are decurrent, somewhat branched, eroded, distinct at the base, nearly of the same color. Spores .00036 by .00015 inch. Fries.

I found this species near Bowling Green on willow stumps. About every ten days the stumps offered me a very excellent dish, better than any meat market could afford. September to November.

PLEUROTUS ULMARIUS. BULL.

THE ELM PLEUROTUS. EDIBLE.

Figure 119 Pleurotus ulmarius. One-third natural size.

Ulmarius, from ulmus, an elm. It takes its name from its habit of growing on elm trees and logs. It appears in the fall and may be found in company with the Oyster mushroom, late in December, frozen solid. This species is frequently seen on elm trees, both dead and alive, on live trees where they have been trimmed or injured in some way. It is often seen on elms in the cities, where the elm is a common shade tree. Its cap is large, thick and firm, smooth and broadly convex, sometimes pale yellow or buff. Frequently the epidermis in the center of the cap cracks, giving the surface a tessellated appearance as in Figure 119. The flesh is very white and quite compact. The gills are white or often becoming tawny at maturity, broad, rounded or notched, not closely placed, sometimes nearly decurrent. The stem is firm and solid, various in length, occasionally very short, inclined to be thick at the base and curved so that the plant will be upright, as will be seen in Figure 119.

The cap is from three to six inches broad. A specimen that measured over ten inches across the cap, was found some thirty feet high in a tree. While it was very large, it was quite tender and made several meals for two families. But this species is not limited entirely to the elm. I found it on hickory, about Chillicothe. There are a few elm logs along my rambles that afford me fine specimens with great regularity. Insects do not seem to infest it as they do the ostreatus and the sapidus. Sometimes, when the plant grows from the top of a log or the cut surface of a stump, the stem will be longer, straight, and in the center of the cap. This form is called by some authors var. verticalis.

For my own use I think the Elm mushroom, when properly prepared, very delicious. Like all tree mushrooms it should be eaten when young. It is easily dried and kept for winter use. Found from September to November.

PLEUROTUS PETALOIDES. BULL.

THE PETALOID PLEUROTUS. EDIBLE.

Figure 120   Pleurotus petaloides.

This species is so called from its likeness to the petals of a flower. Pileus fleshy, spathulate, entire; margin at first involute, finally fully expanded; villous, depressed. The stem is compressed and villous, often channelled, nearly erect. The gills are strongly decurrent, crowded, narrow, and white or whitish. Spores minutely globose, .0003 by .00015.

The plant varies very greatly in form and size. Its chief characteristic is the presence of numerous short white cystidia in the hymenium, which dot the surface of the hymenium, and under an ordinary pocket lens give to the gills a sort of fuzzy appearance. Frequently it will have the appearance of growing from the ground, but a careful examination will reveal a piece of wood of some kind, which serves as a host for the mycelium. I have found this plant but a few times, It seems to be quite rare in our state, especially in the southern part of the state. The plants in Figure 120 were photographed by Prof. G. D. Smith of Akron, Ohio.

PLEUROTUS SAPIDUS. KALCHB.

THE SAPID PLEUROTUS. EDIBLE.

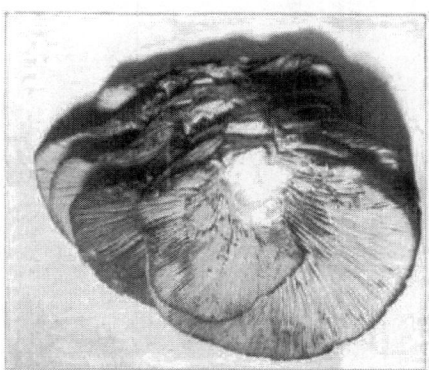

Figure 121 Pleurotus sapidus. One-third natural size, showing imbricated growth. Spores lilac.

Figure 122   Pleurotus sapidus.

Sapidus, savory. This plant grows in clusters whose stems are more or less united at the base as in Figure 121. The caps when densely crowded are often irregular. They are smooth and vary much in color, being whitish, ash-gray, brownish, yellowish-gray.

The flesh is thick and white. The gills are white or whitish, rather broad, running down on the stem, and slightly connected, at times, by oblique or transverse branches. The stem is generally short, solid, several usually springing from a thickened base, white or whitish and either laterally or eccentrically connected with the cap.

This plant is classed with the white-spored species, yet its spores, after a short exposure to the air, really exhibit a pale lilac tint. This can only be seen when the spores are in sufficient quantity and resting on a suitable surface.

The size of the plant varies, the cap being commonly from two to five inches long. It grows in woods and open places, on stumps and logs of various kinds. Its edible quality is quite as good as the Oyster mushroom. The only way by which it can be distinguished from the P. ostreatus is by its lilac-tinted spores. It is found from June to November.

Plate XX. Figure 123   Pleurotus sapidus.

PLEUROTUS SEROTINOIDES. PK.

THE YELLOWISH PLEUROTUS. EDIBLE.

Figure 124   Pleurotus serotinoides. One-third natural size.

Serotinoides, like serotinus, which means late-coming; from its appearing in the winter.

The pileus is fleshy, one to three inches broad, compact, convex or nearly plane, viscid when young and moist, half-kidney-shaped, roundish, solitary or crowded and imbricated, variously colored, dingy-yellow, reddish-brown, greenish-brown or olivaceous, the margin at first involute.

The gills are close, determinate, whitish or yellowish.

The stem is very short, lateral, thick, yellowish beneath, and minutely downy or scaly with blackish points.

The spores are minute, elliptical, .0002 inch long, .0001 inch broad.

There is probably no difference between this and P. serotinus, the European species. It is a beautiful plant. The color and size are quite variable. I found it on Ralston's Run and in Baird's woods on Frankfort Pike. It is found from September to January.

PLEUROTUS APPLICATUS. BATSCH.

LITTLE GRAY PLEUROTUS.

Figure 125   Pleurotus applicatus. Natural size.

Applicatus means lying upon or close to; so named from the sessile pileus. The pileus is one-third of an inch across, when young cup-shaped, dark cinereous, somewhat membranaceous, quite firm, resupinate, then reflexed, somewhat striate, slightly pruinose, villous at the base.

The gills are thick, broad in proportion to the size of the cap, distant, radiating, gray, the margin lighter, sometimes the gills are as dark as the pileus.

Sometimes it is attached only by the center of the pileus; sometimes, growing on the side of a shelving log, it is attached laterally. It is not as abundant as some other forms of Pleurotus. It differs from P. tremulus in absence of a distinct stem.

PLEUROTUS CYPHELLÆFORMIS. BERK.

Cyphellæformis means shaped like the hollows of the ears. The pileus is cup-shaped, pendulous, downy or mealy, upper layer gelatinous, gray, very minutely hairy, especially at the base, margin paler.

The gills are narrow, rather distant, pure white, alternate ones being shorter. These are very small plants, found only in damp places on dead herbaceous plants. They resemble a Cyphella griseo-pallida in habit.

PLEUROTUS ABSCONDENS. PK.

Figure 126 Pleurotus abscondens. Entire plant white.

Abscondens means keeping out of view. It is so called because it persists in growing in places where it is hidden from sight.

The pileus is often two and a half inches broad, delicate-white, strong stringent odor, usually pruinose, margin slightly incurved.

The gills are attached to the stem, rather crowded, very white, somewhat narrow.

The stem is short, solid, pruinose, usually lateral, and curved.

The plant usually grows in hollow stumps or logs, and in this case the stem is always lateral and the plant grows very much as does the P. ostreatus, except that they are not imbricated. Occasionally the plant is found on the bottom of a hollow log and in that case the cap is central and considerably depressed in the center. I have never seen it growing except in a hollow stump or log. Its manner of growth and its delicate shape of white will serve to identify it. It is found from August to November.

PLEUROTUS CIRCINATUS. FR.

Circinatus means to make round, referring to the shape of the pileus.

The pileus is two to three inches broad, white, plane, orbicular, convex at first, even, covered over with silky-pruinose lustre.

The gills are adnate-decurrent, rather crowded, quite broad, white.

The stem is equal, smooth, one to two inches long, stuffed, central or slightly eccentric, rooted at the base.

The form of these plants is quite constant and the round white caps will at first suggest a Collybia. The white gills and its decurrent form will distinguish it from P. lignatilis. It makes quite a delicious dish when well cooked. I found some beautiful specimens on a decayed beech log in Poke Hollow. Found in September and October.

LACTARIUS. FR.

Lactarius means pertaining to milk. There is one feature of this genus that should easily mark it, the presence of milky or colored juice which exudes from a wound or a broken place on a fresh plant. This feature alone is sufficient to distinguish the genus but there are other points that serve to make the determination more certain.

The flesh, although it seems quite solid and firm, is very brittle. The fracture is always even, clean cut, and not ragged as in more fibrous substances.

The plants are fleshy and stout, and in this particular resemble the Clitocybes, but the brittleness of the flesh, milky juice, and the marking of the cap, will easily distinguish them.

Many species have a very acrid or peppery flavor. If a person tastes one when raw, he will not soon forget it. This acridity is usually lost in cooking.

The pileus in all species is fleshy, becoming more or less depressed, margin at first involute, often marked with concentric zones.

The stem is stout, often hollow when old, confluent with the cap.

The gills are usually unequal, edge acute, decurrent or adnate, milky; in nearly all the species the milk is white, changing to a sulphur yellow, red, or violet, on exposure to the air.

LACTARIUS TORMINOSUS. FR.

THE WOOLLY LACTARIUS. POISONOUS.

Figure 127 Lactarius torminosus. Three-fourths natural size. Caps yellowish-red or ochraceous tinged with red, margin incurved.

Torminosus, full of grips, causing colic. The pileus is two to four inches broad, convex, then depressed, smooth, or nearly so, except the involute margin which is more or less shaggy, somewhat zoned, viscid when young and moist, yellowish-red or pale ochraceous, tinged with red.

The gills are thin, close, rather narrow, nearly of the same color as the pileus, but yellower and paler, slightly forked, subdecurrent.

The stem is one to two inches long, paler than the cap, equal or slightly tapering downward, stuffed or hollow, sometimes spotted, clothed with a very minute adpressed down.

The milk is white and very acrid. The spores are echinulate, subglobose, $9–10 \times 7–8\mu$.

This differs from L. cilicioides in its zoned pileus and white milk. Most authorities speak of it as dangerous. Captain McIlvaine speaks of the Russians as preserving it in salt and eating it seasoned with oil and vinegar. They grow in the woods, open places, and in fields. The specimens in Figure 127 were found in Michigan and photographed by Dr. Fischer.

LACTARIUS PIPERATUS. FR.

THE PEPPERY LACTARIUS. EDIBLE.

Figure 128 Lactarius piperatus. One-third natural size.

Piperatus—having a peppery taste. The pileus is creamy-white, fleshy, firm, convex, then expanded, depressed in the center, dry, never viscid, and quite broad.

The gills are creamy-white, narrow, close, unequal, forked, decurrent, adnate, exuding a milky juice when bruised, milky-white, very acrid.

The stem is creamy white, short, thick, solid, smooth, rounded at the end, slightly tapering at the base. Spores generally with an apiculus, .0002 by .00024 inch.

The plant is found in all parts of Ohio, but most people are afraid of it on account of its very peppery taste. Although it can be eaten without harm, it will never prove a favorite.

It is found in open woods from July to October. In its season is one of the very common plants in all of our woods.

LACTARIUS PERGAMENUS. FR.

Pergamenus is from pergamena, parchment. The pileus is convex, then expanded, plane, depressed, wavy, wrinkled, without zones, often repand, smooth, white.

The gills are adnate, very narrow, tinged with straw-color, often white, branched, much crowded, horizontal.

The stem is smooth, stuffed, discolored, not long. The milk is white and acrid. Spores, 8×6. It differs from L. piperatus in its crowded, narrow gills and longer stem. Found in woods from August to October.

LACTARIUS DECEPTIVUS. PK.

DECEIVING LACTARIUS. EDIBLE

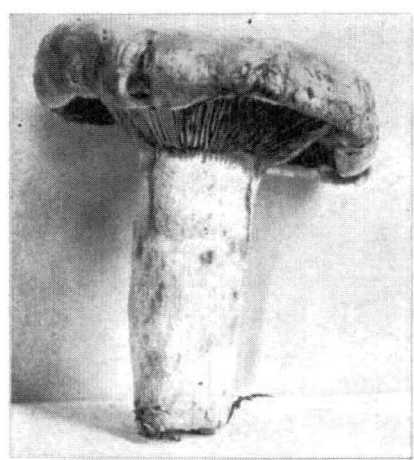

Figure 129   Lactarius deceptivus.

Deceptivus means deceiving.

The pileus is three to five inches broad, compact, at first convex, and umbilicate, then expanded and centrally depressed or subinfundibuliform, obsoletely tomentose or glabrous except on the margin, white or whitish, often varied with yellowish or sordid strains, the margin at first involute and clothed with a dense, soft cottony tomentum, then spreading or elevated and more or less fibrillose.

The gills are rather broad, distant or subdistant, adnate or decurrent, some of them, forked, whitish, becoming cream-colored.

The stem is one to three inches long, equal or narrowed downward, solid, pruinose-pubescent, white. Spores are white, 9–12.7μ. Milk white, taste acrid.

This plant delights in woods and open groves, especially under coniferous trees. It is a large, meaty, acrid white species, with a thick, soft, cottony tomentum on the margin of the pileus of the young plant.

The specimen photographed was sent me from Massachusetts by Mrs. Blackford. It grows in July, August and September. Its sharp acridity is lost in cooking, but like all acrid Lactarius it is coarse and not very good.

LACTARIUS INDIGO. (SCHW.) FR.

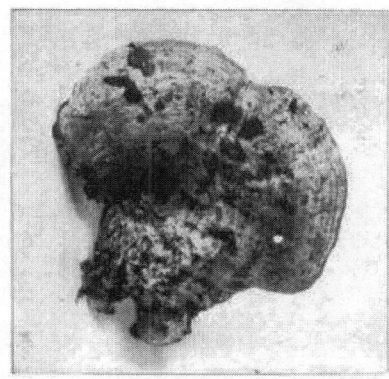

Figure 130   Lactarius indigo. One-third natural size. Entire plant indigo blue.

Figure 131   Lactarius indigo. One-third natural size, showing gills.

This is one of our most striking plants. No one can fail to recognize it, because of the deep indigo blue that pervades the whole plant. I have found it in only one place, near what is known as the Lone-Tree Hill near Chillicothe. I have found it there on several different occasions.

The pileus is from three to five inches broad, the very young plants seem to be umbilicate with the margin strongly incurved, then depressed or funnel-shaped; as the plant ages the margin is elevated and sometimes waved. The entire plant is indigo blue, and the surface of the cap has a silvery-gray appearance through which the indigo color is seen. The surface of the cap is marked with a series of concentric zones of darker shade, as will be seen in Figure 130 especially on the margin; sometimes spotted, becoming paler and less distinctly zonate with age or in drying.

The gills are crowded, indigo blue, becoming yellowish and sometimes greenish, with age.

The stem is one to two inches long, short, nearly equal, hollow, often spotted with blue, colored like the pileus.

It is edible but rather coarse. Found in open woods July and August.

LACTARIUS REGALIS. PK.

Figure 132   Lactarius regalis. Natural size. Caps white, tinged with yellow.

Regalis means regal; so named from its large size. The pileus is four to six inches broad, convex, deeply depressed in the center; viscid when moist; often corrugated on the margin; white, tinged with yellow.

The gills are close, decurrent, whitish, some of them forked at the base.

The stem is two to three inches long and one inch thick, short, equal, hollow. The taste is acrid and the milk sparse, white, quickly changing to sulphur-yellow. The spores are .0003 of an inch in diameter. Peck.

This is frequently a very large plant, resembling in appearance L. piperatus but easily recognized because of its viscid cap and its spare milk changing to yellow, as in L. chrysorrhæus. It grows on the ground in the woods, in August and September. I find it here chiefly on the hillsides. The specimens in Figure 132 were found in Michigan and photographed by Dr. Fischer.

LACTARIUS SCROBICULATUS. FR.

THE SPOTTED-STEMMED LACTARIUS.

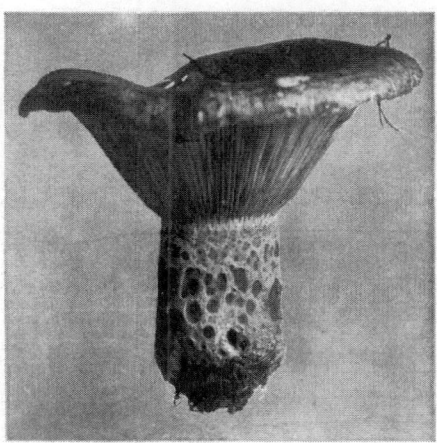

Figure 133 Lactarius scrobiculatus. Natural size. Caps reddish-yellow, zoned. Margin very much incurved, stem pitted.

Scrobiculatus is from scrobis, a trench, and ferro, to bear, referring to the pitted condition of the stem. The pileus is convex, centrally depressed, more or less zoned, reddish-yellow, viscid, the margin very much incurved, downy.

The gills are adnate, or slightly decurrent, whitish, and often very much curled, because of the incurved condition of the cap at first.

The stem is equal, stuffed, adorned often with pits of a darker color.

The spores are white, juice white, then yellowish.

The plant is very acrid to the taste, and solid. Too hot to be eaten. I have found it only a few times on the hills of Huntington township, near Chillicothe. The yellowish hue and markedly incurved margin will identify the plant. Found in August and September.

LACTARIUS TRIVIALIS. FR.

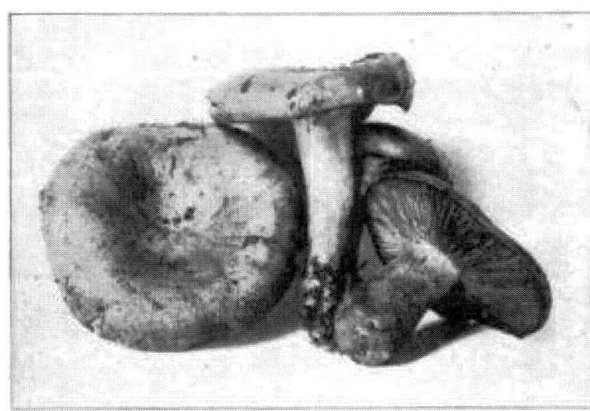

Figure 134 Lactarius trivialis. One-half natural size. Caps light tan with a pinkish hue. Very acrid.

Trivialis means common.

The pileus is three to four inches broad, usually damp or watery, sometimes quite viscid, shining when dry, convex, then expanded, depressed in the center, margin at first incurved, even, smooth; warm, soft tan, rather light, and sometimes a very slight pinkish hue prevails. The flesh is solid and persistent.

The gills are rather crowded, slightly decurrent, at first whitish, then a light yellow, many not reaching to the stem, none forked. The stem is from three to four inches long, of same color as the pileus, often a much lighter shade; tapering from the cap to the base, smooth, stuffed, and finally hollow. The plant is quite full of milk, white at first, then turning yellowish.

The plant is very acrid and peppery. It is quite plentiful along the streams of Ross county, Ohio. It is not poisonous, but it seems too hot to eat. It is found after rains from July to October, in mixed woods where it is damp.

LACTARIUS INSULSUS. FR.

Figure 135 Lactarius insulsus. One-third natural size. Caps yellowish or straw color. Very acrid.

Insulsus, insipid or tasteless. This is a very attractive plant. Quite solid and maintains its form for several days; The pileus is two to four inches broad, convex, depressed in the center, then funnel-shaped, smooth, viscid when moist, more or less zoned, the zones much narrower than L. scrobiculatus, yellowish or straw-color, margin slightly incurved and naked.

The gills are thin, rather crowded, adnate and sometimes decurrent, some of them forked at the base, whitish or pallid. Spores subglobose, rough, $10 \times 8\mu$.

The stem is one to two inches long, equal or slightly tapering downward, stuffed, whitish, generally spotted. Milk, white.

Most authorities class this as an edible plant, but it is so hot and the flesh so solid that I have never tried it. I found two plants which fully answered the description of the European plants. The zones were orange-yellow and brick-red. I have visited the place many times since, but have never been able to find another. It is not an abundant plant with us. Found from July to October, in open woods.

LACTARIUS LIGNYOTUS. FR.

THE SOOTY LACTARIUS. EDIBLE.

Plate XXI. Figure, 136   Lactarius lignyotus.

Natural size. Caps a sooty umber. Flesh mild to the taste.

Lignyotus is from lignum, wood. The pileus is one to four inches in diameter, fleshy, convex, then expanded, sometimes slightly umbonate, often in age slightly depressed, smooth or often wrinkled, pruinosely velvety, sooty umber, the margin in the old plants wavy and distinctly plaited; the flesh white and mild to the taste.

The gills are attached to the stem; unequal; snow-white or yellowish-white, slowly changing to a pinkish-red or salmon color when bruised; distant in old plants.

The stem is one to three inches long, equal, abruptly constricted at the apex, smooth, stuffed, of the same color as the pileus. Milk white, taste mild or tardily acrid. The spores are globose, yellowish, 9–11.3μ.

This is called the Sooty Lactarius and is very easily identified. It will be frequently found associated with the Smoky Lactarius which it greatly resembles. It seems to delight in wet swampy woods. It is said to be one of the best of the Lactarii. The specimens in Figure 136 were collected at Sandusky, Ohio, and photographed by Dr. Kellerman.

LACTARIUS CINEREUS. PK.

Figure 137 Lactarius cinereus.

Cinereus is from cineres, ashes; so called from the color of the plant.

The pileus is one to two and a half inches broad, zoneless, somewhat viscid, floccose-scaly, depressed in the center, margin thin, even, flesh thin and white, mild to the taste, ashy-gray.

The gills are adnate, rather close, sometimes forked (usually near the stem), uneven, white or creamy-white, milk white, not plentiful.

The stem is two to three inches long, tapering upward, loosely stuffed, finally hollow, often floccose at the base.

This plant is quite common from September to November, growing in damp weather on leaves in mixed woods. It has a mild taste. While I have not eaten it I have no doubt of its edibility. The color of the pileus is sometimes quite dark.

LACTARIUS GRISEUS. PK.

GRAY LACTARIUS.

Figure 138   Lactarius griseus.

Griseus means gray.

The pileus is thin, nearly plane, broadly umbilicate or centrally depressed, sometimes infundibuliform, generally with a small umbo or papilla, minutely squamulose tomentose, gray or brownish-gray, becoming paler with age.

The gills are thin, close, adnate, or slightly decurrent, whitish or yellowish.

The stem is slender, equal or slightly tapering upward, rather fragile; stuffed or hollow; generally villose or tomentose at the base; paler than, or colored like, the pileus.

The spores are .0003 to .00035 inch; milk white, taste subacrid. Pileus is 6 to 18 lines broad, stem 1 to 2 inches long, 1 to 3 lines thick. Peck.

It resembles L. mammosus and L. cinereus. It differs from the former in not having ferruginous gills and pubescent stems, and from the latter by its smaller size, its densely pubescent pileus, and its habitat. It grows on mossy logs or in mossy swamps. The base of one of the plants in Figure 138 is covered with the moss in which they grew. These plants were found in Purgatory Swamp, near Boston, by Mrs. Blackford. They grow from July to September.

LACTARIUS DISTANS. PK.

THE DISTANT-GILLED LACTARIUS. EDIBLE.

Distans means distant, so called because the gills are very wide apart.

The pileus is firm, broadly convex or nearly plane, umbilicate or slightly depressed in the center; with a minute, velvety pruinosity; yellowish-tawny or brownish-orange.

The gills are rather broad, distant, adnate or slightly decurrent, white or creamy yellow, interspaces veined; milk white, mild.

The stem is short, equal or tapering downward, solid, pruinose, colored like the pileus.

The spores are subglobose, 9–11μ broad. Peck, N. Y. Report, 52.

I frequently mistake this plant for L. volemus when seen growing in the ground, but the widely separated gills distinguish the plant as soon as it is gathered. The stem is short and round, tapering downward, solid, colored like the pileus. The milk is both white and mild. I find it on nearly every wooded hillside about Chillicothe. It is found from July to September.

LACTARIUS ATROVIRIDUS. PK.

THE DARK-GREEN LACTARIUS.

Figure 139 Lactarius atroviridus. Cap and stem dark green. Cap depressed in center. Gills white.

Atroviridus is from ater, black; viridus, green; so called from the color of the cap and the stem of the plant.

The pileus is convex, plane, then depressed in the center, with an adherent pellicle, greenish with darker scales, margin involute.

The gills are slightly decurrent, whitish, broad, distant; milk white but not copious as in many of the Lactarii.

The stem is quite short, tapering downward, dark green, scaly.

The stem is so short that the cap seems to be right on the ground, hence it is very easily overlooked. It is found only occasionally on mossy hillsides, where there are not too many leaves.

The plant in Figure 139 was found in Haynes' Hollow, near Chillicothe. I have found the plant on top of Mt. Logan. It is found from July to October. I do not know of its edibility. All specimens that I have found I have sent to my Mycological friends. It should be tasted with caution.

LACTARIUS SUBDULCIS. FR.

THE SWEET LACTARIUS. EDIBLE.

Figure 140   Lactarius subdulcis.

Subdulcis means almost sweet, or sweetish.

The pileus is two to three inches broad, rather thin, papillate, convex, then depressed, smooth, even, zoneless, cinnamon-red or tawny-red, margin sometimes wavy.

The gills are rather narrow, thin, close, whitish, often reddish or tinged with red. Spores, 9–10μ.

The stem is stuffed, then hollow, equal, slightly tapering upward, slender, smooth, sometimes villous at the base. The milk is white, sometimes rather acrid and unpleasant to the taste when raw. It needs to be cooked a long time to make it good.

It is likely to be found anywhere, but it does best in damp places. The plants found with us all seem to have red or cinnamon-red gills, especially before the spores begin to fall. They are found growing on the ground, among leaves, or on well-rotted wood and sometimes on the bare ground. Found from July to November.

LACTARIUS SERIFLUUS. FR.

Serifluus means flowing with serum, the watery part of milk.

The pileus is fleshy, depressed in the center, dry, smooth, not zoned, tawny-brown, margin thin, incurved.

The gills are crowded, light-brown, or yellowish, milk scanty and watery.

The stem is solid, equal, paler than the pileus. Spores, 7–8μ.

It differs from L. subdulcis in having a solid stem and perhaps a shade darker color. Found in woods, July to November.

LACTARIUS CORRUGIS. PK.

THE WRINKLED LACTARIUS. EDIBLE.

Figure 141 Lactarius corrugis. Caps wrinkled, tawny-brown. Gills orange-brown.

Corrugis means wrinkled.

The pileus is convex, plane, expanded, slightly depressed in the center; surface of the cap wrinkled, dry, bay-brown; margin at first involute.

The gills are adnexed, broad, yellowish or brownish-yellow, growing paler with age. The stem is rather short, equal, solid, pruinose, of the same color as the pileus. The spores are subglobose, 10–13μ.

This species looks very much like L. volemus, and its only essential difference is in the wrinkled form, and color of the pileus. The milk when dry is very sticky and becomes rather black. It has just a touch of acridity.

Any one determining this species will not fail to note the number of brown cystidia or setæ, in the hymenium, which project above the surface of the gills. They are so numerous and so near the edge of the gills that they give these a downy appearance. The quality of this species is even better than L. volemus, though it is not as abundant here as the latter. Found in thin woods from August to September. The photograph, Figure 141, was made by Prof. H. C. Beardslee.

LACTARIUS VOLEMUS. FR.

THE ORANGE-BROWN LACTARIUS. EDIBLE.

Figure 142   Lactarius volemus. Natural size. Caps golden-tawny. Milk copious, as will be seen where the plant has been pricked.

Volemus from volema pira, a kind of a pear, so called from the shape of the stem. The pileus is broad, flesh thick, compact, rigid, plane, then expanded, obtuse, dry, golden-tawny, at length somewhat wrinkly.

The gills are crowded, adnate or slightly decurrent, white, then yellowish; milk copious, sweet.

The stem is solid, hard, blunt, generally curved like a pear-stem; its color is that of the pileus but a shade lighter. Spores globose, white.

The milk in this species is very abundant and rather pleasant to the taste. It becomes quite sticky as it dries on your hands. This plant has a good record among mushroom eaters, both in this country and Europe.

There is no danger of mistaking it. The plants grow in damp woods from July to September. They are found singly or in patches. They were found quite plentifully about Salem, Ohio, and also about Chillicothe.

LACTARIUS DELICIOSUS. FR.

THE DELICIOUS LACTARIUS. EDIBLE.

Figure 143   Lactarius deliciosus. One-third natural size. Caps light reddish-yellow. Milk orange color.

Deliciosus, delicious. The pileus is three to five inches broad; color varying from yellow to dull orange or even brownish-yellow with mottled concentric zones of deeper color, especially in younger plants, sometimes a light reddish-yellow, without apparent zones (as is the case of those in Figure 143); convex, when expanded becoming very much depressed; funnel-shaped; smooth, moist, some times irregular, wavy; flesh brittle, creamy, more or less stained with orange.

The gills are slightly decurrent in the depressed specimens, somewhat crowded, forked at the stem, short ones beginning at the margin; when bruised exuding a copious supply of milky juice of an orange color; a pale tan-color, turning green in age or in drying. Spores are echinulate, $9–10×7–8\mu$.

The stem is two to three inches or more, equal, smooth, hollow, slightly pruinose, paler than the cap, occasionally spotted with orange, tinged with green in old plants.

The taste of the raw plant is slightly peppery. It grows in damp woods and is sometimes quite common. Its name suggests the estimation in which it is held by all who have eaten it. Like all Lactarii it must be well cooked. The specimens in Figure 143 were gathered on Cemetery Hill close to the pine trees and in company with Boletus Americanus. Found from July to November. I found the plant in a more typical form about Salem, Ohio.

LACTARIUS UVIDUS. FR.

Figure 144   Lactarius uvidus.

Uvidus is from uva, grape, so called because when exposed to the air changes to the color of a grape.

The pileus is two to four inches broad, flesh rather thin, convex, sometimes slightly umbonate, then depressed in the center, not zoned, viscid, dingy pale ochraceous-tan, margin at first involute, naked, milk mild at first then becoming acrid, white changing to lilac.

The gills are thin, slightly decurrent, crowded, shorter ones very obtuse and truncate behind, connected by veins, white, when wounded becoming lilac.

The stem is soon hollow, two to three inches long, viscid, pallid.

The spores are round, 10μ.

Not only the milk changes to a lilac when cut, but the flesh itself. They are found in damp woods during August and September. The plants in Figure 144 were found near Boston, by Mrs. Blackford. These plants grew in Purgatory Swamp. The Sphagnum moss will be seen at the base of the upright plant.

LACTARIUS CHRYSORRHEUS. FR.

YELLOW-JUICED LACTARIUS.

Chrysorrheus from two Greek words; chrysos, yellow or golden; reo, I flow, because the juice soon turns to a golden yellow.

The pileus is rather fleshy, depressed, then funnel-shaped, yellowish-flesh colored, marked with dark zones or spots.

The stem is stuffed, then hollow, equal, or tapering below, paler than the pileus, sometimes pitted.

The gills are decurrent, thin, crowded, yellowish, milk white, then golden-yellow, very acrid.

The milk is white, quite acrid, has a peculiar taste, and changes at once on exposure to a beautiful yellow. This is a common species about Salem, Ohio, and is quite variable in size. Found in woods and groves from July to October. I do not know whether its edible quality has ever been tested. When I found it some years ago I had less faith in mushrooms than I have now.

LACTARIUS VELLEREUS. FR.

THE WOOLY-WHITE LACTARIUS. EDIBLE.

Vellereus from vellus, a fleece. The pileus is white, compact, fleshy, depressed or convex, tomentose, zoneless, margin at first involute, milk white and acrid.

The gills are white or whitish, distant, forked, adnate or decurrent, connected by veins, bow-shaped, milk scanty.

The stem is solid, blunt, pubescent, white, tapering downward. Spores white and nearly smooth, .00019 by .00034 inch.

This species is quite common; and though very acrid to the taste, this acridity is entirely lost in cooking. It will be readily known by the downy covering of the cap. Found in thin woods and wood margins. July to October.

RUSSULA. PERS.

Russula, red or reddish. The beginner will have little difficulty in determining this genus. There is such a strong family likeness that, finding one, he will say at once it is a Russula. The contour of the cap, the brittleness of its flesh and of its stem, the fragile gills, and the failure of any part of the plant to exude a milky or colored juice, the many gay colors—will all help in determining the genus.

Many species of Russula strongly resemble those of the genus Lactarius, in size, shape, and texture. The spores, too, are quite similar, but the absence of the milky juice will mark the difference at once.

The cap may be red, purple, violet, pink, blue, yellow, or green. The colored zones often seen in the Lactarii do not appear here. The beginner will possibly find trouble in identifying species, because of variation of size and color. The spores are white to very pale yellow, generally spiny. The pileus is fleshy, convex, then expanded, and at length depressed. The stem is brittle, stout, and smooth, generally spongy within, and confluent with the cap. The gills are milkless, with acute edge, and very tender.

Captain McIlvaine, in his very valuable book, One Thousand American Fungi, says: "To this genus authors have done special injustice; there is not a single species among them known to be poisonous, and where they are not too strong of cherry bark and other highly flavored substances, they are all edible; most of them favorites." I can testify to the fact that many of them are favorites, though a few are very peppery and it requires some courage to attack them.

They are all found on the ground in open woods, from early summer to late fall.

RUSSULA DELICA. FR.

THE WEANED RUSSULA. EDIBLE.

Delica means weaned, so called because, though it resembles Lactarius vellereus in appearance, it is void of milk.

The pileus is quite large, fleshy, firm, depressed, even, shining, margin involute, smooth, not striated.

The gills are decurrent, thin, distant, unequal, white.

The stem is solid, compact, white, short.

Specimens will be found that resemble Lactarius piperatus and L. vellereus, but they may be easily distinguished because they have no milk in their gills and the taste is mild. They are not equal to most of the Russulas. Found in woods from August to October.

RUSSULA ADUSTA. PERS.

THE SMOKY RUSSULA. EDIBLE.

Figure 145   Russula adusta.

Adusta means burned.

The pileus is fuliginous, cinereous, flesh compact, margin even and inflexed, depressed in the center.

The gills are attached to the stem, decurrent, thin, crowded, unequal, white, not reddening when bruised.

The stem is obese, solid, of the same color as the pileus, not turning red when bruised.

The plant resembles R. nigricans, but can readily be distinguished from it because of the thin, crowded gills and failure to turn red when cut or bruised. The spores are subglobose, almost smooth, 8–9μ; no cystidia. It is found in the woods during August and September. Edible but not first class. It is a plant very widely distributed.

RUSSULA NIGRICANS. FR.

Figure 146   Russula nigricans.

Nigricans means blackish.

The pileus is two to four inches broad, dark grayish-brown, black with advancing age, fleshy, compact, flesh turning red when bruised or convex, flattened, then depressed, at length funnel-shaped, margin entire, without striate, margin at first incurved, young specimens are slightly viscid when moist, even, without a separable pellicle; whitish at first, soon sooty olive, at length becoming broken up into scales and black; flesh firm and white, becoming reddish when broken.

The gills are rounded behind, slightly adnexed, thick, distant, broad, unequal, the shorter ones sometimes very scanty, forked, reddening when touched.

The stem is rather short, thick, solid, equal, pallid when young, then black. The spores are subglobose, rough, 8–9μ.

The plant is quite compact, inodorous, becoming entirely black with age. It is easily distinguished from R. adusta by the flesh becoming reddish when bruised, and by the much thicker, and more distant gills. It is very close to R. densifolia but differs from it in that its gills are more distant and because of its mild taste.

I am pleased to present to my readers, in Figure 146, a photograph of a plant which grew in Sweden in the locality where Prof. Fries did his great work in fungal study and research. It is a typical specimen of this species. It was gathered and photographed by Mr. C. G. Lloyd.

It is found from June to October. Not poisonous, but not good.

RUSSULA FŒTENS. FR.

THE FETID RUSSULA. NOT EDIBLE.

Figure 147   Russula fœtens.

Fœtens means stinking.

The pileus is four to six inches broad, dirty white or yellowish; flesh thin; at first hemispherical, then expanded, almost plane, often depressed in the center; covered with a pellicle which is adnate; viscid in wet weather; widely striate-tuberculate on the margin, which is at first incurved.

The gills are adnexed, connected by veins, crowded, irregular, many forked, rather broad, whitish, becoming dingy when bruised, exuding watery drops at first.

The stem is stout, stuffed, then hollow, concolor, two to four inches long. The spores are small, echinulate, almost round.

I have found the plants very generally diffused over the state. It is very coarse and uninviting. Its smell and taste are bad. Found from July to October. These plants are widely distributed and usually rather abundant.

RUSSULA ALUTACEA. FR.

THE TAN-COLORED RUSSULA. EDIBLE.

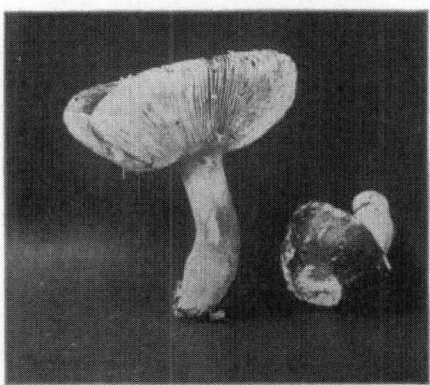

Figure 148   Russula alutacea. Two-thirds natural size. Caps flesh color. Gills broad and yellowish.

Alutacea, tanned leather. The pileus is flesh-color, sometimes red; flesh white; bell-shaped, then convex; expanded, with a viscid covering, growing pale; slightly depressed; even; margin inclined to be thin, striate.

The gills are broad, ventricose, free, thick, somewhat distant, equal, yellow, then ochraceous.

The stem is stout, solid, even; white, though parts of the stem are red, sometimes purple; wrinkled lengthwise; spongy. The spores are yellow.

The taste is mild and pleasant when young, but quite acrid when old. Alutacea will be known mostly by its mild taste, broad, and yellow gills. It is quite common, but does not grow in groups. It is sweet and nutty.

From July to October.

RUSSULA OCHROPHYLLA. PK.

OCHREY GILLED RUSSULA. EDIBLE.

Ochrophylla is from two Greek words meaning ochre and leaf, because of its ochre-colored gills.

The pileus is two to four inches broad, firm, convex, becoming nearly plane or slightly depressed in the center; even, or rarely very slightly striate on the margin when old; purple or dark purplish-red; flesh white, purplish under the adnate cuticle; taste mild.

The gills are entire, a few of them forked at the base, subdistant, adnate at first yellowish, becoming bright, ochraceous-buff when mature and dusted by the spores, the interspaces somewhat venose.

The stem is equal or nearly so, solid or spongy within, reddish or rosy tinted, paler than the pileus. The spores are bright, ochraceous-buff, globose, verruculose, .0004 of an inch broad. Peck.

This is one of the easiest Russulas to determine because of its purple or purplish-red cap, entire gills, at first yellowish, then a bright, ochraceous-buff when mature. The taste is mild and the flavor fairly good.

There is also a plant which has a purplish cap and a white stem, called Russula ochrophylla albipes. Pk. It quite agrees in its edible qualities with the former.

R. ochrophylla is found in the woods, especially under oak trees, in July and August.

RUSSULA LEPIDA. FR.

THE NEAT RUSSULA. EDIBLE.

Figure 149 Russula lepida. Two-thirds natural size. Caps, purplish-red, with more
or less brown.

Lepida, from lepidus, neat.

The pileus firm, solid; varying in color from bright red to dull, subdued purplish with a distinct brown; compact; convex, then depressed, dry unpolished; margin even, sometimes cracked and scaly, not striated.

The gills are white, broad, principally even, occasionally forked, very brittle, rounded, somewhat crowded, connected by veins, sometimes red on the edge, especially near the margin.

The stem is solid, white, usually stained and streaked with pink, compact, even.

The surface is dull, as with a fine dust or plum-like bloom, and thus without polish. Often times the surface will appear almost velvety. The tints of the flesh and the gills will be found uniform. The plant when raw is sweet and nut-like to the taste. This is a beautiful species, the color being averaged under the general hue of dark, subdued red, inclining to maroon. It is simply delicious when properly cooked. Found in woods from July to September.

RUSSULA CYANOXANTHA. FR.

THE BLUE AND YELLOW RUSSULA. EDIBLE.

Cyanoxantha, from two Greek words, blue and yellow, referring to color of the plant.

The pileus is quite variable as to color, ranging from lilac or purplish to greenish; disk yellowish, margin bluish or livid-purple; convex, then plane, depressed in center; margin faintly striate, sometimes wrinkled.

The gills are rounded behind, connected by veins, forked, white, slightly crowded.

The stem is solid, spongy, stuffed, hollow when old, equal, smooth and white.

The color of the cap is quite variable but the peculiar combination of color will assist the student in distinguishing it. It is a beautiful plant and one of the best of the Russulas to eat. The mushroom-eater counts himself lucky indeed when he can find a basketful of this species after "the joiner squirrel" has satisfied his love of this special good thing. It is quite common in woods from August to October.

RUSSULA VESCA. FR.

THE EDIBLE RUSSULA. EDIBLE.

Vesca from vesco, to feed. The pileus is from two to three inches broad; red-flesh-color, disk darker; fleshy; firm; convex, with a slight depression in the center, then funnel-shaped; slightly wrinkled; margin even, or remotely striate.

Gills adnate, rather crowded, unequal, forked, and white.

The stem is firm, solid, sometimes peculiarly reticulated, tapering at the base. The spores are globose, spiny, and white. I frequently found it near Salem, O., in thin chestnut woods and in pastures under such trees. A mushroom lover will be amply paid for the long tramps if he finds a basket full of these dainties. It is mild and sweet when raw. It is found in thin woods and in wood margins, sometimes under trees in pastures, from August to October.

RUSSULA VIRESCENS. FR.

THE GREEN RUSSULA. EDIBLE.

Figure 150 Russula virescens. Two-thirds natural size. Caps pale-green. Gills white.

Virescens, being green. The Pileus is grayish-green; at first globose, then expanded, convex, at last depressed at the center; firm, adorned with flaky greenish or yellow patches, produced by the cracking of the skin; two to four inches broad, margin striate, often white.

The gills are white, moderately close, free or nearly so, narrow as they approach the stem, some being forked, others not; very brittle, breaking to pieces at the slightest touch.

The stem is shorter than the diameter of the cap, smooth, white, and solid or spongy. The spores are white, rough, and nearly globose.

This plant is especially sweet and nutty to the taste when young and unwilted. All Russulas should be eaten when fresh. I have found the plant over the state quite generally. It is a prime favorite with the squirrels. You will often find them half eaten by these little nibblers. Found in open woods from July to September. It is one of the best mushrooms to eat and one that is very easily identified. It is quite common about Chillicothe, Ohio. Its mouldy color is not as prepossessing as the brighter hues of many far less delicious fungi, but it stands the test of use.

RUSSULA VARIATA. BAN.

VARIABLE RUSSULA. EDIBLE.

Pileus is firm, convex becoming centrally depressed or somewhat funnel-form, viscid, even on the thin margin, reddish-purple, often variegated with green, pea-green sometimes varied with purple, flesh white, taste acrid or tardily acrid.

The gills are thin, narrow, close, often forked, tapering toward each end, adnate or slightly decurrent, white.

The stem is equal or nearly so, solid, sometimes cavernous, white. The spores are white, subglobose, .0003 to .0004 of an inch long, .0003 broad. Peck, Rep. State Bot., 1905.

This plant grows in open beech woods, rather damp, and appears in July and August. The caps are often dark purple, often tinged with red, and sometimes the caps contains shades of green. I found the plants plentifully in Woodland Park, near Newtonville, Ohio, in July, 1907. We ate them on several occasions and found them very good. The greenish margin and purplish center will mark the plant.

RUSSULA INTEGRA. FR.

THE ENTIRE RUSSULA. EDIBLE.

Integra, whole or entire. The pileus is three or four inches in diameter, fleshy; typically red, but changing color; expanded, depressed, with a viscid cuticle, growing pale. Margin thin, furrowed and tuberculate. Flesh white, sometimes yellowish above.

The stem is at first short and conical, then club-shaped or ventricose, sometimes three inches long and up to one inch thick; spongy, stuffed, commonly striate; even, and shining white.

The gills are somewhat free, very broad, sometimes three-fourths of an inch; equal or bifid at the stem, rather distant and connected by veins; pallid or white, at length light yellow, being powdered yellow with the spores.

Although the taste is mild it is often astringent. One of the most changeable of all species, especially in the color of the pileus, which, though typically red, is often found inclining to azure-blue, bay-brown, olivaceous, etc. It occasionally happens that the gills are sterile and remain white. Fries.

The spores are spheroid, spiny, pale ochraceous.

R. integra so closely resembles R. alutacea that to distinguish them requires a knowledge of both plants, and even then one may not feel quite sure; however, it matters little as they are equally good. Its powdery gills will help to distinguish R. integra from R. alutacea. Found from July to October.

RUSSULA ROSEIPES. (SECR) BRES.

THE ROSY-STEMMED RUSSULA. EDIBLE.

Figure 151   Russula roseipes. Natural size.

Roseipes is from rosa, a rose; pes, a foot; so called because of its rose-colored or pinkish stem.

The pileus is two to three inches broad, convex, becoming nearly plane, or slightly depressed; at first viscid, soon dry, becoming slightly striate on the margin; rosy-red variously modified by pink, orange or ochraceous hues, sometimes becoming paler with age; taste mild.

The gills are moderately close, nearly entire, rounded behind and slightly adnexed, ventricose, whitish becoming yellow.

The stem is one to three inches long, slightly tapering upward, stuffed or somewhat cavernous, white tinged with red. The spores are yellow, round. Peck, 51 R.

This plant is widely distributed from Maine to the West. It grows best in pine and hemlock woods, but sometimes found in mixed woods. It is found in July and August.

RUSSULA FRAGILIS. FR.

THE TENDER RUSSULA.

Figure 152 Russula fragilis.

Fragilis means fragile.

The pileus is rather small, flesh-color or red, or reddish; thin, fleshy only at the disk; at first convex and often umbonate, then plane, depressed; cuticle thin, becoming pale, viscid in wet weather, margin tuberculate-striate.

The gills are thin, ventricose, white, slightly adnexed, equal, crowded, sometimes slightly eroded at the edge. The spores are minutely echinulate, $8-10 \times 8\mu$.

The stem is stuffed, hollow, shining white.

Quite as acrid as R. emetica, which it resembles in many ways, especially the smaller plants. It can be distinguished by its thinner caps, thinner and crowded gills, more ventricose and often slightly eroded at the edge. It is generally classed among poisonous mushrooms; but Captain Charles McIlvaine in his book says: "Though one of the peppery kind, I have not, after fifteen years of eating it, had reason to question its edibility." I should advise caution. Eat of it sparingly till sure of its effects. Found in woods from July to October.

RUSSULA EMETICA. FR.

THE EMETIC RUSSULA.

Figure 153 Russula emetica. Two-thirds natural size. Caps rose-red to yellow-red. Gills white.

Emetica means making sick, inciting to vomit. The pileus is fleshy, quite viscid, expanded, polished, shining, oval, or bell-shaped when young; its color is very variable from rose-red to a yellow-red or even purple; margin furrowed, flesh white.

The gills are free, equal, broad, distant, white. The spores are round, 8μ.

The stem is stout, solid, though sometimes spongy stuffed, even, white or reddish. The spores are white, round, and spiny.

This species is recognized by its very acrid taste and free gills. A distinct channel will be seen between the gills and the stem. This very pretty mushroom is quite common in most parts of Ohio. I found it in abundance about Salem, Bowling Green, Sidney, and Chillicothe—all in this state.

Captain McIlvaine states that he has repeatedly eaten it and cites a number of others who ate it without bad results, although weight of authority would band it a reprobate. I am glad to report something in its favor, for it is a beautiful plant, yet I should advise caution in its use.

It is found in open woods or in pastures under trees, from July to October. Its viscid cap will distinguish it.

RUSSULA FURCATA. FR.

THE FORKED GILLED RUSSULA. EDIBLE.

Figure 154 Russula furcata. Two-thirds natural size. Caps greenish-umber to
reddish.

Furca, a fork, so called from the forking of the gills. This is not peculiar, however, to this
species. The pileus is two to three inches broad; greenish, usually greenish-umber, sometimes
reddish; fleshy; compact; nearly round, then expanded, depressed in the center; even; smooth;
often sprinkled with a silky luster, pellicle separable, margin at first inflexed, then expanded,
always even, sometimes turned upward. The flesh is firm, white, dry, somewhat cheesy.

The gills are adnate or slightly decurrent, somewhat crowded, broad, narrowed at both ends,
many forked, shining white. The spores, $7–8 \times 9\mu$.

The stem is two to three inches long, solid, white, rather firm, even, equal or tapering
downward. The spores are round and spiny.

I have found it frequently on the wooded hillsides of the state. The taste when raw is mild at
first, but soon develops a slight bitterness which, however, is lost in cooking. Fried in butter they
are excellent. July to October.

RUSSULA RUBRA, FR.

THE RED RUSSULA.

Figure 155   Russula rubra. Two-thirds natural size. Caps bright-vermilion. Gills forked and tinged with red.

Rubra means red, so called from the cap being concolorous, bright vermillion; showy, becoming pale with age, center of the cap usually darker; compact, hard, fragile, convex, expanded, somewhat depressed, dry, no pellicle, often cracked when old. The flesh is white, often reddish under the cuticle.

The gills are adnate, rather crowded, white at first, then yellowish, many forked and with some short ones intermixed, frequently tinged with red at the edge. Spores 8–10μ, cystidia pointed.

The stem is two to three inches long, solid, even, white, often with a faint reddish hue. The spores are nearly round and white.

It is very acrid to the taste, and because of this acridity it is usually thought to be poisonous, but Captain McIlvaine says he does not hesitate to cook it either by itself or with other Russulæ. It is found very generally in the state and is quite plentiful in the woods about Chillicothe, from July to October.

RUSSULA PURPURINA. QUEL & SCHULZ.

THE PURPLE RUSSULA. EDIBLE.

Figure 156   Russula purpurina. Two-thirds natural size. Caps rosy-pink to light-
yellow. Gills yellowish in age.

Purpurina means purple. The pileus is fleshy, margin acute, subglobose, then plane, at length depressed in the center, slightly viscid in wet weather, not striate, often split, pellicle separable, rosy-pink, paling to light-yellow.

Gills are crowded in youth, afterward subdistant, white, in age yellowish, reaching the stem, not greatly narrowed behind, almost equal, not forked.

The stem is stuffed, spongy, very variable, cylindrical, attenuated above, rosy-pink, becoming paler toward the base, color obscure in age. The flesh is fragile, white, reddish under the skin; odor slight and taste mild. The spores white, globose, sometimes subelliptical, $4-8\mu$ long, minutely warted. Peck, 42 Rept., N. Y. State Bot.

This is not a large plant, but it can be readily determined by its red or reddish stem, mild taste and white spores. Found in open woods in July and August.

RUSSULA DENSIFOLIA. GILLET.

Figure 157   Russula densifolia. Two-thirds natural size. Caps whitish, becoming fuliginous gray. Flesh turning red when exposed to the air.

Densifolia has reference to the crowded condition of the gills.

The pileus is from three to four inches broad, fleshy, quite compact, convex, expanded, then depressed, margin inflexed, smooth, not striate, white or whitish, becoming fuliginous, gray, or brownish, quite black in center, flesh red when broken.

The gills are attached to the stem, somewhat decurrent, unequal, thin, crowded, white or whitish, with a rosy tint. Spores, $7–8\mu$.

The stem is short, slightly mealy, white, then gray, at length blackish, smooth, round, turning red or brown on being handled.

It differs from R. nigricans in being much smaller, and in its crowded gills. It differs from R. adusta in flesh turning red when broken. The flesh or substance is white at first, turning red when exposed to the air, then blackish. This plant is not abundant in this state. I found a number of plants on Cemetery Hill, where some shale had been dumped under a large beech tree. Found in July and August.

CANTHARELLUS. ADANSON.

Cantharellus means a diminutive drinking-cup or vase. This genus can be distinguished from all other genera by the character of its gills which are quite blunt on the edge, like folds, polished, and are mostly forked or branched. In some species the gills vary in thickness and number. They are decurrent, folded, more or less thick and swollen. The spores are white. They grow on the ground, on rotten wood, and among moss. They seem to delight in damp shady places.

CANTHARELLUS CIBARIUS. FR.

THE EDIBLE CANTHARELLUS.

Plate XXII. Figure 158   Cantharellus cibarius.

Natural size. Entire plant egg-yellow.

Cibarius means pertaining to food. This plant is frequently spoken of as the Chanterelle. The entire plant is a rich egg-yellow. The pileus is fleshy, at first convex, later flat, three to five inches broad, depressed in the center, finally funnel-shaped; bright to deep yellow; firm, smooth, but often irregular, its margin often wavy; flesh white, the cap has the appearance of an inverted cone.

The gills are decurrent, shallow and fluted, resembling swollen veins, branched, more or less interconnected and tapering downward on the stem, color the same as the pileus.

The stem is solid, variable in length, often curved, tapering towards the base, paler than the pileus and gills.

It grows in woods and rather open places. I found it in great abundance in Stanley's woods, near Damascus, Ohio. I have found it very often about Chillicothe. The plant has a strong prune-like odor; when tasted raw they are peppery and pungent but sweet and quite delicious when cooked. My friends and myself have eaten it and pronounced very good. The plants in Figure 158 were gathered near Columbus, Ohio, and photographed by Dr. Kellerman.

The species is quite common in the state, and is found from June to September.

CANTHARELLUS AURANTIACUS. FR.

FALSE CHANTARELLE.

Figure 159   Cantharellus aurantiacus. One-third natural size. Caps orange-yellow. Gills yellow and forked.

Aurantiacus means orange-yellow. The pileus is fleshy, soft, depressed, downy, the margin strongly incurved when young, in mature plants it is wavy or lobed; color dull yellowish, usually brownish.

The gills are crowded, straight, dark-orange, branched, with a regular bifurcation.

The stem is lighter in color than the pileus, solid at first, spongy, stuffed, hollow, unequal, tapering upward, and somewhat curved.

It is generally labeled poisonous, but some good authorities say it is wholesome. I have never eaten it further than in its raw state. It is easily distinguished from the edible species by its dull orange cap and its orange gills, which are thinner and closer and more regularly forked than those of the Edible Chantarelle. It grows in woods and open places. Found from July to September.

CANTHARELLUS FLOCCOSUS. SCHW.

THE WOOLLY CANTHARELLUS. EDIBLE.

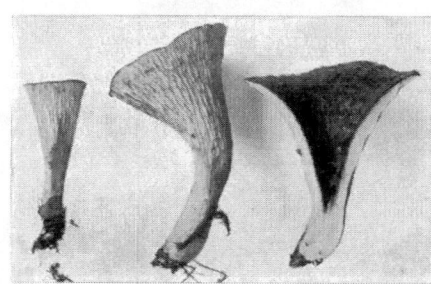

Plate XXIII. Figure 160   Cantharellus floccosus.

Floccosus means floccose or woolly.

The pileus at the top is from one to two inches broad, fleshy, elongated funnel-form or trumpet-shape, floccose-squamose, ochraceous-yellow.

The gills are vein-like, close, much anastomosing above, long decurrent and subparallel below, concolorous.

The stem is very short, thick, rather deeply rooted. The spores are elliptical, $12.5–15 \times 7.6\mu$. Peck, 23 Rep., N. Y.

This plant is funnel-shaped nearly to the base of the stem. It is a small plant, never more than four inches high. I found it in Haynes's Hollow, in rather open woods, on mossy hillsides. July and August.

CANTHARELLUS BREVIPES. PK.

The Short-Stemmed Cantharellus. Edible.

Brevipes is from brevis, short; pes, foot; so called because of its short stem.

The pileus is fleshy, obconic, glabrous, alutaceous, or dingy cream-color, the thin margin erect, often irregular and lobed, tinged with lilac in the young plant; folds numerous, nearly straight in the margin, abundantly anastomosing below; pale umber, tinged with lilac.

The stem is short, tomentose-pubescent, ash-colored, solid, often tapering downward. Spores yellowish, oblong-elliptical, uninucleate, $10–12 \times 5\mu$. Peck, 33d Rep., N. Y.

The plant is small; with us, not more than three inches high and the pileus not more than two inches broad at the top. It differs somewhat in color, in the character of the folds, and materially in the shape of the margin of the pileus. Found occasionally on the hillsides of Huntington Township, near Chillicothe, July to August.

CANTHARELLUS CINNABARINUS. SCHW.

THE CINNABAR CANTHARELLUS. EDIBLE.

Figure 161   Cantharellus cinnabarinus. Cap and stem cinnibar-red, flesh white. Natural size.

Cinnabarinus means cinnabar-red, from the color of the plant.

The pileus is firm, convex, or slightly depressed in the center, often irregular with wavy or lobed margin; glabrous, cinnabar-red, flesh white.

The gills are narrow, distant, branched, decurrent, of the same color as the cap, dull on the edge.

The stem is equal or tapering downward, glabrous, solid, sometimes stuffed, cinnabar-red.

The spores are elliptical, 8–10μ long, 4–5μ broad.

No one will have any difficulty in identifying this plant, since its color suggests the name at once. It is quite common about Chillicothe and throughout the state. It is found frequently with Craterellus cantharellus. It is a very pretty plant, growing in open woods or along the roadside in woods. It will keep for some time after it is gathered. It is found from July to October.

CANTHARELLUS INFUNDIBULIFORMIS. FR.

FUNNEL-SHAPED CANTHARELLUS.

Infundibuliformis means shaped like a funnel.

The pileus is one to two and a half inches broad, somewhat membranaceous, umbilicate, then infundibuliform, usually perforated at the base, and opening into the cavity of the stem, floccosely rugose on the surface, yellowish-gray or smoky when moist, pale when dry, becoming wavy.

The gills are decurrent, thick, distant, regularly forked, straight, yellow or cenereous, at length pruinose.

The stem is two to three inches long, hollow, even, smooth, always yellow, slightly thickened at the base. The spores are elliptical, smooth, $9–10 \times 6\mu$.

They grow on the ground, especially where wood has decayed and become a part of the ground. They also grow on decayed wood. They are found from July to October.

NYCTALIS. FR.

Nyctalis is from a Greek word meaning night.

Pileus symmetrical, in some species bearing large conidia upon its surface.

The gills are adnate or decurrent, thick, soft, margin obtuse.

The stem is central, its substance continuous with the flesh of the pileus. The spores are colorless, smooth, elliptical or globose. Fries.

NYCTALIS ASTEROPHORA. FR.

Figure 162   Nyctalis asterophora.

Asterophora means star-bearing.

The pileus is about one-half inch broad, fleshy; conical, then hemispherical; flocculose and rather mealy, owing to the large, stellate conidia; whitish, then tinged with fawn-color.

The gills are adnate, distant, narrow, somewhat forked, straight, dingy.

The stem is about one-half inch long, slender, twisted, stuffed, white then brownish, rather mealy. The spores are elliptical, smooth, $3 \times 2\mu$. Fries, Hym.

I found, about the last of August, these plants growing on decaying specimens of Russula nigricans, along Ralston's Run, near Chillicothe.

HYGROPHORUS. FR.

Hygrophorus is from two Greek words meaning bearing moisture. So called because the members of this genus may be known from their moist caps and the waxy nature of the gills, which distinguish them from all others. As in the Pleurotus, the gills of some of the species are rounded or notched at the end next to the stem, but of others they are decurrent on it; hence, in some species they are like the gills of Tricholoma in their attachment, in others they run down on the stem as in the Clitocybe. In many of them both cap and stem are very viscid, a characteristic not found in the Clitocybes; and the gills are generally thicker and much farther apart than in that genus. A number of the species are beautifully colored.

HYGROPHORUS PRATENSIS. FR.

THE PASTURE HYGROPHORUS. EDIBLE.

Plate XXIV. Figure 163   Hygrophorus pratensis.

Pratensis, from pratum, a meadow. The pileus is one to two inches broad; when young almost hemispherical, then convex, turbinate or nearly flat, the center more or less convex, as if umbonate; margin often cracked, frequently contracted or lobed; white or various shades of yellow, buffish-reddish, or brownish. Flesh white, thick in the center, thin at the margin. The stem is stuffed, attenuated downwards. The gills are thick, distant, white or yellowish, bow-shaped, decurrent, and connected by vein-like folds. Spores are white, broadly elliptical, .00024 to .00028 inch long.

The pasture hygrophorus is a small but rather stout-appearing mushroom. It grows on the ground in pastures, waste places, clearings, and thin woods, from July to September. Sometimes all white or gray.

Var. cinereus, Fr. Pileus and gills gray. The stem whitish and slender.

Var. pallidus, B. & Br. Pileus depressed, edge wavy, entirely pale ochre.

This species differs mainly from H. leporinus in that the latter is quite floccose on the pileus.

HYGROPHORUS EBURNEUS. BULL.

SHINING WHITE HYGROPHORUS. EDIBLE.

Figure 164   Hygrophorus eburneus.

Eburneus is from ebur, ivory. The pileus is two to four inches broad, sometimes thin, sometimes somewhat compact, white; very viscid or glutinous in wet weather, and slippery to the touch; margin uneven, sometimes wavy; smooth, and shining. When young, the margin is incurved.

The gills are firm, distant, straight, strongly decurrent, with vein-like elevations near the stem. The spores are white, rather long.

The stem is unequal, sometimes long and sometimes short; stuffed, then hollow, tapering downward, punctate above with granular scales. Odor and taste are rather pleasant. It is found in woods and pastures in all parts of Ohio, but it is not plentiful anywhere. I have found it only in damp woods about Chillicothe. August to October.

HYGROPHORUS COSSUS. SOW.

Cossus, because it smells like the caterpillar, Cossus ligniperda.

The pileus is small, quite viscid, shining when dry, white with a yellow tinge, edge naked, very strong-scented.

The gills are somewhat decurrent, thin, distant, straight, firm.

The stem is stuffed, nearly equal, scurvy-punctate upwards. Spores 8×4. Found in the woods. The strong smell will serve to identify the species.

HYGROPHORUS CHLOROPHANUS. FR.

THE GREENISH-YELLOW HYGROPHORUS.

Chlorophanus is from two Greek words, meaning appearing greenish-yellow.

The pileus is one inch broad, commonly bright sulphur-yellow, sometimes scarlet-tinted, not changing color; slightly membranaceous, very fragile, often irregular, with the margin split or lobed, at first convex, then expanded; smooth, viscid, margin striate.

The gills are emarginate, adnexed, quite ventricose, with a thin decurrent tooth, thin, subdistant, distinct, pale-yellow.

The stem is two to three inches long, hollow, equal, round, viscid when moist, shining when dry, wholly unicolorous, rich light-yellow.

The spores are slightly elliptical, $8 \times 5\mu$.

This species resembles in appearance H. ceraceus, but it can be identified by its emarginate gills and somewhat larger form. The plant has a wide distribution, having been found from the New England States through the Middle West. It is found in damp, mossy places from August to October. I have no doubt of its edibility. It has a mild and agreeable taste when eaten in the raw state.

HYGROPHORUS CANTHARELLUS. SCHW.

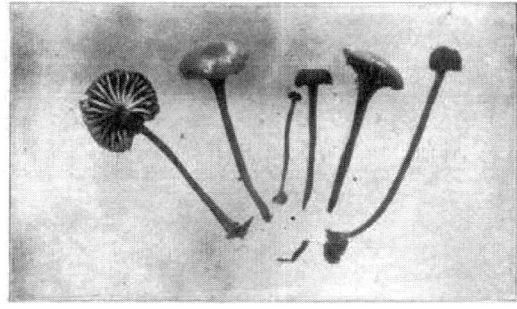

Figure 165 Hygrophorus cantharellus. Natural size. Caps bright red.

Cantharellus means a small vase.

The pileus is thin, convex, at length umbilicate, or centrally depressed, minutely squamulose, moist, bright red, becoming orange or yellow.

The gills are distant, subarcuate, decurrent, yellow, sometimes tinged with vermilion.

The stem is one to three inches long, smooth, equal, sub-solid, sometimes becoming hollow, concolorous, whitish within. Peck.

I have found about Chillicothe a number of the varieties given by Dr. Peck.

Var. flava. Pileus and stem pale yellow. Gills arcuate, strongly decurrent.

Var. flavipes. Pileus red or reddish. Stem yellow.

Var. flaviceps. Pileus yellow. Stem reddish or red.

Var. rosea. Has the pileus expanded and margin wavy scalloped.

Found from July to September.

HYGROPHORUS COCCINEUS. FR.

THE SCARLET HYGROPHORUS. EDIBLE.

Coccineus, pertaining to scarlet. The pileus is thin, convex, obtuse, viscid, scarlet, growing pale, smooth, fragile.

The gills are attached to the stem, with a decurrent tooth, connected by veins, variously shaded.

The stem is hollow and compressed, rather even, not slippery, scarlet near the cap, yellow at the base.

This plant when young is of a bright scarlet, but it soon shades into a light-yellow with advancing age. It is quite fragile and varies very greatly in size in different localities. Found in woods and pastures from July to October.

HYGROPHORUS CONICUS. FR.

THE CONICAL HYGROPHORUS. EDIBLE.

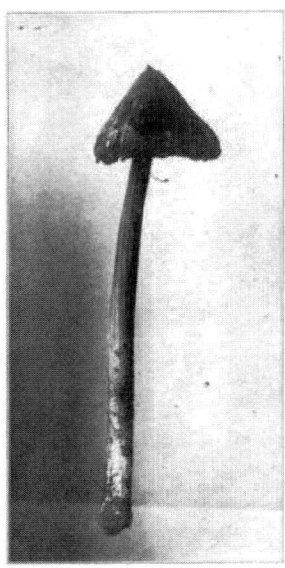

Figure 166   Hygrophorus conicus.

The pileus is one to two inches broad, acutely conical, submembranaceous, smooth, somewhat lobed, at length expanded, and rimose; turning black, as does the whole plant when broken or bruised; orange, yellow, scarlet, brown, dusky.

The gills are free or adnexed, thick, attenuated, ventricose, yellowish with frequently a cinereous tinge, wavy, rather crowded.

The stem is three to four inches long, hollow, cylindrical, fibrillose, striated, colored like the pileus, turning black when handled.

This plant is quite fragile. It can be identified by its turning black when bruised. It sometimes appears early in the spring and continues till late in the fall. It is not abundant but is only occasionally found on the ground in woods and open places.

HYGROPHORUS FLAVODISCUS. FROST.

YELLOW-DISKED HYGROPHORUS. EDIBLE.

Figure 167 Hygrophorus flavodiscus. Natural size. The gluten is shown connecting the margin of the cap to their stem.

Flavodiscus means yellow-disked.

The pileus is one-half to three inches broad, fleshy, convex or nearly plane, glabrous, very viscid or glutinous, white, pale-yellow or reddish-yellow in the center, flesh white.

The gills are adnate or decurrent, subdistant, white, sometimes with a slight flesh-colored tint, the interspaces sometimes venose.

The stem is one to three inches long, solid, subequal, very viscid, or glutinous, white at the top, white or yellowish elsewhere. The spores are elliptical, white, .00025 to .0003 of an inch long, .00016 to .0002 broad.

These mushrooms make a delicious dish. The specimens in the photograph were gathered at West Gloucester, Mass., by Mrs. E. B. Blackford, of Boston. I have found them about Chillicothe. They are very viscid, as the plants in Figure 167 will show. The caps are thick and the margin inrolled. They are found in October and November.

HYGROPHORUS SPECIOSUS. PK.

SHOWY HYGROPHORUS. EDIBLE.

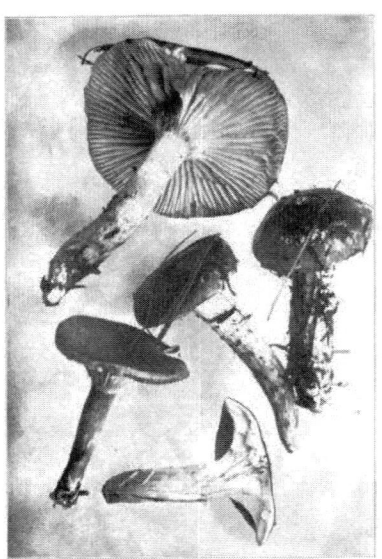

Figure 168　Hygrophorus speciosus.

Speciosus means beautiful, showy; so called from the scarlet color of the umbo. The pileus is one to two inches in diameter, broadly convex, often with small central umbo; glabrous, very viscid or glutinous when moist; yellow, usually bright red or scarlet in the center; flesh white, yellow under the thin, separable pellicle.

The gills are distant, decurrent, white, or slightly tinged with yellow.

The stem is two to four inches long, nearly equal, solid, viscid, slightly fibrillose, whitish or yellowish. The spores are elliptic, .0003 of an inch long, .0002 broad. Peck.

This is a very beautiful and showy plant. It grows in swampy places and under tamarack trees. The specimens in Figure 168 were found in Massachusetts by Mrs. Blackford, and were photographed by Dr. Kellerman. It is found in September and August.

HYGROPHORUS FULIGINEUS. FROST.

SOOTY HYGROPHORUS. EDIBLE.

Figure 169 Hygrophorus fuligineus. Natural size. Specimen on the right is H. caprinus.

Fuligineus means sooty or smoky.

The pileus is one to four inches broad, convex or nearly plane, glabrous, very viscid or glutinous, grayish-brown or fuliginous, the disk often darker or almost black.

The gills are subdistant, adnate or decurrent, white.

The stem is two to four inches long, solid, viscid or glutinous, white or whitish. The spores are elliptic, .0003 to .00035 of an inch long, .0002 broad. Peck, No. 4, Vol. 3.

This species is found frequently associated with H. flavodiscus, which it resembles very closely, save in color. When moist, the cap and stems are covered with a thick coating of gluten, and when the caps are dry this gives them a varnished appearance. I do not find them abundant here. The plants in Figure 169 were found by Mrs. Blackford near West Gloucester, Mass. They are found October and November.

HYGROPHORUS CAPRINUS. SCOP.

THE GOAT HYGROPHORUS. EDIBLE.

Caprinus means belonging to a goat; it is so called from the fibrils resembling goat's hair.

The pileus is two to three inches broad, fleshy, fragile, conical, then flattened and umbonate, rather wavy, sooty, fibrillose.

The gills are very broad, quite distant, deeply decurrent, white, then glaucous.

The stem is two to four inches long, solid, fibrillose, sooty, often streaked or striate, as will be seen in Figure 169, page 212.

The spores are $10 \times 7–8\mu$.

These plants grow in pine woods in company with H. fuligineus and H. flavodiscus. The specimen on the right in Figure 169 was found near West Gloucester, Mass., by Mrs. Blackford. It is found from September till hard frost.

HYGROPHORUS LAURÆ. MORG.

Figure 170 Hygrophorus Lauræ.

This is a beautiful plant, found among leaves, and so completely covered with particles of leaves and soil that it is hard to clean them off. They are very viscid, both stem and cap. They are only occasionally found in our state.

The pileus is two to three inches broad; reddish-brown in the center, shading to a very light tan on the edges; very viscid; convex; margin at first slightly incurved, then expanded.

The gills are adnate, slightly decurrent, not crowded, unequal, yellowish.

The stem is stuffed, tapering downward, whitish, furfuraceous near the cap.

I have found this plant in Poke Hollow, near Chillicothe, on several occasions, also in Gallia county, Ohio. I have not found it elsewhere in this vicinity. While I have not found it in sufficient quantity to try it I have no doubt of its edible qualities. I have found it only about the last of September and the first of October. It grows in rather dense woods on the north sides of the hills, where it is constantly shaded and damp. Named in honor of Prof. Morgan's wife.

HYGROPHORUS MICROPUS. PK.

SHORT-STEMMED HYGROPHORUS. EDIBLE.

Micropus means short-stemmed. The pileus is thin, fragile, convex or centrally depressed, umbilicate; silky, gray, often with one or two narrow zones on the margin; taste and odor farinaceous.

The gills are narrow, close, adnate or slightly decurrent, gray, becoming salmon color with age.

The stem is short, solid or with a slight cavity, often slightly thickened at the top, pruinose, gray, with a white, mycelioid tomentum at the base. The spores are angular, uninucleate, salmon color, .0003–.0004 of an inch long, .00025–.0003 broad. Peck.

This is a very small plant and not frequently found, but widely distributed. I have always found it in open grassy places during damp weather. The caps are thin, often markedly depressed. Its silky appearance and narrow zones on the margin of the cap, together with its rather close gills, broadly attached to the stem, gray at first, then salmon color, will identify the species. July to September.

HYGROPHORUS MINIATUS. FR.

THE VERMILION HYGROPHORUS. EDIBLE.

Figure 171 Hygrophorus miniatus. Cap and stems vermilion-red. Gills yellowish
and tinged with bright-red.

Miniatus is from minium, red lead.

This is a small but a very common species, highly colored and very attractive. The pileus and the stem are bright red and often vermilion. The pileus is at first convex, but, when fully expanded, it is nearly or quite flat, and in wet weather it is even concave by the elevation of the margin, smooth or minutely scaly, often umbilicate. Its color varies from a bright red or vermilion or blood-red to pale orange hues.

The gills are yellow and frequently strongly tinged with red, distant, attached to the stem, and sometimes notched.

The stem is usually short and slender, colored like, or a little paler, than the cap; solid, when young, but becoming stuffed or hollow with age. The spores are elliptical, white, 8μ long.

The Vermilion mushroom grows in woods and in open fields. It is more plentiful in wet weather. It seems to grow best where chestnut logs have decayed. It can be found in such places in sufficient quantities to eat. Few mushrooms are more tender or have a more delicate flavor. There are two other species having red caps, Hygrophorus coccineus and H. puniceus, but both are edible and no harm could come from any mistake. They are found from June to October. Those in Figure 171 were found in Poke Hollow September 29.

HYGROPHORUS MINIATUS SPHAGNOPHILUS. PK.

Plate XXV. Figure 172 Hygrophorus miniatus sphagnophilus.

Natural size.

Sphagnophilus means sphagnum-loving, so called because it is found growing on sphagnum.

The pileus is broadly convex, subumbilicate, red.

The gills are adnate, whitish, becoming yellowish or sometimes tinged with red, occasionally red on the edge.

The stem is colored like the pileus, whitish at the base, both it and the pileus are very fragile.

This is more fragile than the typical form and retains its color better in drying. Peck, 43d Rep.

This is a beautiful plant growing, as Figure 172 shows, on the lower dead portion of the stems of bog moss or sphagnum. It grows very abundantly in Buckeye Lake. The photograph was made by Dr. Kellerman. It is found from July to October. These plants cook readily, have an excellent flavor and because of their color make an inviting dish. I have eaten heartily of them several times.

HYGROPHORUS MARGINATUS. PK.

MARGINED HYGROPHORUS. EDIBLE.

Figure 173   Hygrophorus marginatus.

Marginatus, so called from the frequent vermilion edged gills.

The pileus is thin, fragile, convex, subcampanulate or nearly plane, often irregular, sometimes broadly umbonate, glabrous, shining, striatulate on the margin, bright golden-yellow.

The gills are rather broad, subdistant, ventricose, emarginate, adnexed, yellow, sometimes becoming orange or vermilion on the edge, interspaces venose.

The stem is fragile, glabrous, often flexous, compressed or irregular, hollow, pale-yellow; spores broadly elliptic, .00024–.0003 of an inch long, .00024–.0002 broad. Peck, N. Y., 1906.

This plant has the most beautiful yellow I have ever seen in a mushroom. This bright golden yellow and the orange or vermilion color on the margin or edge of the gills will always characterize the plant.

The specimen in Figure 173 were sent to me by Mrs. Blackford, of Boston, Mass., the last of August. They were not in the best condition when photographed.

HYGROPHORUS CERACEUS. FR.

THE WAX-LIKE HYGROPHORUS. EDIBLE.

Figure 174   Hygrophorus ceraceus. Caps waxy yellow.

Ceraceus is from cera, wax. The pileus is one inch and less broad, waxy-yellow, shining, fragile, thin, occasionally subumbonate, slightly fleshy, slightly striate.

The gills are firmly attached to the stem, subdecurrent, distant, broad, ventricose often connected with veins, almost triangular, yellow.

The stem is one to two inches long, hollow, often unequal, flexuous, sometimes compressed, yellow, occasionally orange at the base, waxy. The spores $8 \times 6\mu$.

This is a very beautiful, fragile plant, usually found growing in the grass. It is easily distinguished by its waxy yellow color. The plants photographed were found on the Cemetery Hill. They are found from August to October.

HYGROPHORUS VIRGINEUS. WULF.

THE IVORY-CAPPED HYGROPHORUS. EDIBLE.

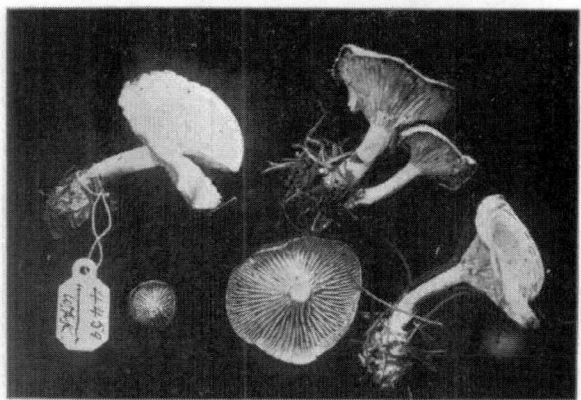

Figure 175 Hygrophorus virgineus. Two-thirds natural size. Entire plant white.

Virgineus, virgin; so called from its whiteness. The pileus is fleshy, convex, then plane, obtuse, at length depressed; moist, sometimes cracked into patches, floccose when dry.

The gills are decurrent, distant, rather thick, often forked.

The stem is curt, stuffed, firm, attenuated at the base, externally becoming even and naked. Spores 12×5–6μ. Fries.

The plant is wholly white and never large. It is easily confounded with H. niveus and sometimes difficult to distinguish from the white forms of H. pratensis. This plant is quite common in pastures, both in the spring and in the fall. I found the specimens in Figure 175 on Cemetery Hill under the pine trees on November 11. They were photographed by Dr. Kellerman.

HYGROPHORUS NIVEUS.

THE SNOW-WHITE HYGROPHORUS. EDIBLE.

Niveus, snow-white. The plant is wholly white. The pileus is scarcely one inch broad, somewhat membranaceous, bell-shaped, convex, then umbilicate, smooth, striate, viscid when moist, not cracked when dry, flesh thin, everywhere equal.

The gills are decurrent, thin, distant, acute, quite entire.

The stem is hollow, thin, equal, smooth. Spores $7 \times 4\mu$. Found in pastures.

HYGROPHORUS SORDIDUS. PK.

THE DINGY HYGROPHORUS. EDIBLE.

Figure 176   Hygrophorus sordidus.

Sordidus means a dirty white, or dingy, referring to the color of the caps, so made by adhering earth.

The pileus is broadly convex or nearly plane, glabrous, slightly viscid, white, but usually defiled by adhering dirt; the margin at first strongly involute, then spreading or reflexed; flesh firm when young, tough when old.

The gills are subdistant, adnate, or decurrent, white or creamy-white.

The stem is five to ten Cm. long, firm, solid, white.

The spores are elliptical, $6.5–7.5 \times 4–5\mu$. Peck.

The specimens I found were clear white, growing among leaves and were especially free from soil. The stems were short and were inclined to be slightly ventricose. Dr. Peck says that this "species is distinguished from H. penarius by its clear white color, though this is commonly obscured by the adhering dirt that is carried up in the growth of the fungus." The young, growing plants were strongly involute but the older plants were reflexed, giving the plants a funnel-shaped appearance and giving the gills a much stronger decurrent appearance. Found October 26th.

HYGROPHORUS SEROTINUS. PK.

LATE HYGROPHORUS.

Figure 177   Hygrophorus serotinus.

Serotinus means late. So called because it is late in the season.

Pileus is fleshy but thin, convex or nearly plane, often with the thin margin curved upward, glabrous or with a few obscure innate fibrils, reddish in the center, whitish on the margin, flesh white, taste mild.

The gills are thin, subdistant, adnate or decurrent, white, the interspaces slightly venose.

The stem is equal, stuffed or hollow, glabrous, whitish. The spores are white, elliptic, .0003 of an inch long, .0002 broad.

Pileus is 8–15 lines broad; stem about 1 inch long, 1.5–2.5 lines thick. Peck.

Some specimens of this species were sent to me from Boston by Mrs. Blackford, but after a careful study of them I was unable to place them. She then sent them to Dr. Peck, who gave them their very appropriate name. Those in Figure 177 were sent me in December, 1907.

They grow a number in the same locality and frequently in close groups or tufts. They seem to delight in oak and pine woods. Dr. Peck observes that this species is similar to Hygrophorus queletii, Bres., both in size and color, but the general characteristics of the plants do not agree. He also says it is similar in size and color to H. subrufescens, Pk., but differs materially in the specific description.

PANUS. FR.

Panus means swelling. The species under this genus are leathery plants, having the stems lateral and sometimes wanting. They dry up but revive with moisture. The gills are simple and thinner than the Lentinus, but with an entire, acute edge. There are a few species which give a phosphorescent light when growing on decayed logs. The genus closely resembles Lentinus but can be readily recognized on account of the smooth edged gills. A number of good authorities do not separate them but give both under the name Lentinus. This genus abounds wherever there are stumps and fallen timber.

PANUS STYPTICUS. FR.

THE STYPTIC PANUS. POISONOUS.

Figure 178   Panus stypticus. Two-thirds natural size. Cinnamon color.

Stypticus means astringent, styptic. The pileus is coriaceous, kidney-shaped, cinnamon-color, growing pale, cuticle breaking up into scales, margin entire or lobed, surface nearly even, sometimes zoned.

The gills are thin, crowded, connected by veins, of same color as cap, determinate, quite narrow.

The stem is lateral, quite short, swollen above, solid, compressed, pruinose, paler than the gills.

It is found very plentifully on decayed logs and stumps, and at times it is quite phosphorescent in its manifestations. It has an extremely unpleasant astringent taste. One might as well eat an Indian turnip as this species. Just a taste will betray it. Found from fall to winter.

PANUS STRIGOSUS. B. & C.

THE HAIRY PANUS. EDIBLE.

Strigosus, covered with stiff hairs. The pileus is sometimes quite large, eccentric, covered with stiff hairs, margin thin, white.

The gills are broad, distant, decurrent, straw-color.

The stem is stout, two to four inches long, hairy like the pileus.

The favorite host of this species is an apple tree. I found a beautiful cluster on an apple tree in Chillicothe. Its creamy whiteness and hairy cap and short hairy stem will distinguish it from all other tree fungi. It is edible when young, but soon becomes woody.

PANUS CONCHATUS. FR.

THE SHELL PANUS.

Conchatus means shell-shaped. The pileus is thin, unequal, tough, fleshy, eccentric, dimidiate; cinnamon, then pale; becoming scaly; flaccid; margin often lobed.

The gills are narrow, forming decurrent lines on the stem, often branched, pinkish, then ochre.

The stem is short, unequal, solid, rather pale, base downy.

This species will frequently be found imbricated and very generally confluent. Its shell-like form, its tough substance, and its thin pileus are its distinguishing marks. The taste is pleasant but its substance very tough. Found from September to frost.

PANUS RUDIS. FR.

Figure 179 Panus rudis.

This is a very plentiful plant about Chillicothe and is found throughout the United States, although it is a rare plant in Europe. It is generally given in American Mycology under the name Lentinus Lecomtei. It grows on logs and stumps. The form of the plant is quite different when growing on the top of a log or a stump, from those springing from the side. Those in the extreme left of Figure 179 grew on the side of the log, while those in the center grew on the top, in which case the plant has usually a funnel-shaped appearance.

The pileus is tough, reddish or reddish-brown, depressed, sinuate, bristling with tufts of hair, the margin quite strongly incurved, cæspitose.

The gills are narrow and crowded, decurrent, considerably paler than the cap.

The stem is short, hairy, tawny; sometimes the stem is almost obsolete.

There is a slight tinge of bitterness in the plant when raw, but in cooking this disappears. When prepared for food it should be chopped fine and well cooked. It can be dried for winter use. It is found from spring to late fall.

PANUS TORULOSUS. FR.

THE TWISTED PANUS. EDIBLE.

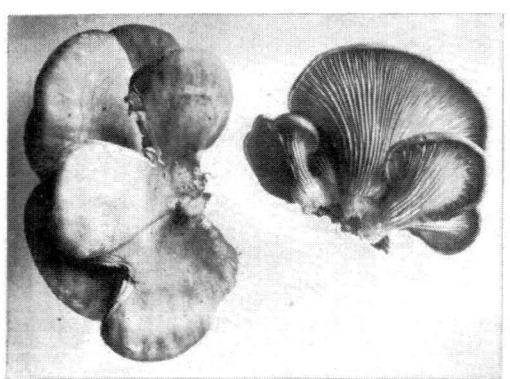

Figure 180 Panus torulosus.

Torulosus means a tuft of hair. The pileus is two to three inches broad, fleshy, then tough, coriaceous; plane, then funnel-shaped, or dimidiate; even; smooth; almost flesh color, varying to reddish-livid, sometimes violet tinted.

The gills are decurrent, rather distant, distinct behind, separate, simple, ruddy, then tan-colored.

The stem is short, stout, oblique, gray, covered with a violaceous down. The spores are $6 \times 3\mu$.

The plant is variable both in form and color. Sometimes shaded very slightly with pink. It is not very common here. I found some very fine specimens growing on a log near Spider Bridge, Chillicothe.

It is edible but quite tough.

PANUS LEVIS. B. & C.

THE LIGHT PANUS. EDIBLE.

Levis, light. Pileus two to three inches broad, orbicular, somewhat depressed, white, covered with a dense mat of hair; margin inflexed and marked by triangular ridges.

The gills are broad, entire, decurrent.

The stem is two to three inches long, attenuated upward, eccentric, lateral, solid, hairy below like the pileus. The spores are white.

This certainly is a very beautiful plant and will hold the attention of the collector. It is not common with us. I have found it only on hickory logs. It is said to be of good flavor and to cook readily.

LENTINUS. FR.

Lentinus means tough. The pileus is fleshy, corky, tough, hard and dry, reviving when moist.

The stem is central or lateral and often wanting, but when present is continuous with the cap.

The gills are tough, unequal, thin, normally toothed, decurrent more or less, margin acute. The spores are smooth, white, orbicular.

All the species, so far as I know, grow on wood. They assume a great variety of forms. This genus is very closely related to Panus in the dry, coriaceous nature of the pileus and the gills, but it can be readily recognized by the toothed margin of the gills.

LENTINUS VULPINUS. FR.

STRONG-SCENTED VULPINUS.

Plate XXVI. Figure 181 Lentinus vulpinus.

One-third natural size.

Vulpinus is from vulpes, a fox.

This is quite a large, massive plant, growing in a sessile and imbricated manner. It has appeared in large quantities for the past four years on an elm, very slightly decayed, but in quite a damp and dark place. The reader will get some idea of the size of the whole plant in Figure 181 if he will consider each pileus to be five to six inches broad. They are built up one on top of another, overlapping each other like shingles on a roof.

The pileus is fleshy but tough, shell-shaped, connate behind, longitudinally rough, costate, corrugate, tan-colored, and the margin is strongly incurved.

The gills are broad, nearly white, flesh-colored near the base, coarsely toothed.

The stem is usually obsolete, yet in some cases it is apparent.

The spores are almost round and very small, .00006 inch in diameter. In all plants which I have found the odor is somewhat strong and the taste is pungent. It grows in the woods in September and October.

LENTINUS LEPIDEUS. FR.

THE SCALY LENTINUS. EDIBLE.

Figure 182 Lentinus lepideus.

Lepideus is from lepis, a scale.

The pileus is fleshy, compact, convex, then depressed, unequal, broken up in dark scales, flesh white, tough.

The gills are sinuate, decurrent, broad, torn, transversely striate, whitish, or with white edges, irregularly toothed.

The stem is stout, central or lateral, tomentose or scaly, often crooked, rooting, whitish, solid, equal or tapering at the base.

This is a peculiar plant, growing sometimes to immense forms. It grows on wood, seemingly to be partial to railroad ties to which its mycelium is very injurious. I found the plant frequently about Salem, Ohio. The specimens in the halftone were found near Akron, Ohio, and photographed by Prof. Smith. As an esculent it almost rivals the Pleuroti. It is found from spring to autumn. I found a beautiful cluster on an oak stump near Chillicothe, while looking for Morels, about the last of April.

LENTINUS COCHLEATUS. FR.

THE SPIRAL-FORMED LENTINUS. EDIBLE.

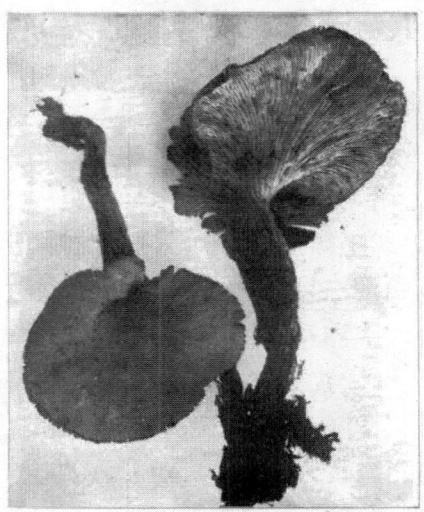

Figure 183 Lentinus cochleatus.

Cochleatus is from cochlea, a snail, from resembling its shell.

The pileus is two to three inches broad, tough, flaccid, irregular, depressed, sometimes funnel-shaped, sometimes lobed or contorted, flesh-color, becoming pale.

The gills are crowded, beautifully serrated, pinkish-white.

The stem is solid, length variable, sometimes central, frequently eccentric, often lateral, smooth. The spores are nearly round, 4μ.

This is a beautiful plant but sparingly found with us. I found a pretty cluster at the foot of a maple stump in Poke Hollow. The serrated form of the gills will attract attention at once. It is found in August and September.

LENZITES. FR.

Lenzites, named after Lenz, a German botanist. The pileus is corky, dimidiate, sessile. The gills are corky, firm, unequal, branched, edge obtuse. It is very common in the woods, sometimes almost covering stumps and logs.

LENZITES BETULINA. FR.

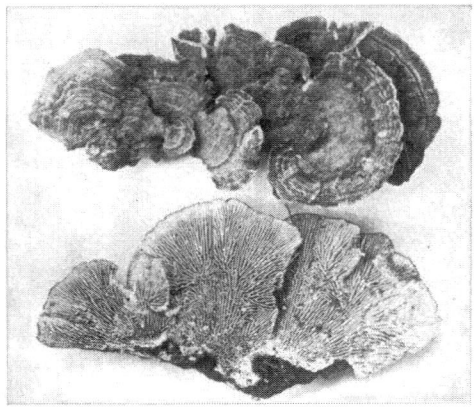

Plate XXVII. Figure 184 Lenzites betulina.

Figure 185 Lenzites betulina.

Betulina, from betula, a birch. This has a somewhat corky, leathery cap, firm and without zones, woolly, sessile, deeply grooved concentrically, margin of the same color.

The gills are radial, somewhat branching, and coming together again, sordid white or tan-color.

This species is wide-spread and is quite variable. It grows in the form of brackets. Figure 185 was photographed by Dr. Kellerman.

LENZITES SEPARIA. FR.

THE CHOCOLATE LENZITES.

The pileus is corky, leathery shells, with the upper surface marked with rough zones of various shades of brown; margin yellowish.

The gills are rather thick, branched, one running into another; yellowish. Stem obsolete. Growing on limbs and branches, especially of the fir tree.

LENZITES FLACCIDA. FR.

FLACCID LENZITES.

Figure 186   Lenzites flaccida. Two-thirds natural size.

Flaccida means limp, flaccid. Pileus is coriaceous, thin, flaccid, unequal, hairy, zoned, pallid, more or less flabelliform, imbricated.

The gills are broad, crowded, straight, unequal, branched, white, becoming pallid. Spores are 5×7.

This is a very attractive plant and quite common. It runs almost imperceptibly into Lenzites betulina. It is found on stumps and trunks.

LENZITES VIALIS. PK.

Pileus is corky, almost woody, firm, zoned.

Gills are thick, firm, serpentine.

Stem, none.

SCHIZOPHYLLUM. FR.

Schizophyllum is from two Greek words, meaning to split, and a leaf.

The pileus is fleshy and arid. The gills are corky, fan-like, branched, united above by the tomentose pellicle, bifid, split longitudinally at the edge. The spores somewhat round and white.

The two lips of the split edge of the gills are commonly revolute. This genus is far removed from the type of Agaricini. It grows on wood and is very common. Stevenson.

SCHIZOPHYLLUM COMMUNE. FR.

Figure 187 Schizophyllum commune.

This is a very common plant, growing in the woods on branches and decayed wood, where it can be found in both winter and summer.

The pileus is thin, adnate behind, somewhat extended, more or less fan-shaped or kidney-shaped, simple, often much lobed, narrowed behind to the point of attachment; whitish, downy, then strigose.

The gills are radiating, gray, then brownish-purple, and sometimes white, branched, split along the edges and rather deeply rolled backwards. The spores are nearly round, 5–6μ.

This is a very common species all over the world. I found it in the winter of 1907 on decayed shade-trees along the streets of Chillicothe. It seems to be partial to maple timber. Some call this S. alneum. It is very easily identified from its purple gills being split.

TROGIA. FR.

Trogia is so called in honor of the Swiss botanist, Trog.

The pileus is nearly membranaceous, soft, quite tough, flaccid, dry, flexible, fibrillose, reviving when moist.

The gills are fold-like, venose, narrow, irregular, crisped.

TROGIA CRISPA. FR.

Crispa means crisp or curled. The pileus tough, cup-shaped, reflexed, lobed, villous, whitish or reddish toward the attachment, often tan-colored.

The gills are quite narrow, vein-like, irregular, more or less branched, blunt on the edge, white or bluish-gray, quite crisped, edge not channeled.

The caps are usually very much crowded and imbricated. It revives during wet weather and is found throughout the year, generally on beech limbs in our woods.

CHAPTER III

THE ROSY-SPORED AGARICS

The spores of this series are of great variety of color, including rosy, pink, salmon-color, flesh-color, or reddish. In Pluteus, Volvaria, and most of Clitopilus, the spores are regular in shape, as in the white-spored series; in the other genera they are generally irregular and angular. There are not so many genera as in the other series and fewer edible species.

PLUTEUS. FR.

Pluteus means a shed, referring to the sheds used to make a cover for besiegers at their work, that they might be screened from the missiles of the enemy.

They have no volva, no ring on the stem. Gills are free from the stem, white at first then flesh-color.

PLUTEUS CERVINUS. SCHÆFF.

FAWN-COLORED PLUTEUS. EDIBLE.

Plate XXVIII. Figure 188 Pluteus cervinus. Natural size.

Cervinus is from cervus, a deer. The pileus is fleshy, bell-shaped, expanded, viscid in wet weather, smooth, except a few radiating fibrils when young, margin entire, flesh soft and white; color of the cap light-brown or fawn-color, sometimes sooty, often more than three inches across the cap.

The gills are free from the stem, broad, ventricose, unequal in length, almost white when young, flesh-colored when mature from the falling of the spores. The stem is solid, slightly tapering upward, firm, brittle, white, spread over with a few dark fibrils, generally crooked. The spores are broadly elliptical. The cystidia in the hymenium on the gills will be of interest to those who have a microscope.

This is a very common mushroom about Chillicothe. It is found on logs, stumps, and especially on old sawdust piles. Note how easily the stem is removed from the cap. This will distinguish it from the genus Entoloma. You cannot get anything in the market that will make a better fry than Pluteus cervinus; fried in butter, it is simply delicious. Found from May to October.

Figure 189 Pluteus cervinus.

PLUTEUS GRANULARIS. PK.

Figure 190 Pluteus granularis.

Pileus is convex, then expanded, slightly umbonate, wrinkled, sprinkled with minute blackish granules, varying in color from yellow to brown.

The gills are rather broad, close, ventricose, free, whitish, then flesh-colored.

The stem is equal, solid, pallid, or brown, usually paler at the top, velvety with a short, close pile.

The spores are subglobose, about .0002 inch in diameter. The plant is two to three inches high, pileus one to two inches broad, stem one to two lines thick. Peck, 38th Rep. N. Y. State Bot.

This is a much smaller species than P. cervinus, but its esculent qualities are quite as good. Found from July to October.

PLUTEUS EXIMIUS. SMITH.

Eximius, choice, distinguished. The pileus is fleshy, bell-shaped when young, expanded, beautifully fringed on the margin, larger than the cervinus.

The gills are free, broad, ventricose, white at first, then rose-colored, flesh white, and firm.

The stem is thick, solid, and clothed with fibers. Dr. Herbst, Fungal Flora of the Lehigh Valley.

I found some beautiful specimens in George Mosher's icehouse. I am very sorry I did not photograph them.

VOLVARIA. FR.

The spores of this genus are regular, oval, rosy-spored. The veil is universal, forming a perfect volva, distinct from the cuticle of the pileus. The stem is easily separable from the pileus. The gills are free, rounded behind, at first white, then pink, soft. Most of the species grow on wood. Some on damp ground, rich mold, in gardens, and in hot-houses. One is a parasite on Clitocybe nebularis and monadelphus.

VOLVARIA BOMBYCINA. (PERS.) FR.

THE SILKY VOLVARIA. EDIBLE.

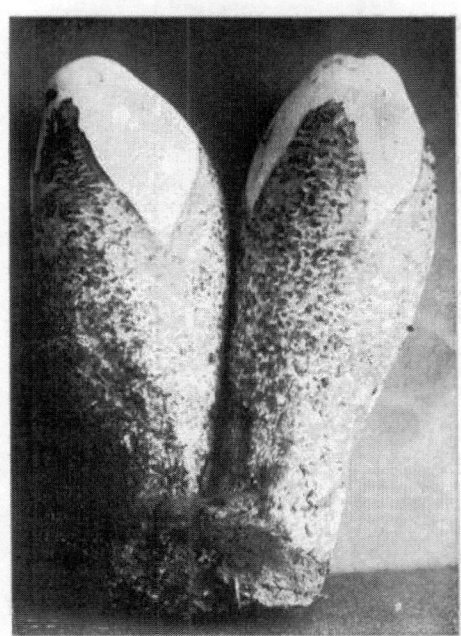

Plate XXIX. Figure 191   Volvaria bombycina.

The egg form of the V. bombycina showing the universal veil or volva bursting at the apex. These are unusually large specimens.

Figure 192   Volvaria bombycina. Two-thirds natural size. Entire plant white and silky.

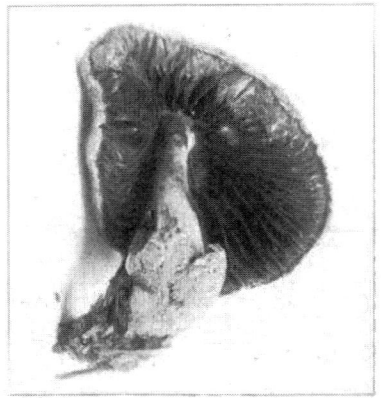

Figure 193 Volvaria bombycina. Two-thirds natural size, showing the gills, which are pink, then dark-brown.

Bombycina is from bombyx, silk. This plant is so called because of the beautiful silky lustre of the entire plant. The pileus is three to eight inches broad, globose, then bell-shaped, finally convex and somewhat umbonate, white, the entire surface silky, in older specimens more or less scaly, sometimes smooth at the apex. The flesh is white and not thick.

The gills are free, very crowded, broad, ventricose, flesh-colored, not reaching the margin, toothed. The stem is three to six inches long, tapering upward, solid, smooth, the tough volva remaining like a cup at the base. The spores are rosy in mass, smooth, and elliptical. The volva is large, membranaceous, somewhat viscid.

The plant in Figure 192 was found August 16th, on a maple tree where a limb had been broken, on North High Street, Chillicothe. Many people had passed along and enjoyed the shade of the trees but its discovery remained for Miss Marian Franklin, whose eyes are trained to see birds, flowers, and everything beautiful in nature.

I have found the plant frequently about Chillicothe, usually solitary; but on one occasion I found three specimens upon one trunk, apparently growing from the same mycelial mass. The caps of two of them were each five inches across. It usually grows on maple and beech. If you will observe a hollow beech, or sugar snag of which one side is broken away, leaving the sheltered yet open nestling place, you are very likely to find snugly enscounced in its decaying heart one or more specimens of these beautiful silky plants. The volva is quite thick and frequently the plant, when in the egg state, has the appearance of a phalloid. Found from June to October.

VOLVARIA UMBONATA. PECK.

THE UMBONATE VOLVARIA.

Figure 194 Volvaria umbonata. Two-thirds natural size. Entire plant white and silky.

Umbonata, having an umbo or conical projection like the boss of a shield. This plant is quite common on the richly manured lawns of Chillicothe. I have found it from June to October. The pileus is white or whitish, sometimes grayish, often smoky on the umbo; globose when young, bell-shaped, plane when fully expanded, umbonate, smooth; slightly viscid when moist, shining when dry, inch to an inch and a half broad. The flesh is white and very soft.

The gills are free, white at first, then from flesh-color to a reddish hue from the rosy-colored spores; some of the gills are dimidiate, somewhat crowded, broader in the middle.

The stem is two inches to two and a half long, tapering from the base up, smooth, cylindrical, hollow and firm. The volva is always present, free, variously torn, white and sometimes grayish.

The entire plant is silky when dry. I have found it growing in my buggy shed. It is not abundant, though quite common. I have never eaten it, but I do not doubt its edibility.

VOLVARIA PUSILLA. PERS.

Figure 195   Volvaria pusilla.

The pileus is explanate, white, fibrillose, dry, striate, center slightly depressed when mature.

The gills are white, becoming flesh-color, from the color of the spores, free, distant.

The stem is white, smooth, volva split to the base into four nearly equal segments. The spores are broadly elliptical, 5–6 mc.

This is the smallest species of the Volvaria. It grows on the ground among the weeds and is apt to escape the attention of the collector unless he knows its habitat. It is quite likely that V. parvula is the same plant as this. Also V. temperata, although it has a different habitat, seems to be very near this species. The plants in Figure 195 were collected in Michigan and photographed by Dr. Fischer. The volva is brown-tipped as shown in the figure given.

VOLVARIA VOLVACEA. BULL.

THE STOVE VOLVARIA.

It is called "The Stove Volvaria" because it has been found in old unused stoves. Pileus fleshy, soft, bell-shaped, then expanded, obtuse, virgate, with adpressed black fibrils. The gills are free, flesh-colored, and inclined to deliquesce. The stem is solid, subequal, white. The volva loose, whitish. The spores are smooth, elliptical.

This is a much smaller plant than the V. bombycina and grows in the ground. It is often found in hot-houses and cellars.

ENTOLOMA. FR.

Entoloma is from two Greek words; entos, within; loma, a fringe, referring to the inner character of the veil, which is seldom even apparent. The members of this genus have rosy spores which are prominently angular. There is neither volva, nor annulus. The gills are attached to the stem or notched near the junction of the gills and the stem. The pileus is fleshy and the margin incurved, especially when young. The stem is fleshy, fibrous, sometimes waxy, continuous with the pileus. It corresponds with Hypholoma, Tricholoma, and Hebeloma. It can always be separated from the rosy-spored genera by the notched gills. The flesh-colored spores and gills distinguish the Entoloma from the Hebeloma, which has ochre-spored ones, and Tricholoma, which has white ones.

All the species, so far as I know, have rather a pleasant odor, and for that reason it is highly necessary that the genus and species should be thoroughly known, as they are all dangerous.

ENTOLOMA RHODOPOLIUM. FR.

THE ROSE-GRAY ENTOLOMA.

Figure 196 Entoloma rhodopolium. Three-fourths natural size.

Rhodopolium is composed of two Greek words, rose and gray.

The pileus is two to five inches broad, hygrophanous; when moist dingy-brown or livid, becoming pale when dry, isabelline-livid, silky-shining; slightly fleshy, bell-shaped when young, then expanded and somewhat umbonate, or gibbous, at length rather plane and sometimes depressed; fibrillose when young, smooth when full grown; margin at first bent inwards and when large, undulated. Flesh white.

The gills adnate, then separating, somewhat sinuate, slightly distant, broad, white, then rose color.

The stem is two to four inches long, hollow; equal when smaller, when larger, attenuated upward; white pruinate at the apex, otherwise smooth; slightly striate, white, often reddish from spores. Spores 8–10×6–8μ. Fries.

The plant is found in mixed woods and is rather common. Captain McIlvaine reports it edible, but I have never eaten any of the Entolomas. Some of them have a bad reputation. Found in September and October.

ENTOLOMA GRAYANUM. PK.

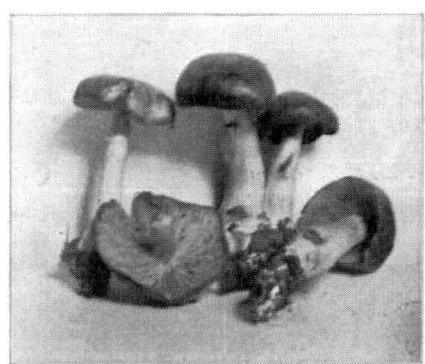

Figure 197   Entoloma grayanum. One-half natural size.

The pileus is convex to expanded, sometimes broadly umbonate, drab in color, the surface wrinkled or rugose, and watery in appearance. The flesh is thin and the margin incurved.

The gills are at first drab in color, but lighter than the pileus, becoming pinkish in age. The spores on paper are very light salmon-color. They are globose or rounded in outline, 5–7 angled, with an oil globule, 8–10μ, in diameter.

The stem is of the same color as the pileus, but lighter, striate, hollow, somewhat twisted, and enlarged below. The above accurate description was taken from Atkinson's Studies of American Fungi. The plants were found near a slate cut on the B. & O. railroad near Chillicothe. Not edible. This species and E. grisea are very closely related. The latter is darker in color, with narrower gills, and has a different habitat.

ENTOLOMA SUBCOSTATUM. ATKINSON N. SP.

Plate XXX. Figure 198   Entoloma subcostatum.

Mature plants showing broad gills and very thin flesh, also fibrous striate stems.

Subcostatum means somewhat ribbed, referring to the gills.

Plants gregarious or in troups or clusters, 6–8 cm. high; pileus 4–8 cm. broad; stems 1–1.5 cm. thick.

The pileus is dark-gray to hair-brown or olive-brown, often subvirgate with darker lines; gills light salmon-color, becoming dull; stem colored as the pileus, but paler; in drying the stems usually become as dark as the pileus.

Pileus subviscid when moist, convex to expanded, plane or subgibbous, not umbonate, irregular, repand, margin incurved; flesh white, rather thin, very thin toward the margin.

Gills are broad, 1–1.5 cm. broad, narrowed toward the margin of the pileus, deeply sinuate, the angles usually rounded, adnexed, easily becoming free, edge usually pale, sometimes connected by veins, sometimes costate, especially toward the margin of the pileus.

Basidia four-spored. Spores subglobose, about six angles, 8–10μ in diameter, some slightly longer in the direction of the apiculus, pale-rose under the microscope.

Stem even, fibrous striate, outer bark subcartilaginous, flesh white, stuffed, becoming fistulose.

Odor somewhat of old meal and nutty, not pleasant; taste similar.

Related to E. prunuloides, Fr., and E. clypeatum, Linn. Differs from the former in dark stem and uneven pileus, differs from the latter in being subviscid, with even stem, and pileus not umbonate and much more irregular, and differs from both in subcostate gills. Atkinson.

The specimens in Plate XXX grew in grassy ground on the campus of the Ohio State University, Columbus, Ohio. They were collected by R. A. Young and photographed by Dr. W. A. Kellerman, and through his courtesy I publish it. The plants were found the last of October, 1906.

ENTOLOMA SALMONEA. PK.

Figure 199   Entoloma salmonea.

Pileus thin, conical or campanulate, subacute, rarely with a minute papilla at the apex, smooth, of a peculiar soft, ochraceous color, slightly tinged with salmon or flesh color.

The gills and stem are colored like the pileus. Peck.

Dr. Peck says, "It is with some hesitation that this is proposed as a species, its resemblance to another species is so close. The only difference is found in its color and in the absence of the prominent cusp of that plant. In both species the pileus is so thin that in well dried specimens, slender, dark, radiating lines on it, mark the position of the lamellæ beneath, although in the living plant these are not visible." The plant in Figure 199 was found in Purgatory Swamp near Boston, by Mrs. Blackford. They are found in August and September.

ENTOLOMA CLYPEATUM. LINN.

THE BUCKLER ENTOLOMA.

Clypeatum, a shield or buckler. The pileus is slightly fleshy, lurid when moist, when dry gray and rather shining, streaked, spotted, campanulate, then expanded, umbonate, smooth, watery.

Gills just reaching the stem, rounded, ventricose, somewhat distant, minutely toothed, dirty flesh-color.

The stem is stuffed, then hollow, equal, round, clothed with small fibers, becoming pale, covered with a minute powdery substance. The flesh is white when dry. This plant will be distinguished usually by the amount of white mycelium at the base of the stem. Dr. Herbst remarks that it is a genuine Entoloma. It is certainly a beautiful plant when fully developed. It is found in woods and in rich grounds from May till September. Label it poisonous until its reputation is established.

CLITOPILUS. FR.

Clitopilus is from clitos, a declivity; pilos, a cap. This genus has neither volva nor ring. It is often more or less eccentric, margin at first involute; stem fleshy, diffused upward into the pileus; the gills are white at first, then pink or salmon-color as the plant matures and the spores begin to fall; decurrent, never notched. The pileus is more or less depressed, darker in the center. The spores are salmon-color, in some cases rather pale, smooth or warted. Clitopilus is closely related to Clitocybe, the latter having white gills, the former pink. It differs from Entoloma just as Clitocybe differs from Tricholoma. It can always be distinguished from Eccilia because the stem is never cartilaginous at the surface. It differs from the genus, Flammula, mainly in the color of the spores.

CLITOPILUS PRUNULUS. SCOP.

THE PLUM CLITOPILUS. EDIBLE.

Figure 200   Clitopilus prunulus.

Prunulus means a small plum; so called from the white bloom covering the plant.

The pileus is two to four inches broad, fleshy, firm; at first convex, then expanded, at length becoming slightly depressed, often eccentric, as will be seen in Figure 200; whitish, often covered with a frost-like bloom, margin often wavy, bending backward.

The gills are strongly decurrent, comparatively few of full length, white, then flesh-color.

The stem is solid, white, naked, striate, short. Spores, 7–8 × 5.

This is one of the most interesting plants because of the various forms it presents.

I have found it in various parts of the state and frequently about Chillicothe. It has a pleasant taste, and an odor reminding you of new meal. It is tender and its flavor is excellent.

Found in woods or open woods, especially where it is damp, and under beech trees, as well as oak. Found from June to October.

The plants in Figure 200 were collected near Ashville, N. C., and photographed by Prof. H. C. Beardslee.

CLITOPILUS ORCELLUS. BULL.

THE SWEET-BREAD CLITOPILUS. EDIBLE.

Figure 201   Clitopilus orcellus.

Orcellus is a diminutive meaning a small cask; from orca, a cask.

The pileus is fleshy, soft, plane, or slightly depressed, often irregular, even when young; slightly silky, somewhat viscid when moist; white or yellowish-white, flesh white, taste and odor farinaceous.

The gills are deeply decurrent, close, whitish, then flesh-color.

The stem is short, solid, flocculose, often eccentric, thickened above. The spores are elliptical, $9–10 \times 5\mu$. Peck, 42d Rep. N. Y.

This plant resembles the Plum mushroom, C. prunulus, very closely in appearance, taste and odor, but it is considerably smaller. It grows in wet weather, in open fields and lawns. It is quite widely distributed in our state, having found it in Salem, Bowling Green, Sidney, and Chillicothe. I frequently find it associated with Marasmius oreades. The specimens in Figure 201 were found near Ashville, N. C., and were photographed by Prof. H. C. Beardslee. Found from July to October.

CLITOPILUS ABORTIVUS. B. AND C.

THE ABORTIVE CLITOPILUS. EDIBLE.

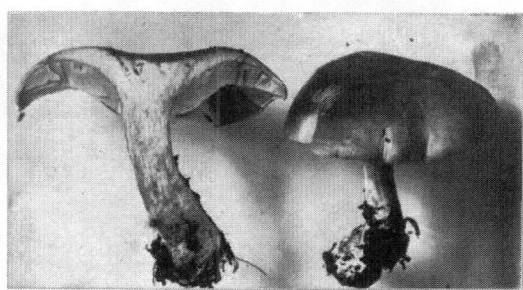

Figure 202   Clitopilus abortivus. Two-thirds natural size, showing the grayish-brown cap and solid stem.

Abortivus means abortive or imperfectly developed; so called from its many irregular and undeveloped forms.

The pileus is fleshy, firm, convex, or nearly plane, regular or irregular, dry, clothed with a minute silky tomentum, becoming smoother with age, gray or grayish-brown, flesh white, taste and odor subfarinaceous.

The gills are slightly or deeply decurrent, at first whitish or pale gray, then flesh-colored. Spores irregular, $7.5-10 \times 6.5\mu$.

The stem is nearly equal, solid, minutely flocculose, sometimes fibrous, striate, paler than the pileus. Peck, 42d Report N. Y.

There are often three forms of this plant; a perfect form, an imperfect form, and an abortive form as will be seen in Figure 203. The abortive forms seem to be more common, especially in this locality. They will be taken at first to be some form of puff-ball. They are found in open woods and in ravines. I found some very fine specimens under beech trees on Cemetery Hill. They are, however, widely distributed over the state and the United States. The specimens in Figure 203 were collected near Ashville and photographed by Prof. Beardslee.

Figure 203   Clitopilus abortivus. Abortive forms. Edible.

CLITOPILUS SUBVILIS. PK.

THE SILKY-CAPPED CLITOPILUS. EDIBLE.

Subvilis means very cheap, insignificant.

The pileus is thin, centrally depressed or umbilicate, with the margin decurved, hygrophanus, dark-brown, striate on the margin when moist, taste farinaceous.

The gills are subdistant, adnate, or slightly decurrent, whitish when young, then flesh-colored.

The stem is slender, brittle, rather long, stuffed or hollow, glabrous, colored like the pileus or a little paler.

The spores are angular, 7.5–10μ. Peck, 42d Rept.

This plant is distinguished from Clitopilus villis by its shining pileus, widely separated gills, and farinaceous taste. Found on Ralston's Run and in Haynes' Hollow, near Chillicothe, from July to October.

CLITOPILUS NOVEBORACENSIS. PK.

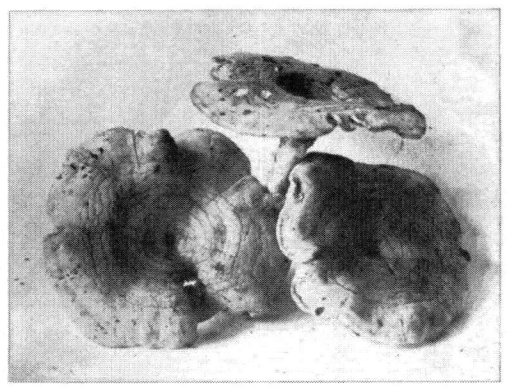

Figure 204   Clitopilus Noveboracensis. Two-thirds natural size.

Noveboracensis, the New York Clitopilus. Pileus thin, convex, then expanded or slightly depressed; dingy-white, cracked in areas or concentrically rivulose, sometimes obscurely zonate; odor farinaceous, taste bitter.

Gills narrow, close, deeply decurrent, some of them forked, white, becoming dingy, tinged with yellow or flesh-color.

Stem equal, solid, colored like the pileus, the mycelium white, often forming white, branching, root-like fibers. Spores globose.

Prof. Beardslee thinks that this species is doubtless identical with C. popinalis of Europe. He has submitted specimens and photographs to European mycologists, who hold to this view.

I found this plant quite abundant on the Huntington Hills after heavy rains in August. Their season is from August to October. The specimens in Figure 204 were found growing among leaves after a heavy rain October 10th. The plants have a tendency to turn blackish if they are bruised in handling them.

Var. brevis. This is so called from its short stem. The margin of the pileus is pure white when moist. Gills attached to the stem or slightly decurrent.

ECCILIA. FR.

Eccilia is from a Greek verb which means "I hollow out"; so called because the hollow cartilaginous stem expands upward into a membranaceous pileus, whose margin at first is incurved. Gills decurrent, attenuated behind.

This genus corresponds with Omphalia and is separated from Clitopilus by the cartilaginous, smooth stem.

ECCILIA CARNEO-GRISEA. B. & BR.

THE FLESH-GRAY ECCILIA. EDIBLE.

Figure 205   Eccilia carneo-grisea. Caps dark-gray or slate color. Gills rosy.

Carneo-grisea means fleshy-gray.

The pileus is one inch or more broad, umbilicate, dark-gray or grayish flesh color, finely striate, margin darkened with micaceous particles.

The gills are distant, adnate, decurrent, rosy, slightly undulate, margin irregularly darkened.

The stem is one to two inches long, slender, smooth, hollow, wavy, same color as the pileus, white tomentose at the base.

Spores irregularly oblong, rough, $7 \times 5\mu$.

It is found from Nova Scotia through the Middle West. It is commonly reported in fir and pine woods but I find it on the hillsides about Chillicothe in mixed woods. It is frequently found here associated with Boletinus porosus.

Found in July, August, and September.

ECCILIA POLITA. PERS.

Polita means having been furbished.

Figure 206　Eccilia polita. Natural size. Caps hair-brown to olive, umbilicate.

The pileus is one inch or more broad, convex, umbilicate, somewhat membranaceous, watery, livid or hair-brown to olive, smooth, shining when dry, finely striate on the margin.

The gills are slightly decurrent, crowded, irregular or uneven, flesh color.

The stem is cartilaginous, stuffed or hollow, lighter in color than the pileus, equal or sometimes slightly enlarged at the base, polished from which the specific name is derived.

This is a larger plant than E. carneo-grisea; and it differs materially in the character of its spores, which are strongly angled and some of them square, $10-12\mu$ in diameter, with a prominent mucro at one angle. It is found in the woods from September to frost.

LEPTONIA. FR.

Leptonia means slender, thin.

The spores are salmon-color and irregular. The pileus is never truly fleshy, cuticle always torn into scales, disk umbilicate, and often darker than the margin which is at first incurved.

The gills are attached to the stem and easily separated in old plants. The stem is rigid, with cartilaginous bark, hollow or stuffed, smooth, shining, often dark-blue, confluent with the cap.

LEPTONIA INCANA. FR.

THE HOARY LEPTONIA.

Incana means hoary or grayish-white.

The pileus is about one inch broad, somewhat membranaceous, convex, then plane, depressed in the center, smooth, with a silky lustre, margin striate.

The gills are attached to the stem, brcad, somewhat distant, white, then greenish.

The stem is hollow, shining, smooth, brownish-green. The spores are very irregular, dull-yellowish, pink, rough, 8–9μ.

It is frequently found in pastures after warm rains. They grow in clusters, and have the odor of mice to a marked degree.

LEPTONIA SERRULATA. PERS.

SAW LEPTONIA.

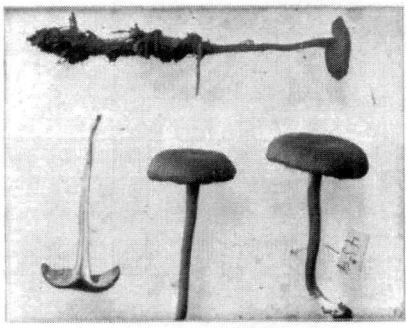

Figure 207   Leptonia serrulata.

Serrulata means saw-bearing, so named from the serrulate character of the gills.

The pileus is dark-blue, flesh thin, umbilicate, depressed, without striate, squamulose.

The gills are attached to the stem, with a dark serrulate edge.

The stem is thin, cartilaginous, paler than the pileus.

NOLANEA. FR.

Nolanea means a little bell, so called from the shape of the pileus.

It is rosy-spored. The stem is cartilaginous and hollow. The pileus is submembranaceous, thin, bell-shaped, papillate, margin straight, pressed close to the stem. The gills are free and not decurrent. They are found growing on the ground in the woods and pastures.

NOLANEA PASCUA. P.

THE PASTURE NOLANEA.

Pascua means pasture.

The pileus is membranaceous, conical, then expanded, slightly umbonate, smooth, striate, watery; when dry, shining like silk.

The gills are nearly free, ventricose, crowded, dirty-grayish.

The stem is hollow, fragile, silky-fibrous, striate. The spores are irregular, 9–10. They are found in pastures in summer and fall, after a rain.

NOLANEA CONICA. PK.

THE CONE NOLANEA.

The pileus is thin, membranaceous, conical, with a minute umbo or papilla, cinnamon-color, striatulate when moist.

The gills are light flesh-color, nearly free.

The stem is slender, straight, hollow.

Found in moist woods.

CLAUDOPUS. SMITH.

Claudopus is from two Greek words: claudos, lame; pus, foot.

The pileus is eccentric or lateral like the Pleuroti. The species were formerly placed in the Pleuroti and Crepidoti, which they very closely resemble, save in the color of the spores. This genus formerly included those plants which have lilac spores, but Prof. Fries limited it to those which have pink spores. The spores in some species are even and in others, rough and angular. The stem is either wanting or very short, hence its name. All are found on decayed wood.

CLAUDOPUS NIDULANS. PERS.

Figure 208   Claudopus nidulans. One-half natural size. Cap yellow or buff. Gills orange-yellow.

Nidulans is from nidus, a nest.

The pileus is sessile, sometimes narrowed behind into a short stem-like base, caps often overlapping one another, kidney-shaped, quite downy, the margin involute, hairy toward the margin, a rich yellow or buff color.

The gills are broad, moderately close, orange-yellow.

The spores are even, $3–5 \times 1\mu$, elongated, somewhat curved, delicate pink in mass. It is quite common in the woods about Chillicothe. A maple log from which I secured the specimen photographed in Figure 208 was completely covered and presented a beautiful sight. It has a rather strong and disagreeable odor. It is edible, but generally tough, and must be chopped very fine and cooked well. It is found in woods, on logs and stumps, from August to November.

CLAUDOPUS VARIABILIS. PERS.

Variabilis, variable or changeable. The pileus is white, thin, resupinate—that is the plant seems to be on its back, the gills being turned upward toward the light, quite downy, even, being fastened in the center to a short downy stem.

The gills are at first white, then of the color of the spores.

It is found on decaying limbs and branches in the woods. It is quite common everywhere.

CHAPTER IV

THE RUSTY-SPORED AGARICS.

The spores are of various shades of ochre yellow, rusty, rusty-brown, brown, yellowish-brown. The hymenophore is never free from the stem in the rusty-spored series, nor is there a volva.

PHOLIOTA. FR.

Pholiota, a scale. The members of this genus have rusty spores. These may be sepia-brown, bright yellowish-brown or light red. There is no volva, but there is a ring which is sometimes persistent, friable, and fugacious. In this respect it corresponds with the Armillaria among the white spored agarics. The pileus is fleshy. The gills are attached to the stem and sometimes notched with a decurrent tooth, tawny or rusty in color on account of the falling of the spores. Many species grow on wood, logs, stumps, and branches of trees, although others grow on the ground.

PHOLIOTA PRECOX. PERS.

THE EARLY PHOLIOTA. EDIBLE.

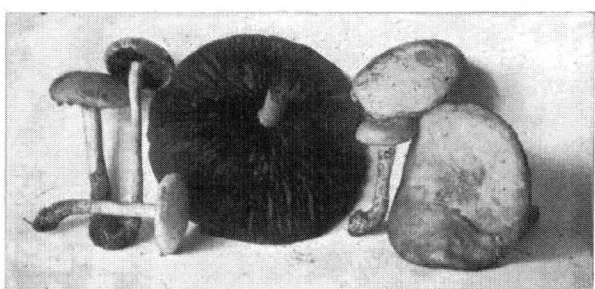

Figure 209 Pholiota precox. Two-thirds natural size. Caps whitish, often tinged with yellow.

Precox, early. Pileus is fleshy, soft, convex, then expanded, at length smooth, even, margin at first incurved; moist but not sticky, whitish, often with slight tinge of yellow or tan-color; when the plant is fully matured it is often upturned and fluted.

The gills are attached to the stem and slightly decurrent by a tooth, moderately broad, crowded, unequal, creamy white, then rusty-brown. Spores brownish, $8-13\times6-7\mu$.

The stem is stuffed, then hollow, often striate above the ring, rather slender, sometimes mealy, skin peeling readily, whitish. The spores are rusty-brown and elliptical. The caps are from one to two inches broad, and the stem is from two to three inches long. The veil is stretched like a drumhead from the stem to the margin of the cap. It varies in manner of breaking; sometimes it separates from the margin of the cap and forms a ring around the stem; again, but little remains on the stem and much on the rim of the cap.

It appears every year on the Chillicothe high school lawn. The gills are creamy-white when the cap first opens, but they soon turn to a rusty-brown. It comes in May. I have never found it after June. I am always delighted to find it for it is always appetizing at that season. Look for them on lawns and pastures and in grain fields.

PHOLIOTA DURA. BOLT.

THE HARD PHOLIOTA. EDIBLE.

Figure 210   Pholiota dura. One-half natural size. Caps tawny tan-color.

Dura, hard; so called because the surface of the cap becomes quite hard and cracked. The pileus is from three to four inches or more broad, very compact, convex, then plane, cuticle often very much cracked, margin even, tawny, tan-color, sometimes quite brown.

The gills are firmly attached to the stem, somewhat decurrent with a tooth, ventricose, livid, then a brown rusty color. Spores elliptical, $8-9\times5-6\mu$.

The stem is stuffed, hard, externally fibrous, thickened toward the apex, sometimes ventricose, often irregularly shaped.

On June 6th, 1904, I found Mr. Dillman's garden on Hickory street, Chillicothe, white with this plant. Some were very large and beautiful and I had an excellent opportunity to observe the irregularity in the form of the stem. Some years previous I found a garden in Sidney, Ohio, equally filled. In the fall of 1905 I was asked to drive out about seven miles from Chillicothe to see a wheat-field, the last of October, that was white with mushrooms. I found them to be of this species.

Only the young plants should be used, as the older ones are a bit tough.

PHOLIOTA ADIPOSA. FR.

THE FAT OR PINEAPPLE PHOLIOTA. EDIBLE.

Figure 211   Pholiota adiposa. Two-thirds natural size. Caps saffron-yellow.

Adiposa is from adeps, fat. The pileus is showy, deep-yellow, compact, convex, obtuse, slightly umbonate, quite viscid when moist, shining when dry; cuticle plain or broken into scales which are dark-brown, the margin incurved; the flesh is saffron-yellow, thick at the center and thinning out toward the margin.

The gills are firmly attached to the stem, sometimes slightly notched, close, yellow, then rust-color with age. Spores elliptical, $7 \times 3\mu$.

The stem is equal, stuffed, tough, thickening at the base, brown below and yellow above, quite scaly.

The beautiful appearance of the tufts or clusters in which the Pineapple Pholiotas grow will attract the attention of an ordinarily unobservant beholder. The scales on the cap seem to contract and rise from the surface and sometimes disappear with age. The caps of mushrooms should not ordinarily be peeled before cooking, but it is better to peel this one.

The ring is slight and the specimens represented here were found on a stump in Miss Effie Mace's yard, on Paint Street, Chillicothe.

PHOLIOTA CAPERATA. PERS.

THE WRINKLED PHOLIOTA. EDIBLE.

Plate XXXI. Figure 212   Pholiota caperata.

Caperata means wrinkled.

The pileus is three to four inches broad, fleshy, varying from a clay to a yellowish color, at first somewhat egg-shaped, then expanded, obtuse, wrinkled at the sides, the entire cap and especially at the center is covered with a white superficial flocci.

The gills are adnate or attached to the stem, rather crowded, this, somewhat toothed on their edges, clay-cinnamon color. Spores elliptical, $12 \times 4.5\mu$.

The stem is four to five inches long, solid, stout, round, somewhat bulbous at the base, white, scaly above the ring, which is often very slight, often only a trace, as will be seen on the left hand plant in Figure 212.

The spores are dark ferruginous when caught on white paper, but paler on dark paper.

The white superficial flocci will mark the plant. It has a wide distribution throughout the states. I found it in a number of places in Ohio and it is quite plentiful about Chillicothe. It is a favorite in Germany and it is called by the common people "Zigeuner," a Gypsy.

It is found in September and October.

PHOLIOTA UNICOLOR. FL. DAN.

Figure 213   Pholiota unicolor. Natural size.

Unicolor means of one color.

The pileus is campanulate to convex, subumbonate, hygrophanous, bay, then ochre, nearly even, never fully expanded.

The gills are subtriangular, adnate, seceding, broad, ochraceous-cinnamon. Spores $9–10 \times 5\mu$.

The stem is stuffed, then hollow, colored as the pileus, nearly smooth, ring thin but entire.

They are a late grower and found on well-decayed logs. They are quite common in our woods. Found in November. The plants in Figure 213 were found on the 24th of November, in Haynes' Hollow.

PHOLIOTA MUTABLIS. SCHAFF.

THE CHANGEABLE PHOLIOTA. EDIBLE.

Mutablis means changeable, variable. The pileus two to three inches broad, fleshy; deep cinnamon when moist, paler when dry; margin rather thin, transparent; convex, then expanded, sometimes obtusely umbonate, and sometimes slightly depressed; even, quite smooth, flesh whitish and taste mild.

The gills are broad, adnate, slightly decurrent, close, pale umber, then cinnamon-color.

The stem is two to three inches long, slender, stuffed, becoming hollow, smooth above or minutely pulverulent, and pale, below slightly scaly up to the ring, and darker at the base, ring membranaceous, externally scaly. The spores are ellipsoid, $9–11 \times 5–6\mu$.

I find this specimen growing in a cæspitose manner on decayed wood. It is quite common here late in the season. I found some very large specimens on Thanksgiving day, 1905, in Gallia County, Ohio. It is one of the latest edible plants.

PHOLIOTA HETEROCLITA. FR.

BULBOUS-STEMMED PHOLIOTA.

Figure 214   Pholiota heteroclita. Natural size. Caps whitish or yellowish.

Heteroclitus means leaning to one side, out of the center.

The pileus is three to six inches broad, compact, convex, expanded, very obtuse, rather eccentric, marked with scattered, innate, adpressed scales, whitish or yellowish, sometimes smooth when dry, viscid if moist.

The gills are very broad, at first pallid, then ferruginous, rounded, adnexed.

The stem is three to four inches long, solid, hard, bulbous at the base, fibrillose, white or whitish; veil apical, ring fugacious, appendiculate. The spores are subelliptical, $8-10 \times 5-6\mu$.

This species has a strong and pungent odor very much like horse-radish. It grows on wood and its favorite hosts are the poplar and the birch. It is found at almost any time in the fall. The specimens in the Figure 214 were found in Michigan and photographed by Dr. Fischer, of Detroit.

PHOLIOTA AUREVELLA. BATSCH.

GOLDEN PHOLIOTA.

Aurevella is from auri-vellus, a golden fleece.

The pileus is two to three inches in diameter, bell-shaped, convex, gibbous, tawny-yellow, with darker scales, rather viscid.

The gills are crowded, notched behind, fixed, very broad, plane, pallid olive, at length ferruginous.

The stem is stuffed, nearly equal, hard, various in length, curved, with rusty adpressed squamules, ring rather distant. On trunks of trees in the fall, generally solitary. Not very common.

PHOLIOTA CURVIPES. FR.

Curvipes, with a curved foot or stem. Pileus is rather fleshy, convex, then expanded, torn into adpressed floccose scales.

The gills are adnate, broad, white, then yellowish, at length tawny.

The stem is somewhat hollow, thin, incurved (from which it derives its name), fibrillose, yellow, as well as is the floccose ring. Spores 6–7×3–4. Cooke.

I found several specimens of this species at different times on one well rotted beech log on Ralston's Run, but was unable to find it on any other log in any woods near Chillicothe. I had trouble to place it till Prof. Atkinson helped me out. I found it from August to November.

PHOLIOTA SPECTABILIS. FR.

THE SHOWY PHOLIOTA.

Spectabilis, of notable appearance, worth seeing. The pileus is compact, convex, then plane, dry, torn into silky scales disappearing toward the margin, golden orange color, flesh yellow.

The gills are adnexed, rounded near the stem, slightly decurrent, crowded, narrow, yellow, then ferruginous.

The stem is solid, three to four inches high, quite thick, tough, spongy, thickened toward the base, even, bulbous, somewhat rooting. Ring inferior. I found the specimens in October and November. It may grow earlier. Found on decayed oak stumps.

PHOLIOTA MARGINATA. BATSCH.

THE MARGINATE PHOLIOTA. EDIBLE.

Figure 215 Pholiota marginata. Two-thirds natural size. Caps honey-colored and tan-colored.

Marginata means edged, margined; so called from the peripheral striæ of the pileus.

The pileus is rather fleshy, convex, then plane, smooth, moist, watery, striate on the margin, honey-colored when moist, tan-colored when dry.

The gills are firmly attached to the stem, crowded, unequal; when mature, of a dark reddish-brown from the shedding of the spores. Spores $7-8 \times 4\mu$.

The stem is cylindrical, smooth, hollow, of the same color as the pileus, covered with a frost-like bloom above the ring, which is distant from the apex of the stem and frequently disappears entirely.

It is quite common, being found on nearly every rotten log in our woods. It comes early and lasts till late in the fall. The caps are excellent when well prepared.

PHOLIOTA ÆGERITA. FR.

Ægerita is the Greek name for the black poplar; so called because it grows on decayed poplar logs. The pileus is fleshy, convex, then plane, more or less checked or rivulose, wrinkled, tawny, edge of the cap rather pale.

The gills are adnate, with a decurrent tooth, rather close, pallid, then growing darker.

The stem is stuffed, equal, silky-white, ring superior, fibrillose, tumid. Spores $10 \times 5\mu$.

Found in October and November, in the woods wherever there are decayed poplar logs.

PHOLIOTA SQUARROSOIDES. PK.

LIKE THE SCALY PHOLIOTA. EDIBLE.

Figure 216 Pholiota squarrosoides. Two-thirds natural size. Caps yellow or yellowish.

Squarrosoides means like Squarrosa. The pileus is quite firm, convex, viscid, especially when moist; at first densely covered with erect papillose or subspinose tawny scales, which soon separate from each other, revealing the whitish or yellowish color of the cap and its viscid character.

The gills are close, emarginate, at first whitish, then pallid or dull cinnamon color.

The stem is equal, firm, stuffed, rough, with thick squarrose scales, white above the thick floccose annulus, pallid or tawny below. The spores are minute, elliptical, .0002 inch long, .00015 inch broad.

They grow in tufts on dead trunks and old stumps, especially of the sugar maple. They closely resemble P. squarrosa. Found late in the fall. Its favorite haunt is the inside of a stump or within the protection of a log.

PHOLIOTA SQUARROSA. MULL.

THE SCALY PHOLIOTA. EDIBLE.

Plate XXXII. Figure 217   Pholiota squarrosa.

Squarrosa means scaly. The pileus is three to four inches broad, fleshy, bell-shaped, convex, then expanded; obtusely umbonate, tawny-yellow, clothed with rich brown scales; flesh yellow near the surface.

The gills are attached to the stem, with a decurrent tooth, at first yellowish, then of a pale olive, changing to rusty-brown in color, crowded, and narrow. The spores are elliptical, $8 \times 4\mu$.

The stem is three to six inches high, saffron yellow, stuffed, clothed with small fibers, scaly like the pileus, attenuated at the base from the manner of its growth. The ring is close to the apex, downy, rich brown, inclining to orange in color.

This is quite a common and showy mushroom. It is found on rotten wood, on or near stumps, growing out from a root underground, and is often found at the foot of trees. Only the caps of the young specimens should be eaten. It is found from August to late frost.

INOCYBE. FR.

Inocybe is from two Greek words meaning fiber and head; so called from the fibrillose veil, concrete with the cuticle of the pileus, often free at the margin, in the form of a cortina. The gills are somewhat sinuate, though they are sometimes adnate, and in two species are decurrent; changing color but not powdered with cinnamon. Spores are often rough but in other specimens are even, more or less brownish rust-color. Stevenson.

INOCYBE SCABER. MULL.

ROUGH INOCYBE. NOT EDIBLE.

Scaber means rough. The pileus is fleshy, conical, convex, obtusely gibbous, sprinkled with fibrous adpressed scales; margin entire, grayish-brown.

The gills are rounded near the stem, quite crowded, pale dingy-brown.

The stem is solid, whitish or paler than the pileus, clothed with small fibers, equal, veiled. The spores are elliptical, smooth, $11 \times 5\mu$.

It is found on the ground in damp woods. Not good.

INOCYBE LACERA. FR.

THE TORN INOCYBE.

Lacera means torn. The pileus is somewhat fleshy, convex, then expanded, obtuse, umbonate, clothed with fibrous scales.

The gills are free, broad, ventricose, white, tinged with red, light-gray. Spores are obliquely elliptical, smooth, $12 \times 6\mu$.

The stem is slender, short, stuffed, clothed with small fibers, naked above, reddish within.

Found on the ground where the soil is clayish or poor. Not good.

INOCYBE SUBOCHRACEA BURTII. PECK.

Figure 218   Inocybe subochracea Burtii. Natural size.

This is a very interesting species. It is thus described by Dr. Peck: "Veil conspicuous, webby fibrillose, margin of the pileus more fibrillose; stem longer and more conspicuously fibrillose. The well developed veil, and the longer stem, are the distinguishing characters of this variety."

The plants are found in mossy patches on the north hillsides about Chillicothe. The pale ochraceous yellow and the very fibrillose caps and stem will attract the attention of the collector at once. The caps are one to two and a half inches broad and the stem is two to three inches long.

INOCYBE SUBOCHRACEA. PECK.

Pileus thin, conical or convex, sometimes expanded, generally umbonate, fibrillose squamulose, pale ochraceous-yellow.

The gills are rather broad, attached, emarginate, whitish, becoming brownish-yellow.

The stem is equal, whitish, slightly fibrillose, solid. Peck.

This is a small plant from one to two inches high whose cap is scarcely over an inch broad. It grows in open groves where the soil is sandy. It is found on Cemetery Hill from June to October.

INOCYBE GEOPHYLLA, VAR. VIOLACEA. PAT.

Figure 219   Inocybe geophylla, var. violacea.

This is a small plant and has all the characteristics of Inocybe geophylla excepting color of cap and gills.

The pileus is an inch to an inch and a half broad, hemispherical at first, then expanded, umbonate, even, silky-fibrillose, lilac, growing paler in age.

The gills are adnexed, lilac at first, then colored by the spores. Spores 10×5.

The stem equal, firm, hollow, slightly violaceous.

This plant grows in September in mixed woods among the dead leaves. Its bright violet color will arrest the attention at once.

INOCYBE DULCAMARA. A. & S.

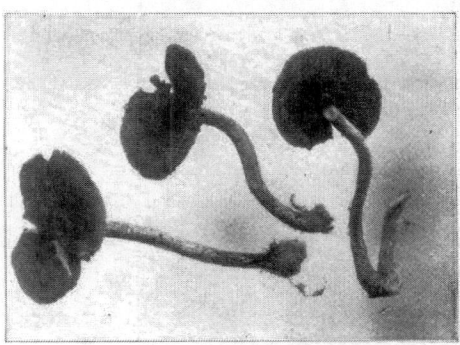

Figure 220   Inocybe dulcamara.

Dulcamara means bitter-sweet. The pileus is an inch to an inch and a half in diameter, rather fleshy, convex, umbonate, pilosely-scaly.

The gills are arcuate, ventricose, pallid olivaceous.

The stem is somewhat hollow, fibrillose and squamulose from the veil, farinaceous at the apex. Spores $8-10\times5\mu$.

Found from July to September, in grassy places.

INOCYBE CINCINNATA. FR.

Figure 221 Inocybe cincinnata. Two-thirds natural size. Caps scaly, dark or grayish-brown.

Cincinnata means with curled hair. This is quite an interesting little plant. It is found on Cemetery Hill, in Chillicothe, under the pine trees and along the walks where there is but little grass. It is gregarious and quite a hardy plant.

The pileus is fleshy, convex, then plane, quite squarrosely scaly, somewhat dark or grayish-brown.

The gills are grayish-brown with a tinge of violet at times; adnexed, rather close, ventricose.

The stem is solid, slender, scaly, somewhat lighter than the pileus. The spores are $8-10\times5\mu$.

This plant seems to be a late grower. I did not find it till about the 15th of October and it continued till the last of November. I had found two other species on the same hill earlier in the season. No Inocybes are good to eat.

INOCYBE PYRIODORA. PERS.

Pyriodora, smelling like a pear. The pileus is one to two inches broad, quite strongly umbonate, at first conical, expanded, covered with fibrous adpressed scales, in old plants the margin turned up, smoky or brown-ochre becoming pale.

The gills are notched at the stem, not crowded, dingy-white, becoming nearly cinnamon-brown, somewhat ventricose.

The stem is two to three inches long, stuffed, firm, equal, pale, apex pruinose, veil very fugacious. Flesh tinged with red.

Common in the woods in September and October. The plant is not edible.

INOCYBE RIMOSA. BULL.

THE CRACKED INOCYBE.

Rimosa, cracked. The pileus is one to two inches broad, shining, satiny, adpressed fibrillose, brown-yellow, campanulate, then expanded, longitudinally cracked.

The gills are free, somewhat ventricose, at first white, brownish-clay color.

The stem is one to two inches high, distant from the pileus, solid, firm, nearly smooth, bulbous, mealy white above. Spores smooth, $10–11 \times 6\mu$.

I. eutheles differs from this species in being umbonate; I. pyriodora in its strong smell. Many plants will often be found in one place in open woods or in cleared places. Their radiately cracked pilei, with the inner substance showing yellow through the cracks, will help to distinguish the species. Found from June to September.

HEBELOMA. FR.

Hebeloma is from two Greek words meaning youth and fringed. Partial veil fibrillose or absent. Pileus is smooth, continuous, somewhat viscid, margin incurved. The gills are notched adnate, edge of different color, whitish. The spores clay-color. All found on the ground.

HEBELOMA GLUTINOSUM. LINN.

Glutinosum, abounding in glue. The pileus is one to three inches broad, light-yellow, the disk darker, fleshy, convex, then plane, covered with a viscid gluten in wet weather; flesh is white, becoming yellow.

The gills are attached to the stem, notched, slightly decurrent, crowded, pallid, light yellow, then clay-color. Spores elliptical, $10–12 \times 5\mu$.

The stem is stuffed, firm, somewhat bulbous, covered with white scales, and mealy at the top. There is a partial veil in the form of a cortina.

Found among leaves in the woods. In wet weather the gluten is abundant. While it is not poisonous it is not good.

HEBELOMA FASTIBILE. FR.

OCHREY HEBELOMA. POISONOUS.

Fastibilis means nauseous, disagreeable; so called from its pungent taste and smell.

The pileus is one to three inches across, convex, plane, wavy, viscid, smooth, pale yellowish-tan, margin involute and downy.

The gills are notched, rather distant, pallid, then cinnamon; lachrymose.

The stem is two to four inches long, solid, subbulbous, white, fibrous scaly, sometimes twisted, often becoming hollow, veil evident. The spores are pip-shaped, $10 \times 6\mu$.

The odor is much the same as in H. crustuliniforme but it differs in having a manifest veil and more distant gills. Found in woods from July to October.

HEBELOMA CRUSTULINIFORME. BULL.

THE RING HEBELOMA. NOT EDIBLE.

Crustuliniforme means the form of a cake or bun.

The pileus is convex, then expanded, smooth, somewhat viscid, often wavy, yellowish-red, quite variable in size.

The gills are notched, thin, narrow, whitish then brown, crowded, edge crenulate, and with beads of moisture.

The stem is solid, or stuffed, firm, subbulbous, whitish, with minute white recurved flecks.

It is found in woods or about old sawdust piles. The plants sometimes grow in rings. September to November.

HEBELOMA PASCUENSE. PK.

Figure 222　Hebeloma pascuense. Natural size. Caps chestnut-color.

Pascuense, pertaining to pastures; referring to its habitat.

The pileus is convex, becoming nearly plane, viscid when moist, obscurely innately fibrillose; brownish-clay, often darker or rufescent in the center, the margin in the young plant slightly whitened by the thin webby veil; the margin of the cap more or less irregular, flesh white, the taste mild, odor weak.

The gills are close, rounded behind, adnexed, whitish, becoming pale ochraceous.

The stem is short, firm, equal, solid, fibrillose, slightly mealy at the top, whitish or pallid.

The spores are pale ochraceous, subelliptical. I found the plants in Figure 222 on Cemetery Hill late in November. It is a very low plant, growing under the pine trees and keeping close to the walks. The whitened margin of the young plant is a very good ear-mark by which to know this species.

PLUTEOLUS. FR.

Pluteolus means a small shed. It is the diminutive of pluteus, a shed or penthouse, from its conical cap.

The pileus is rather fleshy, viscid, conical or campanulate, then expanded; margin at first straight, adpressed to the stem. Stem somewhat cartilaginous, distinct from the hymenophore. Gills free, rounded behind.

PLUTEOLUS RETICULATUS. PERS.

Reticulatus means made like a net; from rete, a net, so called from the net-like appearance of veins on the cap.

The pileus is slightly fleshy, campanulate, then expanded, rugoso-reticulate, viscid, margin striate, pale violaceous.

The gills are free, ventricose, crowded, saffron-yellow, to ferruginous.

The stem is one to two inches long, hollow, fragile, fibrillose, inclined to be mealy at the top, white.

I have found only a few plants of this species in our state. It seems to be rare. The anastomosing veins on the cap and its pale violaceous color will mark the species. I have always found it on decayed wood. Captain McIlvaine speaks of finding it in quantities on the stems of fallen weeds and says it was tender and of fine flavor. September.

GALERA. FR.

Galera means a small cap. The pileus is more or less bell-shaped, margin straight, at first depressed to the stem, hygrophanous, almost even, atomate when dry, more or less membranaceous.

The gills are attached to the stem or with a decurrent tooth, as in Mycena.

The stem is cartilaginous, hollow, confluent with, but different in texture from the cap. The veil is often wanting, but when present is fibrous and fugacious. The spores are ochraceous ferruginous.

GALERA HYPNORUM. BATSCH.

THE MOSS-LOVING GALERA.

Hypnorum means of mosses; from hypna, moss.

The pileus is membranaceous, conic, campanulate, smooth, striate, watery when moist, pale when dry, cinnamon.

The gills are attached to the stem, broad, rather distant, cinnamon-colored, whitish on the edge.

The stem is slender, wavy, same color as the pileus, pruinose at the apex. This plant is very like G. tenera, only much smaller, and of a very different habitat. Found in mosses from June to October.

GALERA TENERA. SCHAEFF.

THE SLENDER GALERA. EDIBLE.

Figure 223 Galera tenera.

Tenera is the feminine form of tener, slender, delicate.

The pileus is somewhat membranaceous, at first cone-shaped, partially expanded, bell-shaped, hygrophanous, ochraceous when dry.

The gills are attached to the stem, crowded, rather broad, ascending, cinnamon-brown, the edges whitish, sometimes slightly serrate.

The stem is straight, hollow, fragile, rather shining; three to four inches long, equal or sometimes inclined to thicken downward, of nearly the same color as the pileus. The spores are elliptical and a dark rust-color, $12–13 \times 7\mu$.

You will frequently meet a variety whose cap and stem are quite pubescent but whose other characteristics agree with G. tenera. Prof. Peck calls it G. tenera var. pilosella.

Found in richly manured lawns and pastures. It is quite common. The caps, only, are good.

GALERA LATERITIA. FR.

THE BRICK-RED GALERA. EDIBLE.

Lateritia means made of brick, from later, a brick; so called because the caps are brick-colored.

The pileus is somewhat membranaceous, cone-shaped, then bell-shaped, obtuse, even, hygrophanous, rather pale yellow when wet, ochraceous when dry.

The gills are almost free, adnexed to the top of the cone, linear, very narrow, tawny or ferruginous.

The stem is three to four inches long, hollow, slightly tapering upward, straight, fragile, white pruinose, whitish. Spores are elliptical, $11–12 \times 5–6\mu$.

This plant resembles G. ovalis, from which it can be distinguished by its linear ascending gills and the absence of a veil.

Found on dung and in richly manured pastures, from July to frost.

GALERA KELLERMANI. PK. SP. NOV.

Figure 224   Galera Kellermani. Showing young plants.

Figure 225   Galera Kellermani. Showing older plants.

Kellermani is named in honor of Dr. W. A. Kellerman, Ohio State University.

The pileus is very thin, subovate or subconic, soon becoming plane or nearly so; striatulate nearly to the center when moist, more or less wavy and persistently striate on the margin when dry, minutely granulose or mealy when young, unpolished when mature, often with a few scattered floccose squamules when young, and sometimes with a few slight fragments of a veil adhering to the margin which appears as if finely notched by the projecting ends of the gills; watery-brown when moist, grayish-brown when dry, a little darker in the center; taste slight, odor faint, like that of decaying wood.

The gills are thin, close, adnate, a delicate cinnamon-brown becoming darker with age. The stem is two and a half to four cm. long, slender, equal, or slightly tapering upward; finely striate, minutely scurvy or mealy, at least when young; hollow, white. The spores are brownish ferruginous with a faint pinkish tint in mass, elliptic, 8–12×6–7μ. Peck.

Dr. Peck says the distinguishing features of this species are its broadly expanded or plane grayish-brown pileus, with its granulose or mealy surface, its persistently striate margin, and its very narrow gills becoming brownish with age. I have seen the plant growing in the culture beds in the greenhouse of the Ohio State University. It is a beautiful plant. Plants of all ages are shown in Figures 224 and 225.

GALERA CRISPA. LONGYEAR.

Figure 226   Galera crispa. Natural size. Cap ochraceous-brown.

Crispa means crisped; the specific name is based on the peculiar character of the gills which are always crisped as soon as the pileus is expanded.

The pileus is 1.5 to 3.5 cm. broad, membranaceous, persistently conico-campanulate, subacute, uneven and somewhat rivulose, ochraceous-brown on disk, lighter toward the margin which becomes crenulate and upturned in older specimens; slightly pruinose at first, rugulose and a little paler when dry.

The gills are adnexed, not crowded, rather narrow, interspersed with anastomosing veins; much crisped; at first nearly white, then becoming ferruginous from the spores.

The stem is 7 to 10 cm. long, tapering from a somewhat bulbous base, yellowish-white, pruinose at base, hollow, fragile. The spores are 8–10μ broad, 12–16μ long. Longyear.

They are found in grass on lawns and in pastures, June and July.

Dr. Peck, to whom specimens were referred, suggested that they may be a variety of G. lateritia, unless the peculiar character of the gills proved to be constant. Prof. Longyear has found the plant frequently in Michigan and it was found by him in the City Park, Denver, Col., in July, 1905.

Its distinguishing characteristic is sufficiently constant to make the recognition of the species a matter of ease. The plants in Figure 226 were photographed by Prof. B. O. Longyear.

GALERA OVALIS. FR.

THE OVAL GALERA.

The pileus is somewhat membranaceous, oval or bell-shaped, even, watery, dusky-rust color, somewhat larger than G. tenera.

The gills are almost free, ventricose, very broad, rust-colored.

The stem is straight, equal, slightly striate, nearly of the same color as the cap, about three inches long. Found in pastures where stock has been. I have found it in the Dunn pasture, on the Columbus pike, Ross County, O.

CREPIDOTUS. FR.

Crepidotus is from a Greek word meaning a slipper. The spores are dark or yellowish-brown. There is no veil. The pileus is excentric, dimidiate or resupinate. The flesh is soft. The stem is lateral or wanting, when present it is continuous with the cap. They generally grow on wood.

CREPIDOTUS VERSUTUS. PK.

Figure 227 Crepidotus versutus. Natural size. Caps pure white.

This is a very modest little plant growing on the underside of rotten logs or bark, thus, no doubt, escaping the attention of many. Sometimes it may be found growing from the side of a log, in which case it grows in a shelving form When growing under the log the upper side of the cap is against the wood and it is said to be resupinate.

The pileus is kidney-form, quite small, thin, pure white, covered with a soft whitish down.

The gills are radiate from the point of attachment of the cap, not crowded, whitish, then ferruginous from the spores.

CREPIDOTUS MOLLIS. SCHAEFF.

SOFT CREPIDOTUS.

The pileus is between subgelatinous and fleshy; one to two inches broad; sometimes solitary, sometimes imbricated; flaccid, even, smooth, reniform, subsessile, pallid, then grayish.

The gills are decurrent from base, crowded, linear, whitish then watery cinnamon. The spores are elliptical, ferruginous, $8-9 \times 5-6\mu$.

This species is widely distributed and quite common on decayed logs and stumps, from July to October.

NAUCORIA. FR.

Naucoria, a nut shell. The pileus is some shade of yellow, convex, inflexed, smooth, flocculent or scaly. The gills are attached to the stem, sometimes nearly free, never decurrent. The stem is cartilaginous, confluent with the cap but of a different texture, hollow or stuffed. The veil is absent or sometimes small traces may be seen attached to the rim of the pileus, in young plants in the form of flakes. The spores are of various shades of brown, dull or bright. They grow on the ground on lawns and rich pastures. Some on wood.

NAUCORIA HAMADRYAS. FR.

THE NYMPH NAUCORIA. EDIBLE.

Hamadryas, one of the nymphs whose life depended upon the tree to which she was attached.

The pileus is one to two inches broad, rather fleshy, convex, expanded, gibbous, even, bay-ferruginous when young and moist, pale yellowish when old.

The gills are attenuated, adnexed, almost free, rusty, slightly ventricose, somewhat crowded.

The stem is hollow, equal, fragile, smooth, pallid, two to three inches long. The spores are elliptical, rust-color, $13-14 \times 7\mu$.

This is quite a common species, often growing alone along pavements, under shade trees, and in the woods. The caps only are good. Found from June to November.

NAUCORIA PEDIADES. FR.

THE TAN-COLORED NAUCORIA. EDIBLE.

Figure 228   Naucoria pediades. Natural size.

Pediades is from a Greek word meaning a plain or a field, referring to its being found on lawns and pastures.

The pileus is somewhat fleshy, convex, then plane, obtuse or depressed, dry, finally opaque, frequently inclined to be minutely rivulose.

The gills are attached to the stem but not adnate to it, broad, subdistant, only a few entire brownish, then a dingy cinnamon.

The stem is pithy or stuffed, rather wavy and silky, yellowish, base slightly bulbous. The spores are of a brownish-rust color, $10–12 \times 4–5 \mu$.

If the small bulb at the base of the stem is examined, it will be found to be formed chiefly of mycelium rolled together around the base. It is found on lawns and richly manured pastures from May to November. Use only the caps. This plant is usually known as semiorbicularis.

NAUCORIA PALUDOSELLA. ATKINSON N. SP.

Plate XXXIII. Figure 229   Naucoria paludosella.

Showing mode of growth, clay-brown scales on the caps.

Paludosella is a diminutive of palus, gen. paludis, a swamp or marsh.

Plants six to eight cm. high; pileus two and a half to three cm. broad; stem three to four mm. thick.

Pileus viscid when moist, convex to expanded, in age somewhat depressed; clay color, darker over center, often with appressed clay brown scales with a darker color.

Gills raw umber to Mars brown (R), emarginate, adnate sometimes with a decurrent tooth, easily becoming free.

Cystidia on sides of gills none, edge of gills with large, hyaline, thin-walled cells, subventricose, sometimes nearly cylindrical, abruptly narrowed at each end with a slight sinus around the middle.

Spores subovate to subelliptical, subinequilateral, smooth, $7-9 \times 4-5\mu$, fuscous ferruginous, dull ochraceous under microscope.

Stem same color as pileus but paler, cartilaginous; floccose from loose threads or, in some cases, abundant threads over the surface; becoming hollow, base bulbous, the extreme base covered with whitish mycelium.

Veil rather thick, floccose, disappearing, leaving remnant on stem and margin of pileus when fresh. Atkinson.

Dr. Kellerman and I found this plant growing on living sphagnum, other mosses and on rotten wood on Cranberry Island, in Buckeye Lake, Ohio. Figure 229 will illustrate its mode of growth, and the older plant with upturned cap will show the conspicuous clay-brown scales of the pileus. The plants are found in September and October.

FLAMMULA. FR.

Flammula means a small flame; so called because many of the species have bright colors. The spores are ferruginous, sometimes light yellow. The cap is fleshy and at first usually inrolled, bright colored; veil filamentous, often wanting. The gills are decurrent or attached with a tooth. The stem is fleshy, fibrous, and of the same character as the cap.

The species of the Flammula are mostly found on wood. A few are found on the ground.

FLAMMULA FLAVIDA. SCHAEFF.

THE YELLOW FLAMMULA.

Flavida means yellow.

The pileus is fleshy, convex, expanded, plane, equal smooth, moist, margin at first inrolled.

The gills are firmly attached to the stem, yellow, turning slightly ferruginous.

The stem is stuffed, somewhat hollow, fibrillose, yellow, ferruginous at the base.

These plants are of a showy yellow, and are frequently found in our woods on decayed logs. They are found in July and August.

FLAMMULA CARBONARIA. FR.

THE VISCID FLAMMULA.

Figure 230   Flammula carbonaria.

Carbonaria is so called because it is found on charcoal or burned earth.

The pileus is quite fleshy, tawny-yellow, at first convex, then becoming plane, even, thin, viscid, margin of the cap at first inrolled, flesh yellow.

The gills are firmly attached to the stem, clay-colored or brown, moderately close.

The stem is stuffed or nearly hollow, slender, rigid, squamulose, pallid, quite short.

The spores are ferruginous-brown, elliptical, $7 \times 3.5\mu$.

I have found this species quite frequently where an old stump had been burned out. It is gregarious. I have only found it from September to November but the specimens in Figure 230 were sent to me in May, from Boston. They were found in great abundance in Purgatory Swamp, where the grass and vegetation had been burned away.

Flammula fusus. Batsch. Fusus means a spindle; so called from the spindle-shaped stem.

The pileus is compact, convex, then expanded, even, rather viscid, reddish-tan, flesh yellowish.

The gills are somewhat decurrent, pallid yellow, becoming ferruginous.

The stem is stuffed, firm, colored like the pileus, fibrillose, striate, attenuated and somewhat fusiform, rooting. The spores are broadly elliptical, $10 \times 4\mu$.

Found on well-decayed logs or on ground made up largely of decayed wood. Found from July to October.

FLAMMULA FILLIUS. FR.

The pileus is two to three inches broad, even, smooth, with rather viscid cuticle, pale orange-red with the disc reddish.

The gills are attached to the stem, arcuate, rather crowded, white, then pallid or tawny-yellow.

The stem is three to five inches long, hollow, smooth, pallid, reddish within. The spores are elliptical, $10 \times 5\mu$.

Found on the ground in the woods from July to October.

FLAMMULA SQUALIDA. PK.

Figure 231 Flammula squalida.

The pileus is one to one and a half inches broad, fleshy, convex, or plane, firm, viscose, glabrous, dingy-yellowish or rufescent, flesh whitish but in color similar to the pileus under the separate cuticle.

The gills are rather broad, adnate, pallid, becoming dark ferruginous.

The stem is one and a half to three inches long, one to two lines thick, slender, generally flexuose, hollow fibrillose, pallid or brownish, pale-yellow at the top when young; spores are brownish-ferruginous, .0003 inch long, .00016 broad. Peck.

It is found in bushy and swampy places. Dr. Peck says it is closely related to F. spumosa. Its dingy appearance, slender habit, more uniform and darker color of the pileus, and darker color of the lamellæ. It grows in groups. The plant in Figure 231 was found in Purgatory Swamp, by Mrs. Blackford. Found in August and September.

PAXILLUS. FR.

Paxillus means a small stake or peg. The spores as well as the entire plant are ferruginous. The pileus, with an involute margin, gradually unfolds. It may be symmetrical or eccentric. The stem is continuous with the hymenophore. The gills are tough, soft, persistent, decurrent, branching, membranaceous, usually easily separating from the hymenophore.

The distinctive features of this genus are the involute margin and the soft, tough, and decurrent gills which are easily separable from the hymenophore. Some grow on the ground, others grow on stumps and sawdust.

PAXILLUS INVOLUTUS. FR.

Involutus means rolled inward. The pileus is two to four inches broad, fleshy, compact, convex, plane, then depressed; viscid when moist, the cap being covered with a fine downy substance, so that when the margin of the cap unrolls the marks of the gills are quite prominent; yellowish or tawny-ochraceous, spotting when bruised.

The gills are decurrent, branched; anastomosing behind, near the stem; easily separating from the hymenophore.

Figure 232   Paxillus involitus

The stem is paler than the pileus, fleshy, solid, firm, thickened upward, brown spotted.

The flesh is yellowish, changing to reddish or brownish when bruised. The spores are rust-colored and elliptical, 8–10μ. It is found on the ground and decayed stumps. When found on the side of a decayed stump or a moss-covered log the stem is usually eccentric, but in other cases it is generally central.

It will be found around swampy places in an open woods. I found quite large specimens around a swamp in Mr. Shriver's woods near Chillicothe, but they were too far gone to photograph. It is edible but coarse. It appears from August to November. Some authors call it the Brown Chantarelle.

PAXILLUS ATROTOMENTOSUS. FR.

Atrotomentosus is from ater, black, and tomentum, woolly or downy.

Figure 233   Paxillus atrotomentosus.

The pileus is three to six inches broad, rust-color or reddish-brown, compactly fleshy, eccentric, convex then plane or depressed, margin thin, frequently minutely rivulose, sometimes tomentose in the center, flesh white, tinged with brown under the cuticle.

The gills are attached to the stem, slightly decurrent, crowded, branched at the base, yellowish-tawny, interspaces venose.

The stem is two to three inches long, stout, solid, elastic, eccentric or lateral, rooting, covered except at the apex with a dark-brown velvety down. The spores are elliptical, $5–6 \times 3–4\mu$.

I found the specimen in Figure 233 at the foot of an old pine tree on hillside at Sugar Grove, Ohio. I found the plant frequently at Salem, Ohio. It grows where the pine tree is a native. It is not poisonous. I do not regard it as very good. Found during August and September.

PAXILLUS RHODOXANTHUS. SCHW.

THE YELLOW PAXILLUS. EDIBLE.

Figure 234 Paxillus rhodoxanthus. Two-thirds natural size. Cap reddish-yellow or chestnut-brown. Gills yellow.

Rhodoxanthus means a yellow rose. The pileus is one to two inches broad, convex, then expanded, cushion-shaped, the epidermis of the cap often cracked showing the yellow flesh, resembling very much Boletus subtomentosus; reddish-yellow or chestnut-brown. The flesh is yellow and the cap dry.

The gills are decurrent, somewhat distant, stout, chrome yellow, occasionally forked at the base; anastomosing veins quite prominent, the cystidia being very noticeable.

The stem is firm, stout, of the same color as the cap, perhaps paler and more yellow at the base. The spores are oblong, yellow, $8–12 \times 3–5\mu$.

This is one of the most troublesome plants whose genus we have to settle. One of my mycological friends advised me to omit it from the genus altogether. It has been placed in various genera, but I have followed Prof. Atkinson and classed it under Paxillus. The plant is widely distributed. I find it frequently about Chillicothe. It is edible. Found in August, September and October. A full discussion of the plant will be found in Prof. Atkinson's book.

CORTINARIUS. FR.

Cortinarius is from cortina, a curtain, alluding to a cobwebby veil seen only in the comparatively young plants. Sometimes, parts of it will seem more substantial, remaining for a time on the margin of the cap or on the stem. The color of the pileus varies and its flesh and that of the stem are continuous. The hymenophore and the gills are continuous. The gills are attached to the stem, frequently notched, membranaceous, persistent, changing color, dry, powdery, with rusty-yellow spores which drop slowly. The veil and gills are the chief marks of distinction. The former is gossamer-like and separate from the cuticle, and the latter are always powdered. It is always essential to note the color of the gills in the young plant, since color is variable and sometimes shows only the slightest trace on the stem, colored from the falling spores.

Most authorities divide the genus into six tribes, from the appearance of the pileus. They are as follows:

I. Phlegmacium, meaning a shiny or clammy moisture. The pileus has a continuous pellicle, viscid when moist, stem dry, veil spider-webby.

II. Myxacium, meaning mucus, slime; so called from the glutinous veil. The pileus is fleshy, glutinous, rather thin; the gills are attached to the stem, slightly decurrent; the stem is viscid, polished when dry, slightly bulbous.

III. Inoloma, meaning a fibrous fringe; from is, genitive inos, a fibre; and loma, a fringe.

The pileus is fleshy, dry, not hygrophanous or viscid, silky with innate scales; the gills may be violaceous, pinkish-brown, yellow at first, then in all cases cinnamon-color from the spores; the stem is fleshy and somewhat bulbous; veil simple.

IV. Dermocybe, meaning a skinhead; from derma, skin, and cybe, a head.

The pileus thin and fleshy, entirely dry, at first clothed with silky down, becoming smooth in mature plants. The gills are changeable in color. The stem is equal or tapering downward, stuffed, sometimes hollow, smooth.

V. Telamonia, meaning a bandage or lint. The pileus is moist, watery, smooth or sprinkled with whitish superficial fibres, the remnants of the web-like veil. The flesh is thin, somewhat thicker at the center. The stem is ringed and frequently scaly from the universal veil, slightly veiled at the apex, hence almost with a double veil. The plants are usually quite large.

VI. Hydrocybe, meaning water-head or moist head. The pileus is moist, not viscid, smooth or sprinkled with a whitish superficial fibril, flesh changing color when dry, and rather thin. The stem is somewhat rigid and bare. Veil thin, fibrillose, rarely forming a ring. Gills also thin.

TRIBE I. PHLEGMACIUM.

CORTINARIUS PURPURASCENS. FR.

THE PURPLISH CORTINARIUS. EDIBLE.

Purpurascens means becoming purple or purplish; so named because the blue gills become purple when bruised.

The pileus is four to five inches broad, bay-brown, viscid, compact, wavy, spotted when old; often depressed at the margin, sometimes bending back; the flesh blue.

The gills are broadly notched, crowded, bluish-tan, then cinnamon-color, becoming purplish when bruised.

The stem is solid, bulbous, clothed with small fibres, blue, very compact, juicy; becoming purplish when rubbed. The spores are elliptical, $10–12 \times 5–6\mu$.

This is one of the delicious mushrooms to eat, the stem cooking tender as readily as the caps. I found it in Tolerton's woods, Salem, Ohio, and in Poke Hollow near Chillicothe. September to November.

CORTINARIUS TURMALIS. FR.

THE YELLOW-TAN CORTINARIUS. EDIBLE.

Turmalis means of or belonging to a troop or a squadron, turma; so called because occurring in groups, and not solitary.

The pileus is two to four inches broad, viscid when wet, ochraceous-yellow, smooth, discoid, flesh soft; veil extending from the margin of the cap to the stem in delicate arachnoid threads, best seen in young plants.

The gills are emarginate, decurrent, depending upon the age of the plant; crowded, somewhat serrated, whitish at first, then brownish-ochraceous-yellow. The remnants of the veil will usually show above the middle of the stem as a zone of minute striæ, darker than the stem.

I found specimens on Cemetery Hill under pine trees. September to November.

CORTINARIUS OLIVACEO-STRAMINEUS. KAUFF. N. SP.

Olivaceo-stramineus means an olive straw-color.

Pileus 4–7 cm. broad, viscid from a glutinous cuticle, broadly convex, slightly depressed in the center when expanded; margin incurved for some time; pale-yellow with an olivaceous tinge, slightly rufous-tinged when old; smooth or silky-fibrillose, disk sometimes covered with minute squamules, shreds of the partial veil attached to the margin when expanded. Flesh very thick, becoming abruptly thin toward the margin, white, dingy-yellowish in age, soon soft and spongy. Gills rather narrow, 7 mm. broad, sinuate-adnexed, whitish at first, then pale cinnamon, crowded, edge serratulate and paler. Stem 6–8 cm. long, with a slight bulb when young, from whose margin arises the dense partial veil; white and very pruinate above the veil, which remains as dingy fibrils stained by the spores; spongy and soft within, becoming somewhat hollow. Veil white with an olive tinge. Spores, $10–12 \times 5.5–6.5\mu$, granular within, almost smooth. Odor agreeable.

Kauffman says this resembles C. herpeticus, except that the gills when young are never violet-tinged.

I found this plant in Poke Hollow, near Chillicothe. It was unknown to me and I sent it to Dr. Kauffman of Michigan University to determine. I found it under beech trees, during October and November.

CORTINARIUS VARIUS. FR.

THE VARIABLE CORTINARIUS. EDIBLE.

Varius—Variable, so called because it varies in stature, its color and habit are unchangeable. The pileus is about two inches broad; compact, hemispherical, then expanded; regular, slightly viscid, thin margin at first incurved, sometimes with fragments of the web-like veil adhering.

The gills are notched, thin, crowded, quite entire, purplish, at length clay-colored or cinnamon.

The stem is solid, short, covered with threads, whitish, bulbous, from one and a half to two and a half inches long.

The plant is quite variable in size but constant in color. It is found in woods. I found specimens at Salem, Ohio, and at Bowling Green, Ohio. September to November.

CORTINARIUS CÆRULESCENS. FR.

THE AZURE-BLUE CORTINARIUS. EDIBLE.

Cærulescens, azure-blue. Pileus fleshy, convex, expanded, even, viscid, azure-blue, flesh soft, not changing color when bruised.

The gills are attached to the stem, slightly rounded behind, crowded, quite entire, at first of a pure dark blue, then rusty from the spores.

The stem is solid, attenuated upward, firm, bright violet, becoming pale, whitish, bulb growing less with age, fibrillose from vein. Spores elliptical. Neither the flesh nor the gills change color when bruised. This fact distinguishes it from C. purpurascens. When young the entire plant is more or less blue, or bluish-purple, and the color never entirely leaves the plant. In age it becomes somewhat spotted with yellow. The flesh is a little tough and needs to be stewed for some time. Found in Whinnery's woods, Salem, Ohio. September to October.

TRIBE II. MYXACIUM.

CORTINARIUS COLLINITUS. FR.

THE SMEARED CORTINARIUS. EDIBLE.

Figure 235 Cortinarius collinitus. One-half natural size. Caps purplish-brown, also showing veil.

Collinitus means smeared. The pileus is at first hemispherical, convex, then expanded, obtuse; smooth, even, glutinous, shining when dry; purplish when young, later brownish; at first incurved.

The gills are attached to the stem, rather broad, dingy-white or grayish-tan when young, then cinnamon.

The stem is solid, cylindrical, viscid or glutinous when moist, transversely cracking when dry, whitish or paler than the cap. The spores are elliptical, $12 \times 6\mu$. I found this species in Tolerton's woods, Salem, Ohio, St. John's woods, Bowling Green, Ohio, also on Ralston's Run near Chillicothe, where the specimens in Figure 235 were found. Both cap and stem are covered with a thick gluten. They grow, with us, in woods among leaves. The young plant has a development peculiar to itself. The cap varies greatly in color. The flesh is white or whitish. The peculiar bluish-white gills of the young plant will attract attention at once. It is found from September to November.

TRIBE III. INOLOMA.

CORTINARIUS AUTUMNALIS. PK.

THE FALL CORTINARIUS. EDIBLE.

Figure 236 Cortinarius autumnalis. Two-thirds natural size. Cap a dull rusty-yellow, also showing bulbous stem.

Autumnalis pertaining to fall. The pileus is fleshy, convex or expanded, dull rusty-yellow, variegated, or streaked with innate rust-colored fibrils.

The gills are rather broad, with a wide, shallow emargination.

The stem is equal, solid, firm, bulbous, a little paler than the pileus.

The height is three to four inches, breadth of pileus two to four inches. Peck.

The plant was named by Dr. Peck because it was found late in the fall. I found the plant on several occasions in September, 1905. It grew very sparingly in a mixed woods on a north hillside.

CORTINARIUS ALBOVIOLACEUS. PERS.

THE LIGHT VIOLET CORTINARIUS. EDIBLE.

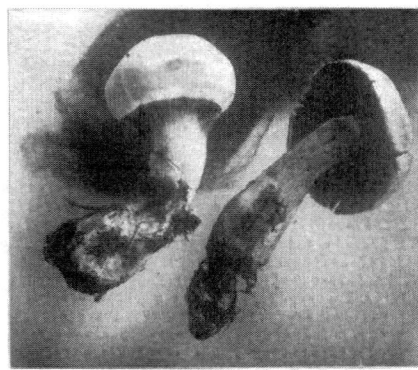

Figure 237   Cortinarius alboviolaceus. The caps are violet.

Alboviolaceus means whitish-violet.

The pileus is two to three inches broad, fleshy, rather thin, convex, then expanded, sometimes broadly subumbonate; smooth, silky, whitish, tinged with lilac or pale violet.

The gills are generally serrulate, whitish-violet, then cinnamon-color.

The stem is three to four inches long, equal or tapering upward, solid, silky, white, stained with violet, especially at the top, slightly bulbous, the bulb gradually tapering into the stem. Spores, 12×5–6μ. Peck's Report.

Sometimes the stem has a median ring-like zone, being violet above the zone and white below. The spider-like veil shows very plainly in the specimen on the left in Figure 237. In the plant on the right is shown the tapering stem from the base to the apex. These plants were found in Poke Hollow, September 21st. They are quite abundant there and elsewhere about Chillicothe. They are very good but not equal in flavor to C. violaceus. They are found in mixed woods. September to frost.

CORTINARIUS LILACINUS. PK.

THE LILAC-COLORED CORTINARIUS. EDIBLE.

The pileus is two to three inches broad, firm, hemispherical, then convex, minutely silky, lilac-color.

The gills are close, lilac, then cinnamon.

The stem is four to five inches long, stout, bulbous, silky-fibrillose, solid, whitish, tinged with lilac. Spores nucleate, $10 \times 6\mu$. Peck.

I have found this plant in but one place near Chillicothe. In Poke Hollow on a north hillside I have found a number of rare specimens. All were identified by Dr. Kauffman of Michigan University. All were found under beech trees within a very small radius. September and October.

CORTINARIUS BOLARIS. FR.

THE COLLARED CORTINARIUS.

The pileus is fleshy, obsoletely umbonate, growing pale, variegated with saffron-red, adpressed, innate, pilose scales.

The gills are subdecurrent, crowded, watery cinnamon.

The stem is two to three inches long, at first stuffed, then hollow, nearly equal, squamose.

Found under beech trees. Only occasionally found here.

CORTINARIUS VIOLACEUS. FR.

THE VIOLET CORTINARIUS. EDIBLE.

Figure 238   Cortinarius violaceus. Two-thirds natural size. Caps dark violet. Stems bulbous. Gills violet.

Violaceus, violet color. The pileus is convex, becoming nearly plane, dry, adorned with numerous persistent hairy tufts or scales; dark violet.

The gills are rather thick, distant, rounded, or deeply notched at the inner extremity; colored like the pileus in the young plant, brownish-cinnamon in the mature plant.

The stem is solid, clothed with small fibres; bulbous, colored like the pileus. The spores are slightly elliptical.

The Violet Cortinarius is a very beautiful mushroom and one easy of recognition. At first the whole plant is uniformly colored, but with age the gills assume a dingy ochraceous or brownish-cinnamon hue. The cap is generally well formed and regular, and is beautifully adorned with little hairy scales or tufts. These are rarely shown in figures of the European plant, but they are quite noticeable in the American plant, and should not be overlooked. The flesh is more or less tinged with violet. Peck. 50th Rep. N. Y. State Bot.

No one can fail to recognize this plant. The web-like veil in the young plant, the bulbous stem, and the violet tinge throughout will readily distinguish it. It grows in rich hilly country. It grows solitary, and in open woods.

TRIBE IV. DERMOCYBE.

CORTINARIUS CINNAMONEUS. FR.

THE CINNAMON CORTINARIUS. EDIBLE.

Figure 239   Cortinarius cinnamoneus. Two-thirds natural size. Caps cinnamon-brown. Stems yellow.

The pileus is thin, convex, nearly expanded, sometimes nearly plane, sometimes slightly umbonate, sometimes the pileus is abruptly bent downward; dry, fibrillose at least when young, often with concentric rows of scales on the margin, cinnamon-brown, flesh yellowish.

The gills are thin, close, firmly attached to the stem, slightly notched, decurrent with a tooth, becoming easily separated from the stem, shining, yellowish, then tawny-yellow.

The stem is slender, equal, stuffed or hollow, thin, clothed with small fibres, yellow, as is also the flesh. The spores are elliptical. This plant is so called because of its color, the entire plant being of a cinnamon-color. Sometimes there are cinnabar stains on the pileus. It seems to grow best under pine trees, but I have found it in mixed woods. My attention was called to it by the little Bohemian boys picking it when they had been in this country but a few days and could not speak a word of English. It is evidently like the European species. There is also a Cortinarius that has blood-red gills. It is var. semi-sanguineus, Fr. July to October.

The plants in Figure 239 were found on Cemetery Hill, Chillicothe, O.

CORTINARIUS OCHROLEUCUS. FR.

THE PALLID CORTINARIA.

Figure 240 Cortinarius ochroleucus. Two-thirds natural size, showing veil and bulbous form of stem.

Ochroleucus, meaning yellowish and white, because of the color of the cap. The pileus is an inch to two and a half inches broad, fleshy; convex, sometimes somewhat depressed in the center, often remaining convex; dry; on the center finely tomentose to minutely scaly, sometimes the scales are arranged in concentric rows around the cap; quite fleshy at the center, thinning out toward the margin; the color is a creamy to a deep-buff, considerably darker at the center.

The gills are attached to the stem, clearly notched, somewhat ventricose; in mature plants, somewhat crowded, not entire, many short ones, pale first, then clay-colored ochre.

Figure 241   Cortinarius ochroleucus. Two-thirds natural size, showing the developed plant.

The stem is three inches long, solid, firm, often bulbous, tapering upward, often becoming hollow, a creamy-buff.

The veil, quite beautiful and strongly persistent, forms a cortina of the same color as the cap but becoming discolored by the falling of the spores. In Figure 240 the cortina and the bulbous form of the stem will be seen.

Found along Ralston's Run. In beech woods from September to November.

TRIBE V. TELAMONIA.

CORTINARIUS MORRISII. PK.

Figure 242   Cortinarius Morrisii.

Morrisii is named in honor of George E. Morris, Ellis, Mass.

Pileus fleshy, except the thin and at length reflexed margin; convex, irregular, hygrophanous, ochraceous or tawny-ochraceous; flesh thin, colored like the pileus; odor weak, like that of radishes.

The gills are broad, subdistant, eroded or uneven on the edge; rounded behind, adnexed, pale-yellow when young, becoming darker with age.

The stem is nearly equal, fibrillose, solid, whitish or pale-yellow and silky at the top, colored like the pileus below and fibrillose; irregularly striate and subreticulate, the double veil whitish or yellowish-white and sometimes forming an imperfect annulus.

The spores are tawny-ochraceous, subglobose or broadly elliptic, nucleate, 8–10μ long, 6–7μ broad. Peck.

Pileus 3–10 cm. broad; stem 7–10 cm. long, 1–2 cm. thick.

They require moist and shady places and the presence of hemlock trees. They are found from August to October. The plants in Figure 242 were found near Boston by Mrs. E. B. Blackford.

CORTINARIUS ARMILLATUS. FR.

THE RED-ZONED CORTINARIUS. EDIBLE.

Armillatus means ringed; so called because the stem is banded with one or more rings, or red

bands. The pileus is two to four inches broad, fleshy, not compact, bell-shaped, then expanded, soon innately fibrillose and torn into scales, smooth when young, reddish-brick-color, margin thin, flesh dingy-pallid.

The gills are very broad, distant, adnate, slightly rounded, pallid, then dark-cinnamon.

The stem is fairly long, solid, bulbous, whitish, with two or three red zones, somewhat fibrillose. The spores $10 \times 6\mu$.

Figure 243 Cortinarius armillatus. Two-thirds natural size, showing the rings on the stem.

This is a very large and beautiful Cortinarius and it has such a number of striking ear marks that it can be easily recognized. The thin and generally uneven margin of the pileus and the one to four red bands around the stem, the upper one being the brightest, will distinguish this species from all others. It is found in the woods in September and October. In quite young specimens the collector will notice two well defined arachnoid veils, the lower one being much more dense. Prof. Fries speaks of them as follows: "Exterior veil woven, red, arranged in 2–4 distant cinnabar zones encircling the stem; partial veil continuous with the upper zone, arachnoid, reddish-white." The specimens in Figure 243 were collected in Michigan and photographed by Dr. Fischer of Detroit. A number of this species form a prize for the table.

CORTINARIUS ATKINSONIANUS. KAUFF.

Figure 244 Cortinarius Atkinsonianus. Caps waxy-yellow, bulbous stem, spider-like veil.

Atkinsonianus is named in honor of Prof. Geo. F. Atkinson.

The pileus is 8 cm. broad, expanded, wax-yellow or gallstone-yellow to clay-colored and tawny (Ridg.), colors very striking and sometimes several present at once; viscid, smooth, even, somewhat shining when dry. Flesh thick, except at margin, bluish-white like the stem, or paler, scarcely or not at all changing when bruised.

The gills are comparatively narrow, 6–8 mm., width uniform except near outer end, adnate, becoming slightly sinuate, purplish to yellow, then cinnamon.

The stem is violaceus-blue, 8 cm. long, 12–15 mm. thick, equal or slightly tapering upward, bulbous by a rather thick, marginate bulb 3 cm. thick, hung with fibrillose threads of the universal veil, which is a beautiful pale-yellow and clothes the bulb even at maturity; violaceous-blue within, solid. Spores $13–15\mu \times 7–8.5\mu$, very tubercular. Kauff.

The specimens in Figure 244 were found in Poke Hollow near Chillicothe. I have found them on several occasions. They are edible and of very good flavor. Found from September to frost. The specimens illustrate the spider-like veil that gives rise to the genus.

CORTINARIUS UMIDICOLA. KAUFF.

Figure 245   Cortinarius umidicola. One-half natural size. Caps pinkish-buff.

Umidicola means dwelling in moist places. Pileus as much as 16 cm. broad (generally 6–7 cm. when expanded), hemispherical, then convex and expanded, with the margin for a long time markedly incurved; young cap heliotrope-purplish with umber on disk, or somewhat fawn-colored, fading very quickly to pinkish-buff, in which condition it is usually found; margin when young with narrow strips of silky fibrils from the universal veil; pileus when old covered with innate, whitish, silky fibrils, hygrophanous; surface punctate, even when young. Flesh of stem and pileus lavender when young but soon fading to a sordid white, thick on disk, abruptly thin towards margin, soon cavernous from grubs. The gills are very broad, as much as 2 cm.; at first lavender, soon very pale-tan to cinnamon; rather distant, thick, emarginate with a tooth; at first plane, then ventricose; edge slightly serratulate, concolorous. Stem as much as 13 cm. long (usually 8 to 10 cm.), 1–2 cm. thick, usually thickened below and tapering slightly upwards, mostly thicker also at apex, rarely attenuate at the base, sometimes curved, always stout, solid, lavender above the woven, sordid white, universal veil, which at first covers the lower part as a sheath, but soon breaks up so as to leave a band-like annulus half way or lower down on the stem. The annulus is soon rubbed off, leaving a bare stem. Cortina violaceous-white. Spores 7–9×5–6, almost smooth. Kauffman.

The specimens in Figure 245 were gathered at Detroit, Michigan, and photographed by Dr. Fischer. They grow in groups in damp places, preferring hemlock trees.

CORTINARIUS CROCEOCOLOR. KAUFF. SP. NOV.

SAFFRON-COLORED CORTINARIUS. (TELAMONIA.)

Croceocolor means saffron-colored.

Pileus 3–7 cm. broad, convex then expanded, saffron-yellow, with dense, dark-brown, erect squamules on disk; whole surface has a velvety appearance and feel, scarcely hygrophanous, even; flesh of pileus yellowish-white, rather thin except on disk, slightly hygrophanous, scissile.

Gills cadmium-yellow (Ridg.), moderately distant, rather thick, emarginate, rather broad, 8–9 mm., width uniform except in front where they taper quickly to a point.

Stem 4–8 cm. long, tapering upwards from a thickened base, i.e., clavate-bulbous, 9–15 mm. thick below, peronate three-fourths of its length by the crome-yellow to saffron veil, paler above the veil, solid, saffron-colored within, hygrophanous, soon dingy; attached to strands of yellowish mycelium. Spores subspheroid to short elliptical, 6.5–8×5.5–6.5μ, echinulate when mature.

Found under beech trees in Poke Hollow near Chillicothe. Found in October.

CORTINARIUS EVERNIUS. FR.

Figure 246   Cortinarius evernius.

Evernius comes from a Greek word meaning sprouting well, flourishing.

The pileus is one to three inches broad, rather thin, between membranaceous and fleshy, at first conical, becoming bell-shaped, and finally expanded, very slightly umbonate, everywhere covered with silky, adpressed veil, usually purplish-bay when smooth, brick-red when dry, then pale ochraceous when old, at length cracked and torn into fibrils, very fragile, flesh thin and colored like the pileus.

The gills are attached to the stem, quite broad, ventricose, somewhat distant, purplish-violet, becoming pale, finally cinnamon.

The stem is three to five inches long, equal or attenuated downwards, often slightly striate, soft, violaceous, scaly from the remains of the white veil. The spores are elliptical, granular, $10 \times 7\mu$.

They grow in damp pine woods. The specimens in the photograph were gathered in Purgatory Swamp near Boston, and sent to me by Mrs. Blackford. They are found in August and September.

TRIBE VI. HYDROCYBE.

CORTINARIUS CASTANEUS. BULL.

THE CHESTNUT-COLORED CORTINARIUS. EDIBLE.

Figure 247   Cortinarius castaneus. Two-thirds natural size.

Castaneus, a chestnut. The pileus one inch or more broad, at first quite small and globose, with a delicate fibrillose veil, which makes the margin appear silvery; dark-bay or dirty-violet, often with a tawny tint; soon expanded, broadly umbonate, pileus often cracked on the margin and slightly upturned.

The gills are fixed, rather broad, somewhat crowded, violet-tinged, then cinnamon-brown, ventricose. Spores, $8 \times 5\mu$.

The stem is one to three inches high, inclined to be cartilaginous, stuffed, then hollow, even, lilac-tinged at the top, white or whitish below the veil, the whole stem beautifully fibrillose, veil white.

This plant is very abundant on Cemetery Hill, growing under pine trees. The caps are small, but they grow in such profusion that it would not be difficult to secure enough for a meal. They compare very favorably with the Fairy Ring mushroom in flavor. They have little or no odor. Found in October and November.

CHAPTER V

PURPLE-BROWN SPORED AGARICS.

AGARICUS. LINN. (PSALLIOTA. FR.)

The pileus is fleshy, but the flesh of the stem is of different texture from that of the pileus, veil universal, concrete with the cuticle of the pileus, and fixed to the stem, forming a ring which soon disappears in some species; the stem is readily separated from the cap and the gills are free from the stem or slightly adnexed, white at first, then pink, afterwards purple-brown.

All the species grow in rich ground, and it includes many of our valuable food mushrooms.

AGARICUS CAMPESTRIS. LINN.

THE MEADOW MUSHROOM. EDIBLE.

Figure 248 Agaricus campestris. Two-thirds natural size.

Campestris, from campus, a field. This is perhaps the widest known of all mushrooms, familiarly known as the "Pink-gilled mushroom." It is the species found in the markets. It is the only species which is sure to respond to the methods of cultivation.

It is the same species which is bought in cans at the store.

In very young plants the pileus is somewhat globular, as will be seen in the small plants in the front row in Figure 248. The edge is connected with the stem by the veil; then round convex, then expanding, becoming almost flat; surface dry, downy, even, quite scaly, varying in color from creamy-white to a light-brown; margin extending beyond the gills, as will be seen in Figure 249 in the one on the extreme right.

The gills, when first revealed by the separation of the veil, are of a delicate pink hue, but with advancing age this generally deepens to a dark-brown or blackish-brown color.

The stem is rather short, nearly equal, white or whitish; the substance in the center is more spongy than the exterior, hence it is said to be stuffed. Sometimes the collar shrivels so much that it is scarcely perceptible, and may disappear altogether in old plants. The spores are brown in mass. The cap of this mushroom is from three to four inches in diameter and the stem from one to three inches long.

This is the first mushroom that yielded to cultivation. It is raised in large quantities, not only in this country, but especially in France, Japan, and China. No doubt other species and genera will be produced in time.

This species grows in grassy places, in pastures, and richly manured grounds, never in the woods. I found it in great abundance in Wood County, in fields which had never been plowed and where the ground was unusually rich. There it seemed to grow in groups or large clusters. Usually it is found singly. Found from August to October. The plants figured here were found near Chillicothe.

Figure 249   Agaricus campestris. Two-thirds natural size.

AGARICUS RODMANI. PK.

RODMAN'S MUSHROOM. EDIBLE.

Figure 250   Agaricus rodmani. Two-thirds natural size.

The pileus is creamy, with brownish spots, firm, surface dry. The mature specimens frequently have the surface of the cap broken into large, brownish scales.

The gills are whitish, then pink, becoming dark-brown; narrow, close and unequal.

The stem is fleshy, solid, short, thick, about two inches long. The collar when well developed exhibits a striking characteristic. It appears as if there were two collars with a space between them. Its spores are broadly elliptical, .0002 to .00025 inch long.

It may be easily distinguished from the common Agaric by the time when found, its thick firm flesh, its narrow gills, which are almost white at first, and its double collar. I have found people eating it, supposing they were eating the common mushroom.

It is found in grassy places and especially between the cobble stones along the gutters in the cities. The specimens in Figure 250 were found in Chillicothe in the gutters. It is a meaty plant and one can soon tell it from its weight alone. It is found through May and June. It is fully as good to eat as the common mushroom. Macadam speaks of finding it in the fall, but I have never succeeded in finding it later than June.

AGARICUS SILVICOLA. VITT.

THE SILVAN AGARIC. EDIBLE.

Figure 251   Agaricus silvicola. One-half natural size.

Silvicola, from silva, woods and colo, to inhabit. The pileus is convex, sometimes expanded or nearly plane, smooth, shining, white or yellowish.

The gills are crowded, thin, free, rounded behind, generally narrowed toward each end, at first white, then pinkish, finally blackish-brown.

The stem is long, cylindrical, stuffed or hollow, white, bulbous; ring either thick or thin, entire or lacerated. Spores elliptical, 6–8×4–5. The plant is four to six inches high. Pileus three to six inches broad. Peck. 36th N. Y. State Bot.

A. silvicola is very closely related to the common mushroom. Its chief differences are in its place of growth, its being slender, and its hollow stem somewhat bulbous at the base. I have found it many times in the woods about Chillicothe, although I have never succeeded in finding more than one or two at a time. I have always put them with edible species and have eaten them when thus cooked with others.

Because of the resemblance which it bears, in its earlier stages, to the deadly Amanita, one can not exercise too great care in identifying it. It grows in the woods and is found from July to October.

AGARICUS ARVENSIS. SCHAEFF.

THE FIELD OR HORSE MUSHROOM. EDIBLE.

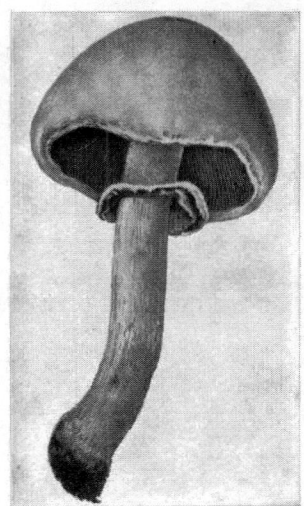

Figure 252   Agaricus arvensis. Two-thirds natural size, showing veil.

Arvensis, pertaining to a field. Pileus is smooth, white or yellowish, convex or conical, bell-shaped, then expanded, more or less mealy. The gills are crowded, free, generally broader toward the stem; at first whitish, then pinkish, finally black-brown.

The stem is stout, equal, slightly thickened at the base, smooth, hollow or stuffed, ring rather large and thick, the upper part membranaceous and white, while the lower or exterior surface is thicker, downy, radically split and yellowish.

The spores are elliptical, .0003 to .0004 inch long.

This plant grows much larger than the common mushroom, and may be distinguished by the collar being composed of two parts closely allied to each other making a double membrane, the lower part being much thicker, softer in texture and split in a stellate manner into broad and yellow rays, as will be seen in Figure 252.

I found it very plentiful in Wood County, Ohio, and in quantities in Dr. Manville's yard in Bowling Green, Ohio. I ate them frequently and gave them to my friends, who all voted them delicious.

When the stem is first cut there exudes from the wound a yellowish liquid which is quite a sure ear mark of this species.

There is a tradition that the spores will not germinate unless they pass through the alimentary canal of the horse or some animal. However this may be, it is found frequently where no trace of the horse can be found. It appears from July to September. I have found it in Fayette County, Ohio, in large rings, resembling the Fairy-Ring Mushroom, only the ring is very large, as well as the mushrooms.

Plate XXXIV. Figure 253   Agaricus arvensis.

AGARICUS ABRUPTUS. PK.

EDIBLE.

Figure 254   Agaricus abruptus.

Abruptus means to break away, referring to the breaking of the veil from the margin of the cap.

The pileus is creamy-white, dry and silky, quite irregular in shape when young, turning yellow when bruised or when the stem is cut.

The gills are slightly pinkish when the veil first breaks, gradually growing a deeper pink, in mature specimens becoming brownish, soft, free from the stem, quite close, unequal.

The stem is creamy-white, much darker toward the base, hollow, rather stiff, quite brittle, frequently found to be split lengthwise, ventricose, tapering toward the cap.

The veil is rather frail, one portion of it often adhering to the cap and another portion forming a ring on the stem.

Through the courtesy of Captain McIlvaine I am able to present an excellent picture of this species. The beginner will have some trouble to distinguish it from A. silvicola. This species, like the A. silvicola, is closely related to the meadow mushroom, but can be readily separated from it. This, too, like the A. silvicola, when seen in the woods at a distance, resembles the Amanita, but a careful glance at the gills will detect the difference.

The gills of the very young plant may appear white, but they will soon develop a pinkish tinge which will distinguish it from the Amanita. It is found in thin woods from July to October.

AGARICUS COMPTULUS. FR.

Comptulus means beautified or luxuriously decked; so called from the silky lustre of its cap.

The pileus is at first convex, then expanded, rather fleshy, thinner at the margin and incurved, usually with an adpressed silky finish to the surface of the cap which gives rise to its specific name.

The gills are free, much rounded toward the margin and the stem, white at first, then grayish, pinkish, purple-brown in old plants.

The stem is hollow, tapering from the base to the cap, slight bulbous, white, then yellowish, fleshy, fibrous. The veil is more delicate than in A. silvaticus, parts of it often found in young plants on the margin of the cap, forming a ring on the stem which soon almost disappears. Spores small, $4-5 \times 2-3\mu$.

The surface of the cap, the rounding of the gills both in front and behind, also the tendency to turn white paper blue or bluish when the flesh of the cap comes in contact with it, will assist in determining this species.

It is found in grassy places in open woods, especially in the vicinity of pine trees, October and November.

AGARICUS PLACOMYCES. PK.

THE FLAT-CAP MUSHROOM. EDIBLE.

Plate XXXV. Figure 255   Agaricus placomyces   **Figure 256**   Agaricus placomyces. Two-thirds natural size.

Figure 257   Agaricus placomyces. Two-thirds natural size.

Placomyces means a flat mushroom. This is one of our prettiest plants.

The pileus is broadly ovate, rather thin, at first convex, but when it is fully expanded it is quite flat, whitish, brown in the center, as will be seen in Figure 256, but it is covered with a persistent brown scale.

The gills are white at first, then pink, turning blackish brown, quite crowded.

The stem is rather long, and slender, cylindrical stuffed, somewhat bulbous at the base, commonly whitish but at times bears yellow stains toward the base, tapering toward the cap. The veil is quite interesting. It is broad and double, loosely joined together by threads, the lower or outer veil breaking first into regular radiating portions. The spores are elliptical, $5–6.5\mu$ long. The caps are two to four inches broad and the stem is three to five inches long.

They are found in lawns or in thin woods. They are much more abundant in hemlock woods though they are frequently found in mixed woods in which there are hemlock trees. The behavior of the veil is very similar to A. arven sis and A. silvicola and indeed this plant seems to be very closely related to these species. It is found from July to September.

AGARICUS CRETACEUS. FR.

THE CHALK AGARIC. EDIBLE.

Cretaceus, relating to chalk.

The pileus is entirely white, fleshy, obtuse, dry; sometimes even, sometimes marked with fine lines around the margin.

The gills are free, remote, quite ventricose, narrowed toward the stem, crowded, white, and only in mature plants do they become brownish. Spores, $5–6\times3.5\mu$.

The stem is two to three inches long, even, smooth, firm, tapering toward the cap, hollow, or stuffed with a fine pith, white.

It is found on lawns and in rich places. I find it more frequently in rich stubble fields. It makes a rare dish. Found in August and September.

AGARICUS SUBRUFESCENS. PK.

THE SLIGHTLY RED MUSHROOM. EDIBLE.

Subrufescens, sub, under; rufescens, becoming red. The pileus is at first inclined to be hemispherical, becoming convex or broadly expanded; silky fibril lose and minutely or obscurely scaly, whitish, grayish, or dull reddish-brown, usually smooth and darker on the disk. Flesh white and unchangeable.

The gills are at first white or whitish, then pink, finally blackish-brown.

The stem is rather long, often somewhat thickened or bulbous at the base, at first stuffed, then hollow, white, the ring is scaly on the under side, mycelium whitish, forming slender branching root-like strings. The spores are elliptical. Peck, 48th Rep. N. Y. State Bot.

The reddish-brown color is due to the coating of fibrils that covers the cap. In the center it does not separate into scales, hence it is smoother and more distinctly reddish-brown than the rest. Its veil resembles that of the A. placomyces, but instead of the lower surface breaking into radial portions it breaks into small floccose flakes or scales.

This species is found about greenhouses, and is frequently found in large clusters.

Dr. McIlvaine says: "This species is now cultivated and has manifest advantages over the market species—it is easier to cultivate, very productive, produces in less time after planting the spawn, is free from attacks of insects, carries better and keeps longer."

Mushroom beds in cellars are becoming quite popular and many are having very good results.

AGARICUS HALOPHILUS. PK.

SEA-LOVING AGARICUS EDIBLE.

Plate XXXVI. Figure 258   Agaricus halophilus.

Showing the globose caps, narrow gills, solid stem, and the peculiar incurved margin. Natural size.

Halophilus is from two Greek words meaning sea and loving, or fond of.

This is a large fleshy plant and does not readily decay. At first it is quite round, then becomes broadly convex. All specimens that I have examined were covered with adpressed scales of a reddish-brown color, becoming grayish-brown when old. The flesh is white, becoming pink or reddish when cut. The margin has a peculiar angular turn, often retaining portions of the rather fragile veil.

The taste is pleasant, and the odor is distinctly that of the seashore.

The gills are quite narrow, as will be seen in Figure 258, much crowded, free, pinkish at first, becoming purplish-brown as the plant matures. The edge of the gills is whitish.

The stem is short, stout, solid, firm, equal, or occasionally slightly bulbous. The ring is rather delicate and in older specimens it is frequently wanting. The spores are broadly elliptical and purplish-brown, 7–8×5–6μ.

The specimens in figure 258 were sent to me from Boston, Mass., by Mrs. Blackford, and on opening the box the odor of the seashore was plainly noticed. The flesh when cut quickly turned to a pinkish or reddish hue and the water in which the plants were prepared for cooking was changed to a faintly pink tinge. These plants were sent me the first of June, but the stems were free from worms and were as easily cooked as the caps. I regard it as one of the very best mushrooms for table use, while also easy to distinguish.

It seems to delight in sandy soil near salt water. This was formerly called Agaricus maritimus.

PILOSACE. FR.

Pilosace is from two Greek words, pilos, felt; sakos, garment.

Hymenophore is distinct from the stem. Gills are free, and at first remote, from the stem. The general and partial veil are both absent, hence it is without ring or volva. This genus seems to have the habit of Agaricus but no ring.

PILOSACE EXIMIA. PK.

Figure 259   Pilosace eximia.

Eximia means choice, distinguished.

The pileus is fleshy, thin, convex or broadly campanulate, at length expanded and subumbonate, smooth, dark sooty-brown.

The gills are close, broad, ventricose, rounded behind, free, dull-red, or brownish-pink, then brown.

The stem is slender, hollow, a little thicker at the base, dull-red. The spores are elliptical, .004 inch long.

These plants are small and quite rare, yet I have found the plants in Haynes' Hollow on three different occasions. Dr. Peck writes that it is a very rare plant. It grows on old stumps and decayed logs. The plants in figure 259 were found in Haynes' Hollow and photographed by Dr. Kellerman.

STROPHARIA. FR.

Stropharia is from the Greek, strophos, a sword belt. The spores are bright purple-brown, brown or slate color. The flesh of the stem and the pileus is continuous. The veil, when ruptured, forms a ring on the stem. The gills are rounded and are not free.

The genus can be distinguished from all the genera of the purple-spored plants except the Agarics by the presence of a ring and by the united flesh of the stem and the cap and by the attachment of the gills. They grow on the ground or are elliptical.

STROPHARIA SEMIGLOBATA. BATSCH.

THE SEMIGLOBOSE STROPHARIA. EDIBLE.

Figure 260   Stropharia semiglobata.

Semiglobata—semi, half; globus, a ball. The pileus is somewhat fleshy at the center, thin at the margin, hemispherical, not expanded, even, viscid when moist.

The stem is hollow, slender, straight, smooth, glutinous, yellowish, veil abrupt.

The gills are firmly attached to the stem, broad, plane, sometimes inclined to be ventricose, clouded with black.

This plant is very common on the Dunn farm on the Columbus Pike, north of Chillicothe, but is found everywhere in grassy places recently manured, or on dung.

This plant has been under the ban for a number of years, but like many others its bad reputation has been outlived. Found from May to November.

STROPHARIA HARDII. ATKINSON N. SP.

Figure 261   Stropharia Hardii.

Hardii is named for the collector and author of this book.

Plant 10 cm. high; pileus 9 cm. broad; stem 1½ cm. thick.

Pileus pale bright ochraceous; gills brownish, near Prout's brown (R); stem pale-yellow tinge.

Pileus convex to expanded, thick at the center, thin toward the margin, smooth; flesh tinged yellow.

Gills subelliptical to subventricose behind, broadly emarginate, adnexed. Basidia 4-spored. Spores suboblong, smooth, $5–9 \times 3–5\mu$, purple-brown under the microscope.

Cystidia not very numerous on side of gills, varying from clavate to subventricose and sublanceolate, the free end more or less irregular when narrow, rarely branching below the apex, and usually with a prominent broad apiculus or with two or several short processes. Similar cells on edge of gills, but somewhat smaller and more regular.

Stem even at the base, tapering to a short root, transversely floccose, scaly both above and below the ring. The ring membranaceous, not prominent but still evident, about 2 cm. from the apex. Atkinson.

The specimens in Figure 261 are very old plants. While the plant was in season I did not photograph it, but when Prof. Atkinson named it I hastened to find some good specimens but only two had survived sufficiently to photograph. They were found October 15, 1906, on Mr. Miller's farm in Poke Hollow near Chillicothe.

STROPHARIA STERCORARIA. FR.

THE DUNG STROPHARIA. EDIBLE.

Stercoraria is from stercus, dung. The pileus is slightly fleshy at the center but thin at the margin; hemispherical, then expanded, even, smooth, discoid, slightly striate on the margin.

The gills are firmly attached to the stem, slightly crowded, broad, white, umber, then olive-black.

The stem is three inches or more long, stuffed with a fibrous pith, equal, ring close to cap, flocculose below the ring, viscid when moist, yellowish.

This species is distinguished from the S. semiglobata by the distinct pithy substance with which the stem is stuffed, also by the fact that the cap is never fully expanded. It is found on dung and manure piles, in richly manured fields, and sometimes in woods.

STROPHARIA ÆRUGINOSA. CURT.

THE GREEN STROPHARIA.

Æruginosa is from ærugo, verdigris. The pileus is fleshy, plano-convex, subumbonate, clothed with a green evanescent slime, becoming paler as the slime disappears.

The gills are firmly attached to the stem, soft, brown, tinged with purple, slightly ventricose, not crowded.

The stem is hollow, equal, fibrillose or squamose below the ring, tinged with blue.

This species is quite variable in form and color. The most typical forms are found in the fall, in very wet weather and in shady woods. This is one of the species from which the ban has not been removed but its appearance will lead no one to care to cultivate its acquaintance further than name it. It is claimed by most writers that it is poisonous. Found in meadows and woods, from July to November.

HYPHOLOMA. FR.

Hypholoma is from two Greek words, meaning a web and a fringe, referring to the web-like veil which frequently adheres to the margin of the cap, not forming a ring on the stem and not always apparent on old specimens.

The pileus is fleshy, margin at first incurved. The gills are attached to the stem, sometimes notched at the stem. The stem is fleshy, similar in substance to the cap.

They grow mostly in thick clusters on wood either above or under the ground. The spores are brown-purple, almost black.

This genus differs from the genus Agaricus from the fact that its gills are attached to the stem and its stem is destitute of a ring.

HYPHOLOMA INCERTUM. PK.

THE UNCERTAIN HYPHOLOMA. EDIBLE.

Plate XXXVII. Figure 262   Hypholoma incertum.

Incertum, uncertain. Prof. Peck, who named this species, was uncertain whether it was not a form of H. candolleanum, to which it seemed to be very closely related; but as the gills of that plant are at first violaceous and of this one white at first, he concluded to risk the uncertainty on a new species.

The pileus is thin, ovate, broadly spreading, fragile, whitish, margin often wavy and often adorned with fragments of the woolly white veil, opaque when dry, transparent when moist.

The gills are thin, narrow, close, fastened to the stem at their inner extremity, white at first, then purplish-brown, edges often uneven.

The stem is equal, straight, hollow, white, slender, at least one to three inches long. The spores are purplish-brown and elliptical. It is found in lawns, gardens, pastures, and thin woods. It is small but grows in such profusion that one can obtain quantities of it. The caps are very tender and delicious. It appears as early as May.

HYPHOLOMA APPENDICULATUM. BULL.

THE APPENDICULATE HYPHOLOMA. EDIBLE.

Appendiculatum, a small appendage. This is so called from the fragments of the veil adhering to the margin of the cap.

The pileus is thin, ovate, expanded, watery, when dry, covered with dry atoms; margin thin and often split, with a white veil; the color when moist dark-brown, when dry nearly white, often with floccose scales on the cap.

The gills are firmly attached to the stem, crowded, white, then rosy-brown, and at length dingy-brown.

The stem is hollow, smooth, equal, white, fibrous, mealy at the apex. The veil is very delicate and only seen in quite young plants.

The plant grows in the spring and the summer and is found on stumps and sometimes on lawns. It is a favorite mushroom with those who know it. The plant can be dried for winter use and retains its flavor to a remarkable degree.

Hypholoma candolleanum, Fr., resembles the H. appendiculatum in many features, but the gills are violaceous, becoming cinnamon-brown and in old plants nearly free from the stem. It has more substance. The caps, however, are very tender and delicious. Found in clusters.

HYPHOLOMA LACHRYMABUNDUM. FR.

THE WEEPING HYPHOLOMA.

Figure 263   Hypholoma lachrymabundum. Two-thirds natural size.

Lachrymabundum—full of tears. This plant is so called because in the morning or in damp weather the edge of the gills retain very minute drops of water. The plant in Figure 263 was photographed in the afternoon yet there can be seen a number of these minute drops.

The pileus is fleshy, campanulate, then convex, sometimes broadly umbonate, spotted with hairy scales; flesh white.

The gills are closely attached to the stem, notched, crowded, somewhat ventricose, unequal, whitish, then brown-purple, distilling minute drops of dew in wet weather or in the morning.

The stem is hollow, somewhat thickened at the base, quite scaly with fibrils, often becoming brownish-red, two to three inches long. The spores are brownish-purple.

Figure 264   Hypholoma lachrymabundum.

I have never found the plant elsewhere than on the Chillicothe high school lawn, and then not in sufficient numbers to test its edible qualities. When I do, I shall try it cautiously, but with full faith that I shall be permitted to try others. Found on the ground and on decayed wood. It often grows in clusters. September to October.

HYPHOLOMA SUBLATERITIUM. SCHAEFF.

THE BRICK-RED HYPHOLOMA. EDIBLE.

Figure 265   Hypholoma sublateritium. Natural size.

Sublateritium is from sub, under, and later, a brick. The pileus is brick-red, with pale yellowish border; the surface is covered with fine silky fibres; fleshy, moist, and firm; the cap is from two to four inches broad; remnants of the veil are often seen on the margin; flesh creamy, firm, and bitter.

The gills are creamy when young, olive when old; attached to the stem at inner extremity, rather narrow, crowded, and unequal.

The stem is creamy when young, lower part slightly tinged with red, hollow or stuffed, having silky fibres on the surface, two to four inches long, often incurved because of position. The spores are sooty-brown and elliptical.

It grows in large clusters around old stumps. It is especially plentiful about Chillicothe. It is not equal to many others of the Hypholomas as an esculent. Sometimes it is bitter even after it is cooked. Captain McIlvaine gives a plausible reason when he says it may be due to the passage of larvæ through the flesh of the plant. It is found from September to early winter.

HYPHOLOMA PERPLEXUM. PK.

THE PERPLEXING HYPHOLOMA. EDIBLE.

Figure 266 Hypholoma perplexum. One-half natural size. Caps brown, with a pale yellow margin.

Perplexum means perplexing; so called because it is quite difficult to distinguish it from H. sublateritium, also from H. fascicularis. From the latter it may be known by its redder cap, its whitish flesh, purple-brown tint of the mature gills and mild flavor. Its smaller size, the greenish and purplish tint of the gills, and the slender hollow stem will aid in distinguishing it from H. perplexum.

The pileus is complex, fleshy, expanded, smooth, sometimes broadly and slightly umbonate, brown with a pale-yellow margin, disk sometimes reddish.

The gills are rounded, notched, easily separating from the stem, pale-yellow, greenish ash-color, finally purplish-brown, thin, quite close.

The stem is nearly equal, firm, hollow, slightly fibrillose, yellowish or whitish above and reddish-brown below. The spores are elliptical and purplish brown.

This plant is very abundant in Ohio. It grows about old stumps, but a favorite habitat seems to be upon old sawdust piles. I have found it after we have had considerable freezing weather.

The plants in the figure were frozen when I found them, the 27th of November. Dr. McIlvaine says in his book, "If the collector gets puzzled, as he will, over one or all of these species, because no description fits, he can whet his patience and his appetite by calling it H. perplexum and graciously eating it."

PSILOCYBE. PERS.

Psilocybe is from two Greek words, naked and head. The spores are purple-brown or slate color. The pileus is smooth, at first incurved, brownish or purple. The stem is cartilaginous, ringless, tough, hollow, or stuffed, often rooting. Generally growing on the ground.

PSILOCYBE FŒNISECII. PERS.

THE BROWN PSILOCYBE.

Figure 267   Psilocybe fœnisecii. One-half natural size.

Fœnisecii means mown hay.

The pileus is somewhat fleshy, smoky-brown or brownish, convex, campanulate at first, then expanded; obtuse, dry, smooth.

The gills are firmly attached to the stem, ventricose, not crowded, brownish-umber.

The stem is hollow, straight, even, smooth, not rooting, white, covered with dust, then brownish.

Quite common in grassy lawns and fields after summer rains. I have never eaten it, but I have no doubt of its esculent qualities.

PSILOCYBE SPADICEA. SCHAEFF.

THE BAY PSILOCYBE. EDIBLE.

Spadicea means bay or date-brown.

The pileus is fleshy, convex-plane, obtuse, even, moist, hygrophanous, bright bay-brown, paler when dry.

The gills are rounded behind, attached to stem, easily separating from it, narrow, dry, crowded, white, then rosy-brown or flesh-color.

The stem is hollow, tough, pallid, equal, smooth, one to two inches long. They grow in dense clusters where old stumps have been or where wood has decayed. The caps are small but very good. They are found from September to frost or freezing weather.

PSILOCYBE AMMOPHILA. MONT.

Figure 268   Psilocybe ammophila. Two-thirds natural size, showing the sand on the base.

Ammophila is from two Greek words; ammos, sand, and philos, loving; so called because the plants seem to delight to grow in sandy soil.

The pileus is small, convex, expanded, umbilicate, at first hemispherical, rather fleshy, yellow, tinged with red, fibrillose.

The gills are smoky in color, with a decurrent tooth, powdered with the blackish spores.

The stem is soft, rather short, hollow, lower half clavate and sunk into the sand, striate. The spores are 12×8.

They are found in August and September. They delight in sandy soil, as the specific name indicates. The plants in the photograph were found near Columbus and photographed by Dr. Kellerman. It is quite common in sandy soil. I do not think it is edible. I should advise great caution in its use.

CHAPTER VI

THE BLACK-SPORED AGARICS.

The genera belonging to this series have black spores. There is an entire absence of purple or brown shades. The genus Gomphidius, placed in this series for other reasons, has dingy-olivaceous spores.

COPRINUS. PERS.

Coprinus is from a Greek word meaning dung. This genus can be readily recognized from the black spores and from the deliquescence of the gills and cap into an inky substance. Many of the species grow in dung, as the name implies, or on recently manured ground. Some grow in flat rich ground, or where there has been a fill, or on dumping grounds; some grow on wood and around old stumps.

The pileus separates easily from the stem. The gills are membranaceous, closely pressed together. The spores, with few exceptions, are black. Most of the species are edible, but many are of such small size that they are easily overlooked.

COPRINUS COMATUS. FR.

THE SHAGGY MANE COPRINUS. EDIBLE.

Figure 269   Coprinus comatus.

Comatus is from coma, having long hair, shaggy. It is so called from a fancied resemblance to a wig on a barber's block. A description is hardly necessary with a photograph before us. They always remind us of a congregation of goose eggs standing on end. This plant cannot be confounded with any other, and the finder is the happy possessor of a rich, savory morsel that cannot be duplicated in any market.

The pileus is fleshy, moist, at first egg-shaped, cylindrical, becoming bell-shaped, seldom expanded, splitting at the margin along the line of the gills, adorned with scattered yellowish scales, tinged with purplish-black, yet sometimes entirely white; surface shaggy.

Figure 270   Coprinus comatus. One-half natural size.

The gills are free, crowded, equal, creamy white, becoming pink, brown, then black, and dripping an inky fluid.

The stem is three to eight inches long, hollow, smooth, or slightly fibrillose, tapering upward, creamy-white, brittle, easily separating from the cap, slightly bulbous at the base. The ring is rarely adherent or movable in young plants, later lying on the ground at the base of the stem or disappearing altogether. The spores are black and elliptical, and are shed in liquid drops.

Found in damp rich ground, gardens, rich lawns, barnyards, and dumping grounds. They often grow in large clusters. They are found everywhere in great abundance, from May till late frost. A weak stomach can digest any of the Coprini when almost any other food will give it trouble. I am always pleased to give a dish of any Coprini to an invalid.

COPRINUS ATRAMENTARIUS. FR.

The Inky Coprinus. Edible.

Figure 271   Coprinus atramentarius. Two-thirds natural size.

Atramentarius means black ink. The pileus is at first egg-shaped, gray or grayish-brown, smooth, except that there is a slight scaly appearance; often covered with a marked bloom, margin ribbed, often notched, soft, tender, becoming expanded, when it melts away in inky fluid.

The gills are broad, close, ventricose, creamy-white in young specimens, becoming pinkish-gray, then black, moist, melting away in inky drops.

The stem is slender, two to four inches in length, hollow, smooth, tapering upward, easily separating from the cap, with slight vestige of a collar near the base when young but soon disappearing. The spores are elliptical, $12 \times 6\mu$., and black, falling away in drops.

Plate XXXVIII. Figure 272   Coprinus atramentarius.

I have found it abundantly all over the state, from May till late frost. In Figure 271 the one in the center will show the spot-like scales; on the others the bloom referred to is quite apparent; the section to the right shows the broad, ventricose gills—cream-white though slightly tinged with pink—also the shape of the stem. The plant at the extreme right has expanded and begun to deliquesce. C. atramentarius is very abundant, growing in rich soil, lawns, filled places, and gardens.

COPRINUS MICACEUS. FR.

THE GLISTENING COPRINUS. EDIBLE.

Figure 273   Coprinus micaceus. Two-thirds natural size.

Micaceus is from micare, to glisten, and refers to the small scales on the pileus which resemble mica scales. The pileus is tawny-yellow, tan or light buff, ovate, bell-shaped; having striations radiating from near the center of the disk to the margin; glistening mica-like scales covering undisturbed young specimens; the margin somewhat revolute or wavy.

The gills are crowded, rather narrow, whitish, then tinged with pinkish or purplish-brown then black.

The stem is slender, fragile, hollow, silky, even, whitish, often twisted, one to three inches long. The spores are blackish, sometimes brown, elliptical, $10 \times 5\mu$.

The Glistening Coprinus is a small but common and beautiful species. One cannot fail to recognize a Coprinus from a photograph. It is somewhat bell-shaped and marked with impressed lines or striations from the margin to or beyond the center of the disk and sprinkled with fugacious micaceous granules all of which show in Figure 273. For eating, this is without doubt the best mushroom that grows. The specimens in Figure 273 grew around an old peach stump in Dr. Miesse's yard, in Chillicothe. You will find them around any stump, especially just before a rain. If you secure a good supply and wish to keep them, partially cook them and warm them for use.

COPRINUS EBULBOSUS. PK.

Figure 274   Coprinus ebulbosus. One-half natural size.

Ebulbosus, without being bulbous. This seems to be the difference between the American and the European plants, the latter being bulbous.

The pileus is membranaceous, at first ovate, bell-shaped, striate, variegated with broad white scales, or white patches; one to two inches broad.

The gills are free, broad, ventricose, grayish-black, soon deliquescing.

The stem is hollow, equal, fragile, smooth, four to five inches long.

Usually found where old stumps have been cut off under the ground, leaving the roots in the ground. It is very abundant. The collector will have no trouble to recognize it from Figure 274. They are found from June to October. Edible, but not as good as C. atramentarius.

COPRINUS EPHEMERUS. FR.

THE EPHEMERAL COPRINUS. EDIBLE.

Ephemerus, lasting for a day. This plant lasts only for a short time. It comes up in the early morning or at night and as soon as the sun's rays touch it it deliquesces into an inky fluid.

The pileus is membranaceous, very thin, oval, slightly covered with bran-like scales, disk elevated, even.

Gills are adnexed, distant, whitish, brown, then black. The stem is slender, equal, pellucid, smooth, from one to two inches high.

When this plant is fully developed it is quite a beautiful specimen, striated from margin to center. Found on dung and dung heaps and in well manured grass plots from May to October. It must be cooked at once. Its chief value is its excellent mushroom flavor.

COPRINUS OVATUS. FR.

THE OVATE COPRINUS. EDIBLE.

Ovatus is from ovum, an egg. It is so called from the shape of the pileus, which is somewhat membranaceous, ovate, then expanded, striate; at first woven into densely imbricated, thick, concentric scales; is bulbous, rooting, flocculose, hollow above, the ring deciduous; gills free, remote, slightly ventricose, for sometime white, then umber-blackish.

This plant is much smaller and less striking than the C. comatus, yet its edible qualities are the same. I have eaten it and found it delicious. It is found in about the same locality in which you would expect to find the C. comatus.

COPRINUS FIMETARIUS. FR.

THE SHAGGY DUNG COPRINUS.

Plate XXXIX. Figure 275   Coprinus fimetarius.

Fimetarius is from fimetum, a dunghill. The pileus is somewhat membranaceous, clavate, then conical, at length torn and revolute; at first rough with floccose scales, then naked; longitudinally cracked and furrowed, even at the apex. The stem is inclined to be scaly, thickened at the base, solid. The gills are free, reaching the stem, at first ventricose, then linear, brownish-black. Fries.

This is quite a variable plant. There are a number of varieties classed under this species. It is said to be of excellent flavor. I have never eaten it.

PANÆOLUS. FR.

Panæolus is from two Greek words, all; variegated. This genus is so called from the mottled appearance of the gills. The pileus is somewhat fleshy, margin even, but never striate. The margin always extends beyond the gills and the gills are not uniform in color. The mottled appearance of the gills is due to the falling of the black spores. The gills do not deliquesce.

The stem is smooth, sometimes scaly, at times quite long, hollow. The veil, when present, is interwoven.

This plant is found on rich lawns recently manured, but principally on dung.

There are only two edible species, P. retirugis and P. solidipes. The other species would not be likely to attract the attention of the ordinary collector.

PANÆOLUS RETIRUGIS. FR.

THE RIBBED PANAEOLUS. EDIBLE.

Plate XL. Figure 276   Panaeolus retirugis.
Natural size, showing portions of the veil on the margin.

Retirugis is from rete, a net; ruga, a wrinkle. The pileus is about one inch in diameter, inclined to be globose, then hemispherical, slightly umbonate, center darker, with united raised ribs, sometimes sprinkled with opaque atoms; veil torn, appendiculate.

The gills are fixed, ascending, broad in middle; and in the expanded forms the gills are separated more and more from the stem and finally appear more or less triangular; cinereous-black, frequently somewhat clouded.

The stem is equal, covered with a frost-like bloom, cylindrical, sometimes tortuous, cartilaginous, becoming hollow, pinkish-purple, always darker below and paler above, bulbous.

The veil in young and unexpanded plants is quite strong and prominent; as the stem elongates it loosens from the stem, and as the cap expands it breaks into segments, frequently hanging to the margin of the cap. By close observation one will sometimes detect a black band on the stem, caused by the falling of the black spores, when the plant is damp, before the pileus has separated from the stem. The spores are black and elliptical.

I have found it a number of times on the Chillicothe high school lawn, especially after it was fertilized in the winter. It is found mostly on dung from June to October. I do not recommend it as a delicacy.

PANÆOLUS EPIMYCES. PK.

Figure 277 Panæolus epimyces. Note black spores in central foreground. Note also huge masses of abortive stuff upon which it grows.

Epimyces is from epi, upon; myces, a mushroom; so called because it is parasitic on fungi. There are a number of species of mushrooms whose habitat is on other mushrooms or fungus growths; such as Collybia cirrhata, C. racemosa, C. tuberosa, Volvaria loveiana and the species of Nyctalis.

The pileus is fleshy, at first subglobose, then convex, white, silky, fibrillose, flesh white or whitish, soft.

The gills are rather broad, somewhat close, rounded behind, adnexed, dingy-white, becoming brown or blackish, with a white edge.

The stem is short, stout, tapering upwards, strongly striate and minutely mealy or pruinose; solid in the young plant, hollow in the mature, but with the cavity small; hairy, or substrigose at the base. The spores elliptical and black, .0003 to .00035 of an inch long, .0002 to .00025 broad. Peck.

The plants are small, about two thirds to an inch broad and from an inch to an inch and a half high. It is referred to this genus because of its black spores. It has other characteristics which would seem to place it better among Hypholomas. It is not common. Found in October and November. The specimens in Figure 277 were found in Michigan and photographed by Dr. Fisher.

PANÆOLUS CAMPANULATUS. LINN.

BELL-SHAPED PANAEOLUS.

Campanulatus is from campanula, a little bell.

The pileus is an inch to an inch and a quarter broad, oval or bell-shaped, sometimes slightly umbonate, smooth, somewhat shining, grayish-brown, sometimes becoming reddish-tinted, the margin often fringed with fragments of the veil.

The gills are attached, not broad, ascending, variegated with gray and black.

The stem is three to five inches long, hollow, slender, firm, straight, often covered with frost-like bloom and often striate at the top, the veil remaining only a short time. The spores are subellipsoid, $8–9 \times 6\mu$.

The gills do not deliquesce. It is widely distributed and is found in almost any horse pasture.

Captain McIlvaine says in his book that he has eaten it in small quantities, because larger could not be obtained, and with no other than pleasant effect. I have found it about Chillicothe quite frequently but have never eaten it. It is found from June to August.

PANÆOLUS FIMICOLUS. FR.

THE DUNG PANAEOLUS.

Fimicolus is from fimus, dung; colo, to inhabit. The pileus somewhat fleshy, convex-bell-shaped, obtuse, smooth, opaque; marked near the margin with a narrow brown zone; the stem is fragile, elongated, equal, pallid, covered with frost-like bloom above; the gills are firmly attached to the stem, broad, variegated with gray and brown. Fries.

The plant is very small and unimportant. It is found on dung, as its name indicates, from June to September. The caps appear lighter in color when dry than when wet.

PANÆOLUS SOLIDIPES. PK.

THE SOLID FOOT PANAEOLUS. EDIBLE.

Plate XLI. Figure 278   Panaeolus solidipes.

Solidipes is from solidus, solid; pes, foot; and is so called because the stem of the plant is solid. The pileus is two to three inches across; firm; at first hemispherical, then subcampanulate or convex; smooth; white; the cuticle at length breaking up into dingy-yellowish, rather large, angular scales. The gills are broad, slightly attached, whitish, becoming black. The stem is five to eight inches long and two to four lines thick, firm, smooth, white, solid, slightly striate at the top. The spores are very black with a bluish tint. Peck. 23d Rep. N. Y. State Bot.

This is a large and beautiful plant and easily distinguished because of its solid stem, growing on dung. Sometimes minute drops of moisture will be seen on the upper part of the stem. The plant is said to be one of the best of mushrooms to eat.

PANÆOLUS PAPILIONACEUS. FR.

THE BUTTERFLY PANAEOLUS.

Figure 279   Panæolus papilionaceus. Natural size.

Papilionaceus is from papilio, a butterfly.

The pileus is about an inch broad, somewhat fleshy, at first hemispherical, sometimes subumbonate, the cuticle breaking up into scales when dry, as will be seen in the photograph, pale-gray with a tinge of reddish-yellow especially on the disk, sometimes smooth.

The gills are broadly attached to the stem, quite wide, at length plane, blackish or with varying tints of black.

The stem is three to four inches long, slender, firm, equal, hollow, powdered above, whitish, sometimes tinged with red or yellow, slightly striate at the top, as will be seen in the photograph with a glass, generally stained with the spores.

The specimens in Figure 279 were found in a garden that had been strongly manured. It is usually found on dung and on grassy lawns during May and June. Captain McIlvaine in his book speaks of this mushroom producing hilarity or a mild form of intoxication. I should advise against its use.

ANELLARIA. KARST.

Anellaria is from anellus, a little ring. This genus is so called because of the presence of a ring on the stem.

The pileus is somewhat fleshy, smooth, and even. The gills are adnexed, dark slate-colored, variegated with black spores. The stem is central, smooth, firm, shining, ring persistent or forming a zone around the stem.

ANELLARIA SEPARATA. KARST.

Separata means separate or distinct.

The pileus is somewhat fleshy, bell-shaped, obtuse, even, viscid, at first ochraceous, then dingy-white, shining, smooth, wrinkled when old.

The gills are firmly attached to the stem, broad, ventricose, thin, crowded, clouded, cinereous, margin nearly white, slightly deliquescent.

The stem is long, straight, shining, white, thickened downward, ring distant, top somewhat striate, bulbous at the base. The spores are broadly elliptic-fusiform, black, opaque, $10 \times 7\mu$.

It is found on dung from May to October. It is not poisonous.

BOLBITIUS. FR.

Bolbitius is from a Greek word meaning cow-dung, referring to its place of growth.

The pileus is membranaceous, yellow, becoming moist; gills moist but not deliquescing, finally losing their color and becoming powdery; stem hollow and confluent with the hymenophore. As the generic name implies the plant usually grows on dung, but sometimes it is found growing on leaves and where the ground had been manured the year before. The spores are of a rusty-red color.

BOLBITIUS FRAGILIS. (L.) FR.

Fragilis means fragile.

The pileus is membranaceous, yellow, then whitish, viscid, margin striate, disk somewhat umbonate.

The gills are attenuated, adnexed, nearly free, ventricose, yellowish, then pale cinnamon.

The stem is two to three inches long, naked, smooth, yellow. The spores are rust-colored, 7×3.5, Massee. 14–15×8–9μ. Saccardo.

This species is much more delicate and fragile than B. Boltoni. I find it often in dairy pastures. It is well flavored and cooks readily. Found from June to October.

BOLBITIUS BOLTONI. FR.

BOLTON'S BOLBITIUS. EDIBLE.

The pileus is somewhat fleshy, viscid, at first smooth, then the margin sulcate, disk darker and slightly depressed.

The gills are nearly adnate, yellowish, then livid-brown.

The stem is attenuated, yellowish, ring fugacious. This is rather common in dairy pastures and is found from May to September.

PSATHYRELLA. FR.

Psathyrella is from a Greek word meaning fragile. The members of this genus are mebranaceous, striated, margin straight, at first pressed to the stem, not extending beyond the gills. Gills adnate or free, sooty-black, not variegated. The stem is confluent with, but different in character from, the spore-bearing surface. Veil inconspicuous and generally absent.

PSATHYRELLA DISSEMINATA. PERS.

THE CLUSTERED PSATHYRELLA. EDIBLE.

Figure 280   Psathyrella disseminata. Natural size.

Disseminata is from dissemino, to scatter. Pileus is about a half inch across, membranaceous, ovate, bell-shaped, at first scurvy, then naked; coarsely striated, margin entire; yellowish then gray. Gills adnate, narrow, whitish, then gray, finally blackish. Stem one to one and a half inches long, rather curved, mealy then smooth, fragile, hollow. Massee.

This is a very small plant, growing on grassy lawns, and very common on old trunks, and about decaying stumps.

A cluster about two yards square shows itself at intervals all summer on the Chillicothe High School lawn. The grass shows itself to be greener and thriftier there on account of fertilization by the mushroom. The entire plant is very fragile and soon melts away. I have eaten the caps raw many times and they have a rich flavor. They are found from May till frost.

PSATHYRELLA HIRTA. PK.

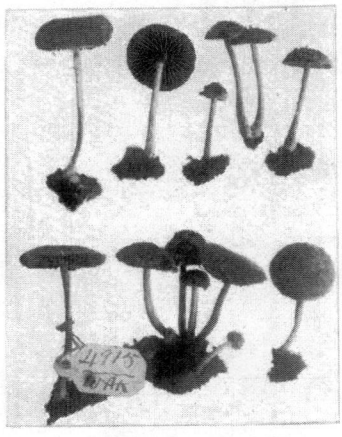

Figure 281   Psathyrella hirta.

Hirta means hairy, rough or shaggy.

Pileus thin, hemispherical or convex, adorned when young with erect or spreading tufts of white, easily determined and quickly evanescent hairs; hygrophanous, brown or reddish-brown and slightly striatulate when moist, pale grayish-brown or dingy-whitish when dry, flesh subconcolorous; lamellæ broad, moderately close, adnate and often furnished with a decurrent tooth, at first pallid, becoming blackish-brown or black; stem flexuose, squamose, hollow, shining, white; spores elliptical, black, .0005 to .00055 inch long, .00025 to .0003 broad.

Subcæspitose; pileus 4 to 6 lines broad; stem 1 to 2 inches long to 1 1-5 lines thick. The specimens in Figure 281 were found in the greenhouse at the State University. When quite young tufts of white hair were very conspicuous. They are scarcely observed in mature specimens. The plants were photographed by Dr. Kellerman.

GOMPHIDIUS. FR.

Gomphidius is from a Greek word meaning a wooden bolt or peg.

The hymenophore is decurrent on the stem. The gills are decurrent, distant, soft, somewhat mucilaginous; edge acute, pruinate with the blackish fusiform spores; veil viscoso-floccose, forming an imperfect ring around the stem.

A small, but distinct, genus, with great difference among species; intermediate by its habits between Cortinarius and Hygrophorus.

GOMPHIDIUS VISCIDUS. FR.

VISCID GOMPHIDIUS.

The pileus is two to three inches broad, viscid, convex, then depressed round the disk, obtusely umbonate, margin acute, reddish-brown to yellowish-brown in the center, the margin liver-color, flesh yellowish-brown.

The gills are decurrent, distant, somewhat branched, firm, elastic, rather thick, purple-brown with an olive tinge.

The stem is two to three inches high, subequal or slightly ventricose; pale yellowish-brown, fibrillose, firm, solid, slimy from the remains of the veil, which form an obsolete filamentose ring.

The spores are elongato-fusiform, $18–20 \times 6\mu$.

Its favorite habitat is under pine and fir trees. Its taste is sweet and it has the mushroom smell. It is edible, but not first-class.

Found in September and October.

CHAPTER VII

POLYPORACEAE. TUBE-BEARING FUNGI.

In this family the cap has no gills on the upper surface, but, instead, there are small tubes or pores. This class of plants may be naturally divided into two groups: The perishable fungi with the pores easily separating from the cap and from each other, which may be called Boletaceæ; and the leathery, corky, and woody fungi, with pores permanently united to the cap and with each other, making the family Polyporaceæ.

In each group the spores are borne on the lining of the pore. A spore print may be made in the same manner as from mushrooms having gills. The color of the spores does not enter into the classification as in the case of the Agaricini.

The distinctive characteristics of these genera may be stated as follows:

Pores compacted together and forming a continuous stratum		1
Pores each a distinct tube, standing closely side by side		Fistulina
1.	Stem central, and stratum of spores easily separable from the cap	Boletus
1.	Stratum of tubes not separating easily, cap covered with coarse scales	Strobilomyces
	Stratum of tubes separating, but not easily; tubes arranged in distinct, radiating lines. In Boletinus porosus the tubes do not separate from the cap	Boletinus
	Stratum of pores not separable from cap; plant soft when young, but becoming hard, corky, stipitate, shelving	Polyporus

BOLETUS. DILL.

Boletus, a clod. There are very many species under this genus and the beginner will experience much trouble in separating the species with any degree of assurance. The Boletus is distinguished from the other pore-bearing fungi by the fact that the stratum of tubes is easily separable from the cap. In the Polyporus the stratum of tubes cannot be separated.

Nearly all Boleti are terrestrial and have central stems. They grow in warm and rainy weather. Many are very large and ponderous; fleshy and putrescent, decaying soon after maturity. It is important to note whether the flesh changes color when bruised and whether the taste is pleasant or otherwise. When I first began to study the Boleti there were but few species that were thought to be edible, but the ban has been removed from very many, even from the most wicked, Boletus Satanus.

BOLETUS SCABER. FR.

THE ROUGH-STEMMED BOLETUS. EDIBLE.

The pileus is from two to five inches in diameter, rounded convex, smooth, viscid when moist, minutely woolly, velvety or scaly, color from nearly white to almost black, the flesh white.

The tubes are free from the stem, white, long, mouths minute and round.

The stem is solid, tapering slightly upward, long, dingy-white; roughened with blackish-brown or reddish dots or scales, this being the most pronounced characteristic by which to distinguish the species; three to five inches long. The spores are oblong fusiform and brown.

Prof. Peck has described a number of varieties under this species, most of which depend on the color of the cap. All are edible and good.

Figure 282   Boletus scaber. Two-thirds natural size.

This is a common plant, usually found in woods and shady waste places, from June to October. Photographed by Prof. H. C. Beardslee.

BOLETUS GRANULATUS. L.

THE GRANULATED BOLETUS. EDIBLE.

Figure 283   Boletus granulatus. One-half natural size.

The pileus is two to three inches broad, hemispherical, then convex; at first covered with a brownish gluten, then turning yellowish; flesh thick, yellowish, does not turn blue; margin involute at first.

The tubes are adnate; at first white, then light yellow; the margin distilling a pale watery fluid which when dry gives the granulated appearance.

The stem is short, one to two inches high, thick, solid, pale yellow above, white below, granulated. The spores are spindle-shaped, rusty-yellow.

This plant grows abundantly in pine regions, but I have found it where only a part of the trees were pine. The brownish gluten, always constant on the pileus, and the gummy juice drying upon the stem, like granules of sugar, will be strong features by which to identify the species.

They are found from July to October.

BOLETUS BICOLOR. PK.

THE TWO-COLORED BOLETUS. EDIBLE.

The pileus is convex, smooth or merely downy, dark red, fading when old, often marked with yellow; flesh yellow, slowly changing to blue when bruised.

The tubes are bright yellow, attached to the stem, the color changing to blue when bruised.

The stem is solid, red, generally red at the top, one to three inches long.

The spores are pale, rusty-brown color.

Found in woods and open places, from July to October.

BOLETUS SUBTOMENTOSUS. L.

THE YELLOW-CRACKED BOLETUS. EDIBLE.

Figure 284   Boletus subtomentosus. One-half natural size.

Subtomentosus, slightly downy. The pileus is from three to six inches broad, convex, plane;

yellowish-brown, olive or subdued tan color; cuticle soft and dry, with a fine pubescence; the cracks in the surface become yellow. The flesh is creamy white in mature specimens, changing to blue, and at length leaden, on being bruised.

The tube surface is yellow or yellowish green, becoming bluish when bruised; opening of tubes large and angular.

The stem is stout, yellowish, minutely roughened with scurvy dots or faintly striped with brown. The spores are a rusty-brown.

The cracks in the cap become yellow, on which account this species is called the Yellow-cracked Boletus. The taste of the flesh is sweet and agreeable. Palmer compares it with the taste of a walnut. The plant should not be feared because the flesh turns blue when bruised. I first found this species in Whinnery's woods, Salem, Ohio. The specimens in Figure 284 grew near Chillicothe and was photographed by Dr. Kellerman. July to August.

BOLETUS CHRYSENTERON. FR.

THE RED-CRACKED BOLETUS. EDIBLE.

Figure 285 Boletus chrysenteron. One-half natural size. Caps yellowish to red. Flesh yellow.

Chrysenteron means gold or golden within. The pileus is two to four inches broad, convex, becoming more flattened, soft to the touch, varying from light to yellowish-brown or bright brick-red, more or less fissured with red cracks; the flesh yellow, changing to blue when bruised or cut, red immediately beneath the cuticle.

The tube surface is olive-yellow, becoming bluish when bruised, tube-openings rather large, angled and unequal in size.

The stem is generally stout, straight, yellowish, and more or less streaked or spotted with the color of the cap. The spores are light brown and spindle-shaped. This species will be easily distinguished from B. subtomentosus because of its bright color and the cracks in the cap turning red, whence the name of the "Red-cracked Boletus."

The cap of this species strongly resembles Boletus alveolatus, but the latter has rose-colored spores and a red pore surface, while the former has light brown spores and an olive-yellow pore surface. Tolerton's and Bower's woods, Salem, Ohio, July to October.

BOLETUS EDULIS. BULL.

THE EDIBLE BOLETUS.

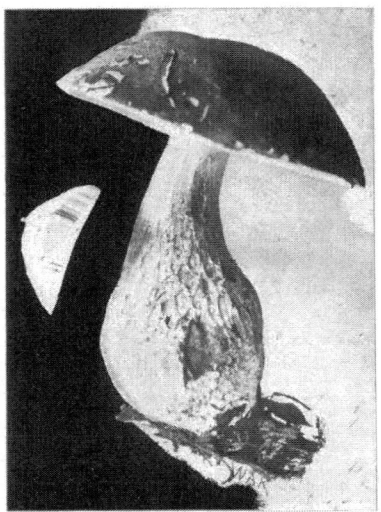

Plate XLII. Figure 286   Boletus edulis.

Pileus light brown, tubes yellowish or greenish-yellow. Stem bulbous and faintly reticulate. Natural size.

This is quite a large and handsome plant and one rather easily recognized. The firm caps of the young plant and the white tubes with their very indistinct mouths, and the mature plants with the tubes changing to a greenish yellow with their mouths quite distinct, are enough to identify the plant at once.

The pileus is convex or nearly plane; variable in color, light brown to dark brownish-red, surface smooth but dull, cap from three to eight inches broad. The flesh is white or yellowish, not changing color on being bruised or broken.

The tube-surface is whitish in very young plants, at length becoming yellow and yellowish-green. Pore openings angled. The tubes depressed around the stem, which is stout, bulbous, often disproportionately elongated; pale-brown; straight or flexuous, generally with a fine raised network of pink lines near junction of cap, sometimes extending to the base. The taste is agreeable and nutty, especially when young. Woods and open places. July and August. Common about Salem and Chillicothe, Ohio.

It is one of our best mushrooms. Captain McIlvaine says: "Carefully sliced, dried, and kept where safe from mold, it may be prepared for the table at any season."

BOLETUS SPECIOSUS. FROST.

THE HANDSOME BOLETUS. EDIBLE.

Figure 287 Boletus speciosus. Natural size. Cap red or deep scarlet. Tubes bright lemon-yellow.

Speciosus means handsome.

The pileus is three to six inches broad, at first very thick, subglobose, compact, then softer, convex, glabrous or nearly so, red or deep scarlet. The flesh is pale yellow or bright lemon-yellow, changing to blue where wounded.

The tubes are adnate, small, subrotund, plane, or slightly depressed around the stem; bright lemon-yellow, becoming dingy-yellow with age, changing to blue where bruised.

The stem is two to four inches long, stout, subequal or bulbous, reticulated, bright lemon-yellow without and within, sometimes reddish at the base. The spores are oblong-fusiform, pale, ochraceous-brown, $10–12.5 \times 4–5\mu$.

The young specimen can be recognized by the whole plant's being of a vivid lemon-yellow except the surface of the cap. The plant quickly turns to green, then blue, wherever touched. It has a wide distribution in the Eastern and Middle states. The plant in Figure 287 was found in Haynes' Hollow by Dr. Chas. Miesse and photographed by Dr. Kellerman.

As an edible it is among the best. Found from August to October.

BOLETUS CYANESCENS. BULL.

Figure 288   Boletus cyanescens.

Cyanescens is from cyaneus, deep blue, so called the moment you touch it, it turns a deep blue.

Pileus is two to four inches across, convex, then expanded, sometimes nearly plane, frequently wavy, covered with an appressed tomentum; opaque, pale-buff, grayish-yellow, or yellowish, flesh thick, white, quickly changing to a beautiful azure-blue where cut or wounded.

The tubes are quite free, openings small, white, then pale-yellow, round, changing color the same as the flesh.

The stem is two to three inches long, ventricose, hoary with fine hair, stuffed at first, then becoming hollow, colored like the pileus.

The spores are subelliptical, $10–12.5 \times 6–7.5\mu$.

The specimens in Figure 288 were found on rather steep wooded hillsides, Sugar Grove, Ohio. They were all solitary. I have found a few specimens about Chillicothe. They are widely distributed in the Eastern states.

Captain McIlvaine says in his book the caps make an excellent dish cooked in any way. I have never tried them. Found on hilly ground in August and September.

BOLETUS INDECISUS. PK.

THE UNDECIDED BOLETUS. EDIBLE.

Indecisus means undecided; so called because it favors very closely Boletus felleus. There is a difference in the style of the two plants by which, after continued tasting, the student can readily separate them.

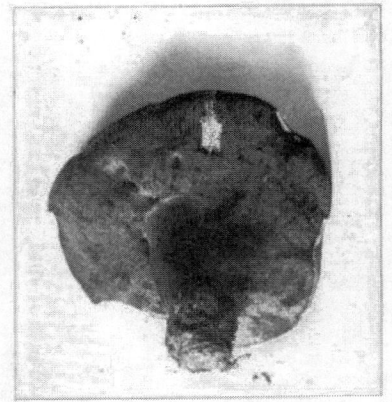

The pileus is three to four inches broad, dry, slightly downy, convex, ochraceous-brown, plane, often irregular on the margin, sometimes wavy, flesh white, and unchangeable, taste mild or sweet.

The tube surface is nearly plane and firmly set against the stem, grayish, becoming tinged with flesh color in age, changing to a brown when bruised; the mouths small and nearly round. The stem is covered with a fine mealy substance, straight or flexuous, sometimes reticulated above. The spores are oblong, brownish flesh color, $12.5–15 \times 4\mu$.

Figure 289   Boletus indecisus. One-half natural size.

The B. indecisus can be readily told from B. felleus by its sweet taste and brownish spores. It is my favorite of all the Boleti, indeed I think it equals the best of mushrooms. Its favorite habitat is under beech trees in the open. It is widely distributed from Massachusetts to the west. Found in July and August.

BOLETUS EDULIS. BULL.—VAR. CLAVIPES. PK.

CLUB-FOOTED BOLETUS. EDIBLE.

Figure 290   Boletus edulis, var. clavipes. Two-thirds natural size. Note confluent caps on right.

Clavipes means club-footed. Pileus fleshy, convex, glabrous, grayish-red or chestnut-color. Flesh white, unchangeable. The tubes at first concave or nearly plane, white and stuffed, then convex, slightly depressed around the stem, ochraceous-yellow. Stem mostly obclavate, inversely club-shaped, and reticulate to the base. The spores oblong-fusiform, $12–15 \times 4–5\mu$. Peck. 51st Rep.

The club-footed Boletus is very closely related to B. edulis. It differs, perhaps, in a more uniform color of the cap, and in having tubes less depressed around the stem, and less tinted with green when mature. The stem is more club-shaped and more completely reticulated.

The pileus in the young plant is much more highly colored and fades out in age, but the margin does not become paler than the disk as is often the case with B. edulis. The specimens in Figure 290 were found in Michigan and photographed by Dr. Fischer. They are quite as good as B. edulis.

BOLETUS SULLIVANTII. B. & M.

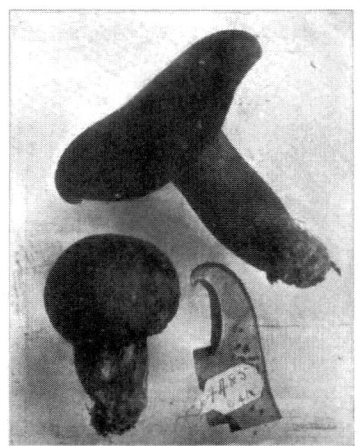

Figure 291　Boletus sullivantii.

Sullivantii is named in honor of Professor Sullivant, an early Ohio botanist.

The pileus is three to four inches broad, hemispherical at first, glabrous, reddish-tawny or brown, brownish when dry, cracked in squares.

The tubes are free, convex, medium size, angular, longer toward the margin, their mouths reddish.

The stem is solid, violaceous at the thickened base, red-reticulated at the apex, expanded into the pileus.

The spores are pallid to ochraceous, oblong-fusiform, $10–20\mu$ long. Peck's Boleti in U. S.

This species is very close to Boletus scaber and Boletus edulis. It differs from B. scaber in its reticulated stem and from B. edulis in its larger tubes. The specimens in Figure 291 were found by Hambleton Young near Columbus, and were photographed by Dr. Kellerman.

BOLETUS PARVUS. PK.

Parvus means small; so named from the smallness of the plant.

The pileus is one to two inches broad, convex, becoming plane, often slightly umbonate, subtomentose, reddish. Flesh yellowish-white, slowly changing to pinkish when bruised.

The tubes are nearly plane, adnate, their mouths rather large, angular, at first bright-red, becoming reddish-brown.

The stem is equal or slightly thickened below, red, from one to two inches long. The spores are oblong, $12.5 \times 4\mu$.

They are found in thin woods, July and August.

BOLETUS EXIMIUS. PK.

THE SELECT BOLETUS. EDIBLE.

Figure 292   Boletus eximius. Two-thirds natural size.

Eximius means select.

The pileus at first is very compact, nearly round, somewhat covered with a mealy substance, purplish-brown, or chocolate color, sometimes with a faint tinge of lilac, becoming convex, soft, smoky red, or pale-chestnut, flesh grayish or reddish-white.

The tube surface is at first concave or nearly plane, stuffed, colored nearly like the pileus, becoming paler with age and depressed around the stem, the mouths minute, round.

The stem is stout, generally short, equal or tapering upward, abruptly narrowed at the base, minutely branny, colored like or a little paler than the cap, purplish-gray within.

The spores are subferruginous, $12.5-15 \times 5-6\mu$. This plant is found in open woods where there are beech trees. I found it frequently on Cemetery Hill, Chillicothe. It is widely distributed, being found from the east to the west. July and August.

BOLETUS PALLIDUS. FROST.

THE PALLID BOLETUS. EDIBLE.

Pallidus, pale. The pileus is convex, becoming plane or centrally depressed, soft, smooth, pallid or brownish-white, sometimes tinged with red. Flesh is white. Tubes plane or slightly depressed around the stem, nearly adnate, very pale or whitish-yellow, becoming darker with age, changing to blue where wounded, the mouths small. The stem is equal or slightly thickened toward the base, rather long, smooth, often flexuous; whitish, sometimes streaked with brown, often tinged with red within. Spores pale ochraceous-brown. Pileus two to four inches broad. Stem three to five inches long. Peck, Boleti of the U. S.

This species is very good, tender, and appetizing. I found it quite abundant in the woods of Gallia County and near Chillicothe, Ohio.

BOLETUS ALVEOLATUS. B. AND C.

THE ALVEOLATE BOLETUS.

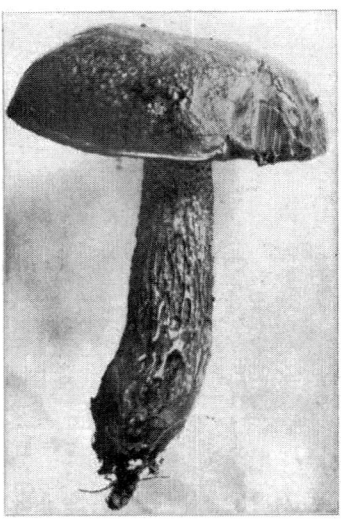

Figure 293 Boletus alveolatus.

Alveolatus is from alveolus, a small hollow, referring to the pitted form of the pore-surface, which is one of the characters of this species. The pileus is convex, smooth, polished, usually rich crimson or maroon, sometimes varied with paler yellowish tints; substance solid, changing to blue on being fractured or bruised, three to six inches broad.

The tube-surface reaches the stem proper, undulate with uneven hollows, maroon, the tubes in section being yellow beyond their dark red mouths.

The stem is usually quite long, covered with depressions or pitted dentations, with intermediate coarse net-work of raised ridges, red and yellow. The spores are yellowish-brown. I found this species in the woods near Gallipolis, Ohio, also near Salem, Ohio. The bright color of its cap will command the attention of any one passing near it. It has been branded as a reprobate, but Captain McIlvaine gives it a good reputation. Found in the woods, especially along streams, August and September. Photographed by Prof. H. C. Beardslee.

BOLETUS FELLEUS. BULL.

THE BITTER BOLETUS.

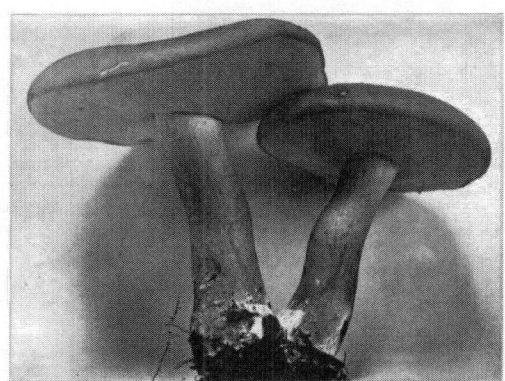

Figure 294   Boletus felleus. Natural size.

Felleus is from fel, gall, bitter. The pileus is convex, nearly plane, at first rather firm in substance, then becoming soft and cushion-like, smooth, without polish, varying in color from pale ochre to yellowish or reddish-brown or chestnut, flesh white, changing to flesh-color when bruised, taste exceedingly bitter, cap three to eight inches in diameter.

The tube-surface is white at first, becoming dull pinkish with age or upon being cut or broken; rounded upward as it reaches the stem, attached to the stem, mouths angular.

The stem is variable, tapering upward, rather stout, quite as smooth as the cap and a shade paler in color, toward the apex covered with a net-work which extends to the base, often bulbous.

The flesh is not poisonous but intensely bitter. No amount of cooking will destroy its bitterness. I gave it a thorough trial, but it was as bitter after cooking as before. It is a common Boletus about Salem, Ohio. I have seen plants there eight to ten inches in diameter and very heavy.

They grow in woods and wood margins, usually about decaying stumps and logs, sometimes in the open fields. July to September.

BOLETUS VERSIPELLIS. FR.

THE ORANGE-CAP BOLETUS. EDIBLE.

Versipellis is from verto, to change, and pellis, a skin. The pileus is two to six inches in diameter, convex, orange-red, dry, minutely woolly or downy, then scaly or smooth, margin containing fragments of the veil, flesh white or grayish.

The tube-surface is grayish-white, tubes long, free, mouths minute and gray.

The stem is equal or tapering upward; solid, white with scaly wrinkles; three to five inches long; and is frequently covered with small reddish or blackish dots or scales. The spores are oblong spindle-shaped.

This plant can be easily distinguished by the remnant of the veil which adheres to the margin of the cap and is of the same color. It is frequently turned under the margin adhering to the tubes. It is a large and imposing plant found in sandy soil and especially among the pines. I found it in J. Thwing Brooke's woods, Salem, Ohio. August to October.

Figure 295   Boletus versipellis. Natural size.

BOLETUS GRACILIS. PK.

THE SLENDER-STEMMED BOLETUS. EDIBLE.

Figure 296   Boletus gracilis. Two-thirds natural size.

Gracilis means slender, referring to the stem.

The pileus is one to two inches broad, convex, smooth or minutely tomentose, the epidermis frequently cracked as in the illustration; ochraceous-brown, tawny, or reddish brown; flesh white.

The tube surface is convex to plane, depressed around the stem, nearly free, whitish, becoming flesh-colored.

The stem is long and slender, equal or slightly tapering upward, usually curved; pruinose or mealy. The spores are subferruginous, .0005 to .0007 inch long, .0002 to .00025 inch broad.

This is quite a pretty plant, but at first sight it will not be taken for a Boletus. They are not plentiful in our woods. I find them only occasionally and then sparsely. They are found in July and August, the months for the Boleti. They grow in leaf mold in mixed woods, especially among beech timber.

BOLETUS STRIÆPES. SECR.

Striæpes means striate stem.

The pileus is convex or plane, soft, silky, olivaceous, the cuticle rust-color within, flesh white, yellow next the tubes, sparingly changing to blue.

The tubes are adnate, greenish, their mouths minute, angular, yellow.

The stem is firm, curved, marked with brownish-black striations, yellow, and brownish-rufescent at the base.

The spores are $10–13 \times 4\mu$. Peck, Boleti of the U. S.

I found some beautiful specimens in a mixed woods on the Edinger hillside, near Chillicothe. I located them here, but observing that this species was not common I sent some to Prof. Atkinson, who placed them under this species. August.

BOLETUS RADICANS. PERS.

The pileus is convex, dry, subtomentose, olivaceous-cinereus, becoming pale-yellowish, the margin thin, involute. Flesh pale-yellow, taste bitterish.

The tubes are adnate, their mouths large, unequal; lemon-yellow.

The stem is two to three inches long, even, tapering downward and radiating, flocculose with a reddish bloom, pale-yellow, becoming naked and dark with a touch.

The spores are fusiform, olive, $10–12.5 \times 5\mu$. Peck, Boleti of the U. S.

I found these specimens in the same locality with the B. striæpes.

The olivaceous cap with its peculiar involute margin and its radiating stem will greatly assist in its determination. August.

BOLETUS SUBLUTEUS. PK.

THE YELLOW BOLETUS. EDIBLE.

Subluteus is from sub, under, nearly; luteus, yellow.

Pileus is two to three inches broad, convex, becoming plane, quite viscid when moist, dull yellowish to reddish brown, frequently more or less streaked. The flesh is whitish or dull yellow.

The tube surface is plane or convex, the tubes set squarely against the stem, being small, nearly round, yellowish or ochraceous, becoming darker in age.

The stem is rather long, nearly equal, about the color of the cap, dotted both above the ring and below it; the ring is membranaceous, quite variable and persistent, usually collapsing as a narrow ring on the stem. The spores are ochraceous-brown, oblong or elliptical, 8–10×4–5.

Figure 297 Boletus subluteus. Natural size.

Prof. Atkinson has made a careful study of both the American and the European plants called in this country B. luteus and B. subluteus, and has come to the conclusion that they should all be called B. luteus. In distinguishing the two we usually say those having much gluten and dotted above the ring are B. luteus, and those dotted both above and below the ring are B. subluteus. The specimens in Figure 297 were collected at the State Farm at Lancaster, Ohio, and photographed by Dr. Kellerman. They are found in July and August.

BOLETUS PARASITICUS. BULL.

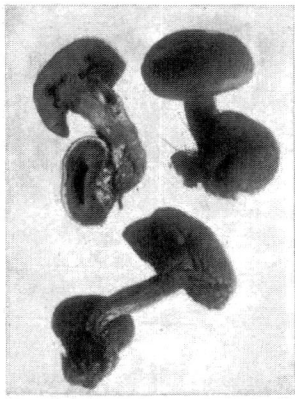

Figure 298 Boletus parasiticus.

Parasiticus means a parasite; so called because it grows on a Scleroderma. It is a small plant and quite rare.

The pileus is one to two inches broad, convex, or nearly plane, dry, silky, becoming glabrous, soon tessellately cracked, grayish or dingy yellow. Tubes decurrent, medium size, golden yellow.

The stem is equal, rigid, incurved, yellow within and without. The spores are oblong-fusiform, pale-brown, $12.5-15 \times 4\mu$. Peck.

The tubes are rather large and unequal, and inclined to run down upon the stem.

This plant was found near Boston, Mass., by Mrs. E. B. Blackford and photographed by Dr. Kellerman. Captain McIlvaine says it is edible but not of good flavor. It is found in July and August.

BOLETUS SEPARANS. PK.

THE SEPARATING BOLETUS. EDIBLE.

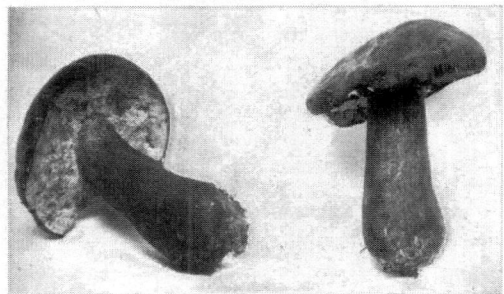

Figure 299 Boletus separans. One-half natural size.

Separans, separating, alluding to the tubes sometimes separating from the stem by the expansion of the pileus.

The pileus is convex, thick, smooth, subshining, often pitted or corrugated; brownish-red or dull-lilac, sometimes fading to yellowish on the margin; flesh white and unchangeable.

Tubes at first are nearly plane, adnate, white and stuffed, then convex, depressed around the stem, ochraceous-yellow or brownish-yellow and sometimes separating from the stem by the expansion of the pileus.

The stem is equal or slightly tapering upward; reticulated, either wholly or in upper part only; colored like the pileus or a little paler, sometimes slightly furfuraceous. Spores subfusiform, brownish-ochraceous. Peck, Boleti of U. S.

The specimens in Figure 299 were found at Londonderry, about fifteen miles east of Chillicothe, in a grassy woods near a stream. The taste is agreeable when raw and quite good when cooked. This might appropriately have been called the lilac Boletus, for that shade of color is usually present in it, somewhere. August to October.

BOLETUS AURIPES. PK.

YELLOW-STEMMED BOLETUS. EDIBLE.

Auripes is from aureus, yellow or golden; pes, foot; so called from its yellow stem.

The pileus is three to four inches broad, convex, nearly smooth, yellowish-brown, the flesh often cracking in areas in old plants; flesh yellow at first, fading to a lighter color, in age.

The tubes are nearly plane, their mouths small, nearly round, at first stuffed, yellow.

The stem is two to four inches long, nearly equal, often reticulated, solid, a bright yellow on the surface and a light yellow within. The spores are ochraceous-brown, tinged with green, $12 \times 5\mu$.

Figure 300 Boletus auripes. One-half natural size. Caps yellowish-brown. Tube surface and stem yellow.

The whole plant, except the upper surface of the cap, is a golden yellow, and even the surface of the cap is more or less yellow. It favors one form of the B. edulis. It is sometimes found in mixed woods, especially if there are mountain laurels in the woods (Kalmia latifolia). It is found in July and August.

BOLETUS RETIPES. B. AND C.

THE BEAUTIFUL-STEMMED BOLETUS. EDIBLE.

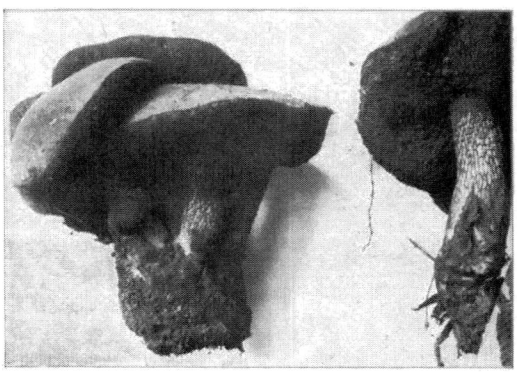

Figure 301 Boletus retipes. Natural size.

Retipes is from rete, a net; pes, a foot; so called from the delicate net-work seen on the stem.

The pileus is convex, dry, powdered with yellow, sometimes rivulose or cracked in areas. The tubes are adnate, yellow.

The stem is subequal, cespitose, reticulate to the base, pulverulent below. The spores are greenish-ochraceous, $12–15 \times 4–5\mu$. Peck, Boleti.

B. retipes is very close to B. ornatipes, but its manner of growth, its pulverulent cap, and its greenish-ochraceous spores will at once distinguish it. I have found them on Ralston's Run, a number from the same mycelial cluster, as in Figure 301. The caps only are good. The specimens in the figure were found near Ashville, N. C., and photographed by Prof. H. C. Beardslee.

BOLETUS GRISEUS. FROST.

THE GRAY BOLETUS.

Griseus means gray. The pileus is broadly convex, firm, dry, almost smooth, gray or grayish black. The flesh is whitish or gray.

The tubes are attached to the stem and slightly depressed around the stem, nearly plane, their mouths being small, nearly round, white or whitish.

The stem is slightly unequal, tapering downward, distinctly reticulated, whitish or yellowish, sometimes reddish toward the base. The spores are ochraceous-brown, $10–14 \times 4–5\mu$. Peck.

This plant, with us, grows singly and it is infrequently found. I have found it always in beech woods along Ralston's Run. It is found in August and September.

Figure 302   Boletus griseus. Two-thirds natural size.

BOLETUS NIGRELLUS. PK.

The Blackish Boletus. Edible.

Figure 303   Boletus nigrellus. Two-thirds natural size.

Nigrellus is a diminutive of niger, black. The entire plant is blackish except the pore surface.

The pileus is three to six inches broad, rather broadly convex or nearly plane, dry, blackish. The flesh is soft and unchangeable.

The tube-surface is rather plane, adhering to the stem, sometimes slightly depressed around the stem, the mouths being small, nearly round; whitish, becoming flesh-colored, changing to black or brown when wounded.

The stem is equal, short, even, black or blackish. The spores are dull flesh-color, $10–12 \times 5–6\mu$.

When I first found this specimen I was inclined to call it B. alboater, but its flesh-colored tubes served to distinguish it. I found the specimens in Figure 303 on Edinger's Hill, near Chillicothe. The taste is mild and fairly good. August and September.

BOLETUS AMERICANUS. PK.

Figure 304   Boletus Americanus. One-half natural size.

This species will attract the attention of the collector because of its very viscid cap. I found the specimens in Figure 304 growing on Cemetery Hill, near Chillicothe, in company with Lactarius deliciosus. They were growing near and under pine trees, both in dense groups and separately. The caps were very viscid, yellow with a slight tinge of red. The stem is covered with numerous reddish-brown dots.

The pileus is one to three inches broad, thin; at first rather globose, convex, then expanded, sometimes broadly umbonate; very viscid when moist, especially on the margin; yellow or becoming dingy or streaked with red in age.

The tube-surface is nearly plane and the tubes join squarely against the stem; quite large, angular, pale yellow, becoming a dull ochraceous.

The stem is slender, equal or tapering upward, firm, with no trace of a ring; yellow, often brownish toward the base, covered with numerous brown or reddish-brown quite persistent granular dots; yellow within. The spores are oblong, ochraceous-ferruginous, $9–11 \times 4–5\mu$.

The veil is only observed in the very young specimens. Only caps are good to eat. The specimens were photographed for me by Dr. Kellerman.

BOLETUS MORGANI. PK.

MORGAN'S BOLETUS. EDIBLE.

Morgani is named in honor of Prof. Morgan.

The pileus is one and a half to two inches broad, convex, soft, glabrous, viscid; red, yellow, or red fading to yellow on the margin; flesh white, tinged with red and yellow, unchangeable.

The tube-surface convex, depressed around the stem, tubes rather long and large, bright yellow, becoming greenish-yellow.

The stem is elongated, tapering upward, pitted with long and narrow depressions, yellow, red in the depressions, colored within like the flesh of the pileus. The spores are olive-brown, 18–22μ, about half as broad. Peck.

Figure 305 Boletus Morgani. One-half natural size.

This plant is found in company with B. Russelli, which it resembles very closely. Its smooth, viscid cap and white flesh will distinguish it. Its stem is much more rough in wet weather than in dry. The peculiar color of the stem will help to identify the species. I found it frequently on Ralston's Run, near Chillicothe. It is found in many of the states of the Union. July and August.

BOLETUS RUSSELLI. FROST.

RUSSELL'S BOLETUS. EDIBLE.

Figure 306 Boletus Russelli. One-half natural size.

The cap is thick, hemispherical or convex, dry, covered with downy scales or bundles of red hairs, yellowish beneath the tomentum, often cracked in areas. The flesh is yellow and unchangeable.

The tubes are subadnate, often depressed around the stem, rather large, dingy-yellow, or yellowish-green.

The stem is very long, equal or tapering upward, roughened by the lacerated margins of the reticular depressions, red or brownish red. The spores are olive-brown, $18–22 \times 8–10\mu$.

The pileus is one and a half to four inches broad, the stem is three to seven inches long, and three to six lines thick. This is distinguished from the other species by the dry squamulose pileus and the color of the stem. The latter is sometimes curved at the base. Peck.

I have found this species frequently in the woods and open places about Chillicothe. It is one of the easiest of the Boleti to determine. The plants here have a bright brownish-red pileus, with a shade lighter color on the stem; the latter quite rough and tapering toward the cap. They are usually solitary. The plants in Figure 306 were collected in Michigan and photographed by Dr. Fischer.

BOLETUS VERMICULOSUS. PK.

Figure 307 Boletus vermiculosus. One-half natural size.

Vermiculosus means full of small worms. The pileus is broadly convex, thick, firm, dry; smooth, or very minutely tomentose; brown, yellowish-brown or grayish-brown, sometimes tinged with red. The flesh is white or whitish, quickly changing to blue where wounded. The tubes are plane or slightly convex, nearly free, yellow; their mouths small, round, brownish-orange, becoming darker or blackish with age, changing promptly to blue where wounded.

The stem is nearly equal, firm, even, paler than the pileus. The spores are ochraceous-brown, $10–12 \times 4–5\mu$. Peck.

The plant represented in Figure 307 grew under the beech trees on Cemetery Hill. I found it frequently in the woods, from July to September.

BOLETUS FROSTII. RUSSELL.

Frostii is named in honor of Mr. Frost, a noted mycologist.

The pileus is three to four inches broad; convex, polished, shining, blood-red; the margin is thin, the flesh scarcely changing to blue.

The tubes are nearly free, greenish-yellow, becoming yellowish-brown with age, their mouths blood-red or cinnabar-red.

The stem is two to four inches long, three to six lines thick, equal or tapering upward, distinctly reticulated, firm, blood-red. The spores are $12.5-15 \times 5\mu$. Peck, Boleti of U. S.

This is a beautiful plant. It is not plentiful, yet it is found frequently on some of our hillsides. The plants in Figure 308 were found in Hayne's Hollow near Chillicothe, and photographed by Dr. Kellerman. The plant is found in New England and through the Middle West. I have had beautiful plants sent me from Vermont. It is not edible, so far as I know. Found in August and September.

Figure 308   Boletus Frostii. Caps blood-red and shining. Natural size.

BOLETUS LURIDUS. SCHAEFF.

THE LURID BOLETUS.

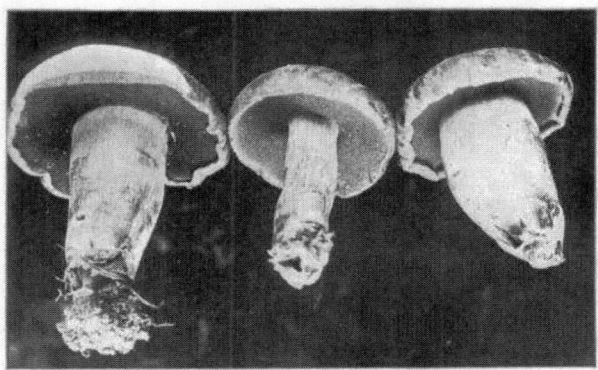

Figure 309   Boletus luridus. One-half natural size.

Luridus means pale-yellow, sallow. The pileus is convex, tomentose, brown-olivaceous, then somewhat viscous, sooty. The flesh is yellow, changing to blue when wounded. Tubes free, yellow, becoming greenish, their mouths round, vermilion, becoming orange. The stem is stout, vermilion, somewhat orange at the top, reticulate or punctuate. The spores are greenish-gray, $15 \times 9\mu$.

The lurid Boletus, though pleasant to the taste, is reputed very poisonous. Boletus rubeolarius, Pers., having a short, bulbous, scarcely reticulated stem, is regarded as a variety of this species. The red-stemmed Boletus, B. erythropus, Pers., is also indicated by Fries as a variety of luridus. It will be seen on the right in Figure 309. It is smaller than B. luridus, has a brown or reddish-brown pileus and a slender cylindrical stem, not reticulated but dotted with squamules. Peck, Boleti of the U. S. The plant is quite abundant in our woods. Found in July and August.

BOLETUS CASTANEUS. BULL.

THE CHESTNUT BOLETUS. EDIBLE.

Figure 310   Boletus castaneus. One-half natural size.

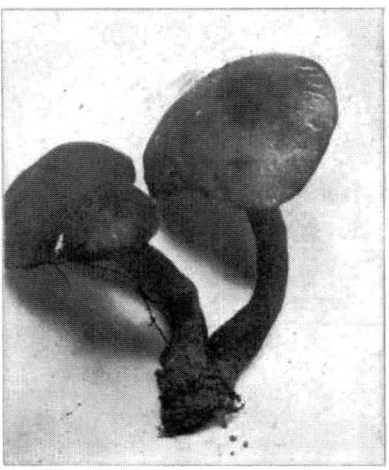

Figure 311   Boletus castaneus.

Castaneus, pertaining to a chestnut. The pileus is dry, convex, then expanded, minutely velvety; cinnamon or reddish-brown, from one to three inches in diameter; the flesh white, not changing when bruised, cap frequently turned upward.

The tube-surface is white, becoming yellow, tubes small and short, free from the stem.

The stem is equal or tapering upward, colored and clothed like the cap, short and not always straight; when young it is spongy in the center but becomes hollow with age. The spores are pale-yellow, oval or broadly elliptical, which is a feature to distinguish the species.

I found a number of specimens in James Dunlap's woods, near Chillicothe, Ohio. A great majority seemed to be attacked by the parasitic fungi, Sepedonium chrysospermum.

The caps are very fine eating. Care should be taken to use only young specimens. Found in open woods from June to September.

BOLETUS SATANUS. LENZ.

SATANIC BOLETUS.

Pileus convex, smooth, somewhat gluey, brownish-yellow or whitish; flesh whitish, becoming reddish or violaceous where wounded. Tubes free, yellow, their mouths bright red, becoming orange-colored with age. The stem thick, ovate-ventricose, marked above with red reticulations. Peck, Boleti of U. S.

Hamilton Gibson and Captain McIlvaine seem to give his Satanic majesty a good reputation, but I would say "Be cautious." His looks always deterred me. Found in woods from June to September.

STROBILOMYCES. BERK.

Strobilomyces is from two Greek words meaning a pine-cone and a fungus. The hymenophore is even, tubes not easily separable from it, large and equal. It is of a brownish-gray color, its shaggy surface more or less studded with deep-brown or black woolly points, each at the center of a scale-like segment. The tubes beneath are covered at first with a veil which breaks and is often found on the rim of the cap. It is a plant that will quickly attract attention.

STROBILOMYCES STROBILACEUS. BERK.

THE CONE-LIKE BOLETUS. EDIBLE.

Figure 312   Strobilomyces strobilaceus. Two-thirds natural size.

Strobilaceus, cone-like. This is especially emphasized from the fact that both the genus and the species are named from the fancied resemblance of the cap to a pine cone. It is ever readily recognized because of this character of the cap.

The pileus is convex, rough with dark umber scales drawn into regular cone-like points tipped with dark-brown; margin veiled, flesh grayish-white, turning red when bruised, and finally black.

Pore-surface grayish-white in young specimens, and usually covered with the veil; tubes attached to the stem, angular, turning red when bruised.

The stem is equal or tapering upward, furrowed at the top, covered with a woolly down. Spores dark-brown, $12–13 \times 9\mu$. Found at Londonderry. Common in woods. August to September.

BOLETINUS. KALCHB.

Boletinus is a diminutive of Boletus.

Hymenium composed of broad radiating lamellæ, connected by very numerous and narrow anastomosing branches or partitions, forming large angular pores. Tubes somewhat tenacious, not easily separable from the hymenophore and from each other, adnate or subdecurrent, yellowish. Peck.

BOLETINUS PICTUS. PK.

THE PAINTED BOLETINUS. EDIBLE.

Pictus, painted. This plant seems to delight in damp pine woods, but I have found it only occasionally about Chillicothe, under beech trees. It is readily recognized by the red fibrillose tomentum which covers the entire plant when young. As the plant expands the reddish tomentum is broken into scales of the same color, revealing the yellowish color of the pileus beneath. The flesh is compact, yellow, often changing to a dull pinkish or reddish tint where wounded.

Figure 313 Boletinus pictus.

The tube-surface is at first pale yellow, but becomes darker with age, often changing to pinkish, with a brown tinge where bruised.

The stem is solid, equal, and covered with a cottony layer of mycelium-threads like the pileus, though often paler. The spores are ochraceous, $15-18 \times 6-8\mu$. The plants are two to four inches broad, and one and a half to three inches high. Found from July to October.

BOLETINUS CAVIPES. KALCHB.

HOLLOW-STEMMED BOLETINUS. EDIBLE.

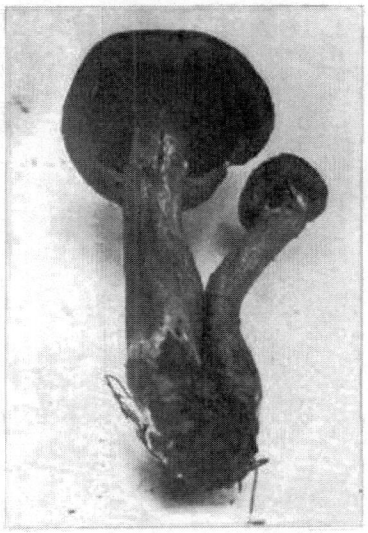

Figure 314 Boletinus cavipes.

Cavipes is from two Latin words meaning a hollow stem.

The pileus is broadly convex, rather tough, flexible, soft, subumbonate, fibrillose-scaly, tawny-brown, sometimes tinged with reddish or purplish, flesh yellowish. The tubes are slightly decurrent, at first pale-yellow, then darker and tinged with green, becoming dingy-ochraceous with age. The stem is equal or slightly tapering upward, somewhat fibrillose or floccose, slightly ringed, hollow, tawny-brown or yellowish-brown, yellowish at the top and marked by the decurrent dissepiments of the tubes, white within. Veil whitish, partly adhering to the margin of the pileus, soon disappearing. The spores are $8-10 \times 4\mu$. Peck, in Boleti of the U. S.

This plant grows in New York and the New England states, under pine and tamarack trees. The caps are convex, covered with a tawny-brown fibrillose tomentum. The stems of those I have seen are hollow from the first. The plants in Figure 314 were sent me from Massachusetts by Mrs. Blackford.

BOLETINUS POROSUS. (BERK.) PK.

Figure 315 Boletinus porosus. Two-thirds natural size. Caps nut-brown, yellowish-brown or olivaceous.

These form a small but interesting species, not usually exceeding three and a half inches in diameter nor more than two inches in height.

The cap is somewhat fleshy, nut-brown, or yellowish-brown, shading to olivaceous in color in most of the specimens which I have found; when fresh and moist, somewhat sticky and shining. The margins are thin, rather even, and inclined to be involute; the shape of the cap is more or less irregular, in many cases almost kidney-shaped.

The stem is laterally attached, tough, and gradually expands into the pileus which it resembles in color; it is markedly reticulated at the top by the decurrent walls of the spore-tubes. The spore-surface is yellow, the tubes arranged in radiating rows, some being more prominent than others, the partitions often assuming the form of gills which branch and are connected by cross partitions of less prominence. The stratum of tubes, while soft, is very tenacious, not separating from the flesh of the pileus.

The odor and taste of all the specimens found were pleasant. Found in damp woods in July and August. When a sufficient number can be found they make an excellent dish.

It is found in abundance about Chillicothe.

FISTULINA. BULL.

Fistulina means a small pipe; so called because the tubes stand close together and separate easily one from another.

The hymenophore is fleshy and hymenium inferior. When first seen springing from a stump or root it looks like a large strawberry. It soon develops into the appearance of a big red tongue. When young the upper side is quite velvety and peach-colored, later it becomes a livid red and loses its velvety appearance. The under surface is flesh-colored and is rough like the surface of a tongue, owing to the fact that the tubes are free from one another. When it is moist it is very viscid, making your hands quite blood-stained in appearance.

FISTULINA HEPATICA. FR.

THE LIVER FUNGUS. EDIBLE.

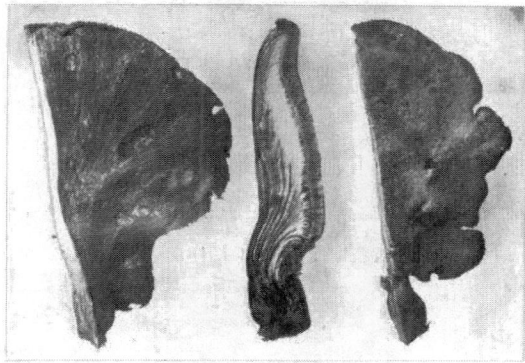

PLATE XLIII. FIGURE 316   FISTULINA HEPATICA. Beefsteak mushroom

This is a beautiful plant, quite common where there are chestnut stumps and trees. I have found it on chestnut oak, quite large specimens, too. It is one of my favorite mushrooms; one cannot afford to pass it by. Its beautiful color will attract attention at once, and having once eaten it well prepared, one will never pass a chestnut stump without examining it.

The pileus is fan-shaped or semicircular, red-juicy, flesh when cut somewhat mottled like beet-root and giving forth a very appetizing odor; the cap is moist and somewhat viscid, the color varying from a red (somewhat beefy) to a reddish-brown in older plants; while the spore surface varies from strawberry-pink through a light-and dark-tan to an almost chestnut-brown.

In young plants the color is much richer and more vivid than in those of greater maturity. The spore surface resembles nothing so much as a very fine sponge, the spore-tubes being short, crowded, yet distinct.

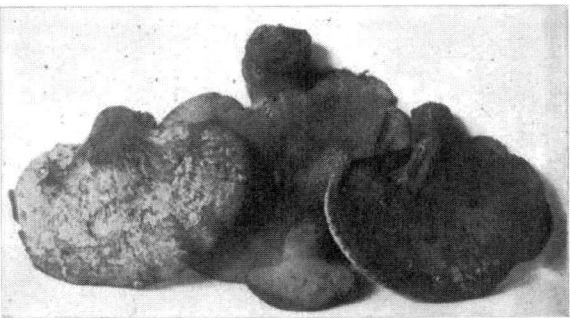

Figure 317   Fistulina hepatica. One-half natural size.

The marked peculiarity of its mode of growth is in the attachment of the stem; somewhat thick, fleshy, and juicy, coming from the side of the pileus like the handle of a fan, it looks as if some one had taken hold of the cap and given it a partial twist to the right or to the left, as may be seen in Figure 317. Another peculiarity I have noticed in this species consists of the nerve-like lines, or veinlets, radiating from the stem and streaking the upper surface of the cap. The taste, when raw, is slightly but pleasantly acid. Its favorite habitat seems to be injured places on chestnut trees, and about chestnut stumps. It is known as Liver Fungus, Beefsteak Fungus, Oak-Tongue, Chestnut-Tongue, etc. It is found from July to October.

I have found it plentiful about Chillicothe on chestnut stumps, and quite generally over the state. I found some very fine specimens on the chestnut oaks, about Bowling Green, Ohio.

When properly prepared it is equal to any kind of meat. It is one of our best mushrooms.

FISTULINA PALLIDA. B. AND RAV.

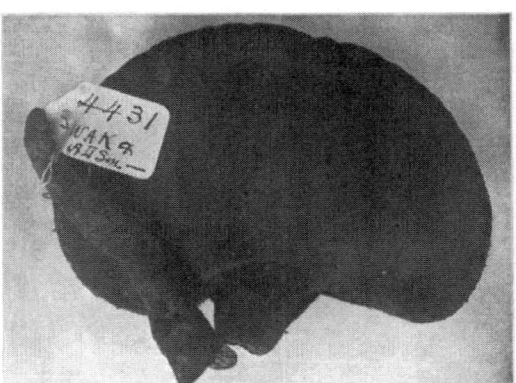

Figure 318   Fistulina pallida. Natural size.

Pallida means pale. Pileus kidney-shaped, pallid-red, fawn or clay-color, thick at the base and thinning toward the margin, which is often crenate and inflexed; pulverulent, firm, flexible, tough; flesh white.

The tubes are long and slender, mouths somewhat enlarged, whitish, the tube surface a pale cream-color and minutely mealy, pores not decurrent but ending with the beginning of the stem.

The stem is uniformly attached to the concave margin of the cap; attenuated downward; whitish below, but near the cap it changes to the same tint. The peculiar manner of attachment of the stem will serve to identify the species, which I have found several times near Chillicothe. The specimen in the illustration was found on the State farm, and photographed by Dr. Kellerman.

POLYPORUS. FR.

Polyporus is from two Greek words meaning many and pores. In this genus the stratum of the pores is not easily separated from the cap. Most of the species under this genus are tough and corky. Many grow on decayed wood, a few on the ground, but even these are inclined to be tough. Very few of those growing on wood have a central stem and many have apparently no stem at all.

POLYPORUS PICIPES. FR.

THE BLACK-FOOTED POLYPORUS.

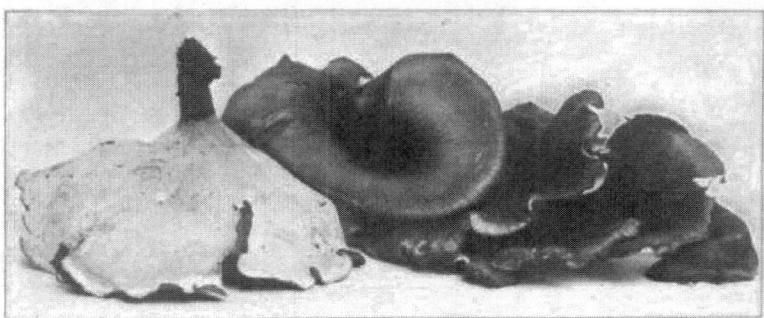

Figure 319 Polyporus picipes. Two-thirds natural size. Note the black stem, which gives name to the species.

Picipes is from pix, pitch or black, and pes, foot.

The pileus is fleshy, rigid, coriaceous, tough, even, smooth, depressed either behind or in the center; livid with a chestnut-colored disk.

The pores are decurrent, rounded, small, tender, white, finally reddish-gray.

The stem is eccentric and lateral, equal, firm; at first velvety, then naked; punctate with black dots, becoming black.

The stem at the base is pitch-black, as will be seen in Figure 319. The margin of the cap is very thin and the caps are irregularly funnel-form. This plant is widely distributed over the United States and is quite common about Chillicothe. Found in damp woods on decayed logs from July to November. When very young and tender it can be eaten.

POLYPORUS UMBELLATUS. FR.

THE SUN-SHADE POLYPORUS. EDIBLE.

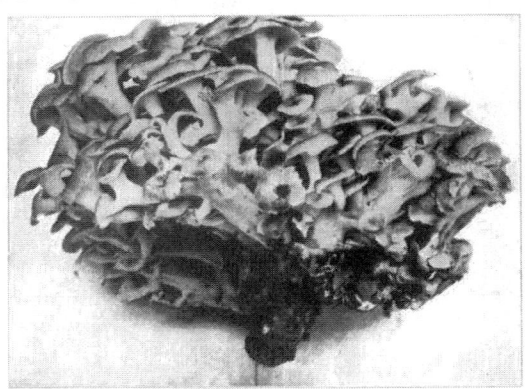

Plate XLIV. Figure 320   Polyporus umbellatus.

Umbellatus is from umbella, a sun-shade. Very much branched, fibrous-fleshy, toughish. The pileoli are very numerous, one-half to one and a half inches broad, sooty, dull-red, united at the base. Pores are minute and white. White pileoli have sometimes occurred. Fries.

The tufts, as will be observed from Figure 320, are very dense, and there seems to be no limit to their branching. Notice that every cap is depressed or umbilicate. The specimen in Figure 320 was collected near Mammoth Cave, Kentucky, by Mr. C. G. Lloyd, Cincinnati, and through his courtesy I have used his print. I have found the plant about Chillicothe and Sidney, Ohio. It is found on decayed roots on the ground, or on stumps. When the caps are fresh they are quite good. May to November.

POLYPORUS FRONDOSUS. FR.

THE BRANCHED POLYPORUS. EDIBLE.

Figure 321   Polyporus frondosus. One-fifth natural size.

Frondosus, full of leafy branches. The tufts are from six inches to over a foot broad, very much branched, fibrous-fleshy, toughish.

The pileoli are very numerous, one-half to two inches broad, sooty-gray, dimidiate, wrinkled, lobed, intricately recurved. Flesh white. Stems, growing into each other, white.

The pores are rather tender, very small, acute, white, commonly round, but in oblique position, gaping open and torn. Fries.

The specimen in Figure 321 was found near Chillicothe. When tender it is very good. Found on stumps and roots from September till the coming of frost.

We are told that in the Roman markets this mushroom is frequently sold as an article of food.

POLYPORUS LEUCOMELAS. FR.

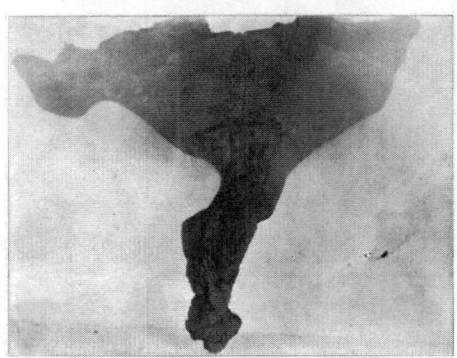

Figure 322   Polyporus leucomelas.

Leucomelas is from two Greek words, leucos, white, and melas, black.

The pileus is two to four inches broad, fleshy, somewhat fragile, irregularly shaped, silky, sooty-black; flesh soft, reddish when broken.

The pores are rather large, unequal, ashy or whitish, becoming black when drying.

The stem is one to three inches long, stout, unequal, somewhat tomentose, sooty-black, becoming black internally. The pileus and stem become black in places.

The spores are cylindric-fusoid, pale-brown, $10-12 \times 4-5\mu$.

They are usually found in pine woods. The caps are often deformed and are easily broken. The pores resemble those of a Boletus. The plant is quite widely distributed. The one in Figure 322 was found in Massachusetts by Mrs. Blackford, and I photographed it after it was partially dry. It is probably the same as P. griseus, P.

POLYPORUS BERKELEYI. FR.

BERKELEY'S POLYPORUS. EDIBLE.

The pileoli are fleshy, tough, becoming hard and corky, many times imbricated, sometimes growing very large, with many in a head; subzonate, finally tomentose; the plant very much branched, alutaceous.

The stem is short or entirely wanting, arising from a long and thick caudex.

The pore surface is very large, the pores are large and irregular, angular, pale-yellowish.

I have seen some very large specimens of this species. The natural size of the specimen in Figure 323 is two and one-fourth feet across. When young it is edible, but not equal to P. sulphureus. It is found growing on the ground near trees and stumps, and is a widely distributed plant.

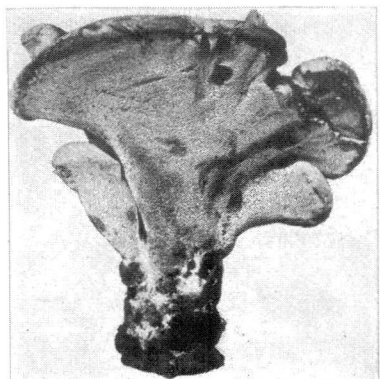

Figure 323   Polyporus Berkeleyi. One-fifth natural size.

Plate XLV. Figure 324   Polyporus Berkeleyi. Reduced. Natural size being 2½ feet across.

POLYPORUS GIGANTEUS. FR.

THE GIANT POLYPORUS. EDIBLE.

Giganteus is from gigas, a giant. The pileoli are very numerous, imbricated, fleshy, tough, somewhat coriaceous, flaccid, somewhat zoned; color a grayish-brown in young specimens, the deep cream pore surfaces tipping the pilecli, rendering it a very attractive plant; this cream-color is quickly changed to black or deep-brown by touching it.

The pores are minute, shallow, round, pallid, at length torn.

The stem is branched, connate from a common tuber.

This is a large and certainly a very attractive plant, being very often two to three feet across. When young and tender it is edible. Found growing on decayed stumps and roots, it is somewhat common in our state. I have found some quite large specimens about Chillicothe. It is easily distinguished by its pore surface turning black or dark-brown to the touch. When young and tender it makes a good stew, but it must be well cooked.

POLYPORUS SQUAMOSUS. FR.

THE SCALY POLYPORUS.

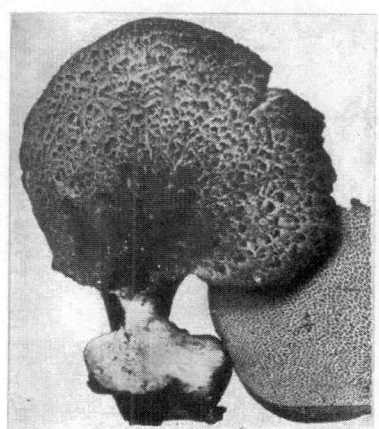

Figure 325   Polyporus squamosus. Natural size.

Squamosus means abounding in scales. The pileus is from three to eighteen inches broad, fleshy, fan-shaped, expanded, flattened, somewhat ochraceous, variegated, with scattered, brown, adpressed scales.

The stem is eccentric and lateral, blunt, reticulated at apex, blackish at the base.

The pores are thin, variable; at first minute, then broad, angular and torn; pallid. Spores are white and elliptical, $14 \times 6\mu$.

It is found from Massachusetts to Iowa, and grows very large. Specimens have been reported seven feet in circumference and attaining a weight of 40 pounds.

The specimen in Figure 325 was found by Mr. C. G. Lloyd in the woods at Red Bank, near Cincinnati. It is quite a common plant in Europe.

It is tough, but it is prepared for eating by being cut fine and stewed for a half hour or more.

In Figure 325 the angular and torn pores are obvious, as well as the scales which give rise to its name. Found on trunks and stumps from May to November.

POLYPORUS SULPHUREUS. FR.

THE SULPHUR-COLORED POLYPORUS. EDIBLE.

Plate XLVI. Figure 326   Polyporus sulphureus.

Sulphureus, pertaining to sulphur, so called from the color of the tube-bearing surface. In mature specimens the growth is horizontal, spreading fan-like from the stem, undulating with radiating flutings. The upper surface is salmon, orange, or orange-red; flesh cheesy, light-yellow, the edge being smooth and unevenly thickened with nodule-like prominences. In young specimens the ascending, under yellow surface outwardly exposed.

The pore surface is a bright sulphur-yellow, which is more persistent than the color of the cap; pores very minute, short, often formed of inflexed masses.

The stem is short, a mere close attachment for the spreading growth. The taste is slightly acid and mucilaginous when raw. The spores are elliptical and white, $7–8 \times 4–5\mu$.

It grows on decayed logs, on stumps, and on decayed places in living trees. The mycelium of this species will frequently be found in the hearts of trees and remain there for years before the tree is injured sufficiently for the mycelium to come to the surface. It may take months, or a century, to accomplish this.

When this plant is young and tender it is a prime favorite with all who know it. It is found from August to November. Its favorite host is an oak stump or log.

POLYPORUS FLAVOVIRENS. B. & RAV.

Figure 327   Polyporus flavovirens. Two-thirds natural size.

Flavovirens means yellowish-green or olivaceous.

The pileus is quite large, three to six inches broad, convex, expanded funnel-form or repand, fleshy, tomentose, yellowish-green or olivaceous; frequently the pileus is cracked when old; flesh white.

The pores are not large, toothed, white or whitish, decurrent upon the stem which is tapering.

This plant is very common on the oak hillsides about Chillicothe. The plants in Figure 327 were found by Miss Margaret Mace on the Governor Tiffin farm, about twelve miles north of Chillicothe, growing in large groups under oak trees. It is edible though often tough. It is found in August and September. It is very abundant in this region.

POLYPORUS HETEROCLITUS. FR.

THE BOUQUET POLYPORUS. EDIBLE.

Figure 328   Polyporus heteroclitus. One-fourth natural size. The Pileoli bright orange.

Heteroclitus is from two Greek words; one of two and to lean, referring to its habit of growth, leaning apparently upon the ground or the base of a tree or stump. It is cæspitose and coriaceous. The pileoli are two and a half inches broad, orange and sessile, expanded on all sides from the radical tubercle, lobed, villous, zoneless.

The pores are irregularly shaped and elongated, golden yellow. Fries.

The specimen in Figure 328 was found by Mr. Beyerly at Richmond Dale, Ohio. It was over a foot in diameter and eight inches high, growing in many cæspitose layers, on the ground under an oak tree, from a radical tubercle. The flesh was juicy and tender, breaking easily. The radical tubercle from which it grew was filled with a milky juice. The flesh was somewhat lighter in color than the outside pilei, which extended horizontally from the tubercle. It is a very showy and attractive plant, and as Captain McIlvaine remarks, it looks like a "mammoth dahlia" in bloom. When young and tender it is good, but in age it becomes rank. This plant was found July 1st. It grows in the months of June and July.

POLYPORUS RADICATUS. SCHW.

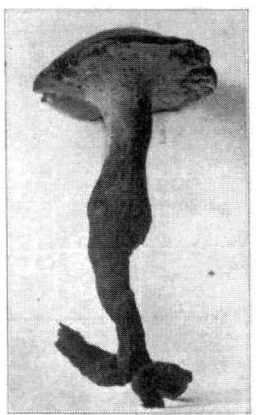

Figure 329   Polyporus radicatus. One-third natural size.

Radicatus, from the long root the plant has. The pileus is fleshy, quite tough, cushion-shaped, slightly depressed, pale sooty, somewhat downy.

The pores are decurrent, quite large, obtuse, equal, white.

The stem is very long, often eccentric, tapering downward, sometimes ventricose as in Figure 329, rooting quite deep, black below.

It is found on the ground in the woods and in old clearings beside old trees and stumps.

The blackish or brown pileus, which is more or less tomentose, with a black stem more or less deformed, will serve to distinguish the species. Found from September to November.

POLYPORUS PERPLEXUS. PK.

Figure 330   Polyporus perplexus. Two-thirds natural size.

The pileus is spongy-fleshy, fibrous, sessile, commonly imbricated, and somewhat confluent, irregular, hairy-tomentose to setose-hispid, grayish-tawny, or ferruginous, the margin subacute, sterile, the substance within tawny-ferruginous, somewhat zonate.

The pores are two to three lines long, unequal, angular, the dissepiments becoming brownish-ferruginous with age or where bruised. The spores are ferruginous, broadly elliptical, .00024 to .0003 inch long and about .0002 broad. Peck.

This is very abundant on beech logs, growing quite large, massive, imbricated, and confluent, the pileoli being often two to four inches broad. It is very closely related to P. cuticularis and P. hispidus. It can be easily distinguished from P. cuticularis by means of its straight margin, and from P. hispidus by its small size and smaller pores. Found from September to November.

POLYPORUS HISPIDUS. FR.

Pileus is very large, eight to ten inches broad and three to four inches thick, compact, spongy, fleshy but fibrous, dimidiate, with occasionally a very short stem; generally very hairy, but sometimes smooth; the pileus is often marked with concentric lines which seem to indicate arrested vegetation; brown, blackish, yellowish or reddish brown, below pale-yellow or rich sienna-brown, margin paler.

The pores are minute, round, inclined to separate, fringed, paler. The spores are yellowish, apiculate, $10 \times 7\mu$. Often found on living trees, the plant gains entrance to the living stem through the bark, by means of a wound made by some agency, as a bird or a boring insect; soon a mass of mycelium is formed, and from this the fruiting body is produced.

POLYPORUS CUTICULARIS. FR.

Pileus is quite thin, spongy, fleshy, then dry; plane, hairy-tomentose, ferruginous, then blackish-brown; margin fibrous, fimbriate, internally loose and parallel, fibrous.

The pores are long, quite small, pale, then ochraceous; pores longer than the thickness of the flesh. The spores are yellow or ochraceous, very abundant, $7 \times 4–5\mu$. The hairs on the pileus are three-cleft.

This is very frequent in beech woods about Chillicothe. Found in September and October.

POLYPORUS CIRCINATUS. FR.

THE ROUND POLYPORUS. EDIBLE.

Circinatus is from circinus, a pair of compasses, hence means rounded like a circle.

The pileus is three to four inches across, with a double cap, one cap within another, both being compact, thick, round, plane, zoneless, velvety, rusty-yellow to reddish-brown, the flesh being of the same color. The upper cap is pliable, compact, soft, and covered with a soft tomentum, the lower cap, contiguous with the stem, is woody and corky.

The pores are decurrent, extending down the stem, entire, rather small, dusky-gray.

The stem is short and rather thick, often swollen, covered with a reddish-brown tomentum.

This is an odd but handsome species and easily determined because of its double cap. It is said to prefer fir woods, but I have frequently found it in oak woods. It grows on the ground, and when young and fresh the pilei are said to be good. I have never found more than one specimen at a time and never in a condition to eat, though good authorities say it is edible when young and tender. Found in September and October.

POLYPORUS ADUSTUS. FR.

Adustus means scorched, so called from the blackish color of margin.

The pileus is often imbricated; fleshy, tough, firm, thin, villous, ash-color; margin straight, blackish.

The pores are minute, round, obtuse, whitish, soon ashy-brown.

It is abundant everywhere on fallen beech or on beech stumps. It is very close to P. fumosus if it is not identical with it. It is found from August to late fall.

POLYPORUS RESINOSUS.

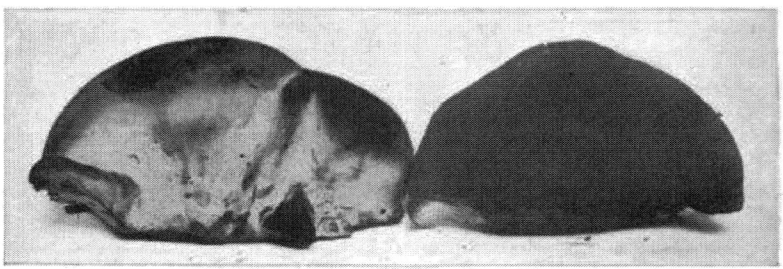

Figure 331   Polyporus resinosus. One-fourth natural size.

Pileus from three to six, and frequently eight, inches long; rich-brown, varying from bright cinnamon to red, handsomely marked with delicate pencilings radiating from the axis of growth; the color of the pileus seems to form a binding about the edge of the light-gray pore surface, which is closely punctured with minute elliptical pores.

The color of the pore surface readily changes to brown upon slight pressure. The whole plant is full of a brownish juice which exudes freely upon pressure. The plant is shelving and imbricated upon the side of a log, without any apparent stem.

Taken altogether the Polyporus resinosus presents one of the handsomest specimens of fungus growth that one will be likely to find in a long day's tramp. When fresh and growing it has rather a pleasant taste.

It is found during October and November, growing on decayed logs, being partial to the beech. Its abundance is equal to its beauty.

POLYPORUS LUCIDUS. FR.

Figure 332   Polyporus lucidus. One-third natural size.

The pileus is two to three or more inches broad, usually very irregular, brownish-maroon, with a distinct double zone of duller dark-brown and tan. Cap glazed especially in the center, wrinkled.

The spore surface is a very light grayish-brown in the young plant, changing to almost a tan in older ones, pores labyrinthiform.

The stem is irregular, knotted and swollen with protuberances somewhat resembling buds, from which develop the caps which in some cases appear as if stuck on the stem like barnacles on a stick. Contrary to most mushrooms the upper surface of the cap and the stem are of nearly the same color, the stem being usually of a more brilliant red. The stem has a distinct root extending into the ground several inches. The whole plant is almost indescribably irregular. It is quite an attractive plant when seen growing among the weeds and beside stumps. The plants in Figure 332 I found growing among Datura stramonium beside old stumps in a pasture. I have found the same species growing on oak stumps. It is known as Ganoderma Curtisii, Berk., G. pseudo-boletus, Merrill. It is found from August till late fall.

POLYPORUS OBLIQUUS. PERS.

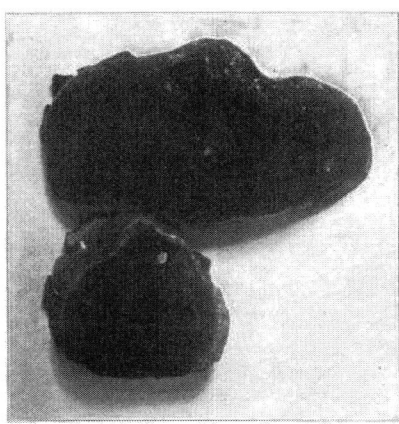

Figure 333   Polyporus obliquus. Two-thirds natural size.

Obliquus means slanting, oblique. This species is widely circumfused, usually hard, quite thick, uneven, pallid, elegant chocolate-brown, then blackish; conversely encircled crested border.

The pores are long, very minute, obtuse, slightly angular. It grows on dead branches of iron-wood and wild cherry. The deep chocolate-brown and the oblique form of its pores will serve to identify the species.

It grows, with us, in the spring. I gathered this specimen in June. In the fall I visited the same trunk, but found they had begun to decay. It is sometimes called Poria obliqua.

POLYPORUS GRAVEOLENS. FR.

Figure 334   Polyporus graveolens.

Graveolens means strong scented. Corky or woody and extremely hard, very closely imbricated and connate, forming a subglobose polycephalous mass, Figure 334. Pileoli innumerable, inflexed and appressed, plicate, brown.

Pores concealed, very minute, round, pale-brown, the dissepiments thick and obtuse. Morgan.

This is a very interesting plant because of its peculiar mode of growth. It is found in woods or clearings on dead logs or on standing dead trees. In some parts of the state it is quite common. From the illustration, Figure 334, it will be seen that the plant consists of an innumerable number of pileoli forming a subglobose or elongated mass. They are frequently three to six inches in diameter and several inches long. I have seen them very much elongated on standing trees. When it is young and growing it is shiny in appearance and has a reddish and sometimes a purplish tint. The inner substance is ferruginous but covered with a hard brown crust. The pores are brown, and when examined with the glass are seen to be lined with a very fine pubescence. The imbricated form of the pileoli show very plainly in the illustration.

POLYPORUS BRUMALIS. FR.

THE WINTER POLYPORUS.

Figure 335   Polyporus brumalis.

Brumalis is from bruma, which means winter; so called because it appears late, in cold weather. The specimens in Figure 335 were found in December.

The pileus is from one to three inches broad, nearly plane, slightly depressed in the center; somewhat fleshy and tough; dingy-brown, clothed with minute scales, becoming smooth, pallid.

The pores are oval, slightly angular, slender, acute, denticulate, white, $5-6 \times 2\mu$.

The stem is short, thin, slightly bulbous at the base, hirsute or squamulose, pale, central.

It usually occurs singly but frequently you will find several in a group. Found on sticks and logs, they are quite hard to detach from their hosts. Too tough to eat. It equals Polyporus polyporus. (Retz) Merrill.

POLYPORUS RUFESCENS. FR.

THE RUFESCENT POLYPORUS.

Rufescens, becoming red. The pileus is flesh-colored, spongy, soft, unequal, hairy or woolly.

The pores are large, sinuose and torn, white or flesh-colored.

The stem is short, irregular, tuberous at the base. Spores elliptical, $6 \times 4–5\mu$.

Rather common about Chillicothe on the ground about old stumps.

POLYPORUS ARCULARIUS. BATSCH.

Figure 336 Polyporus arcularius. Two-thirds natural size, showing dark brown and depressed center; also dark brown stems.

The pileus is dark-brown, minutely scaly, depressed in the center, margin covered with stiff hairs.

The tube surface is of a dingy cream color, openings oblong, almost diamond-shaped, resembling the meshes of a net, the meshes being smaller on the margin, shallow, simply marked out at the top of the stem.

The stem is dark-brown, minutely scaly, mottled, with a ground work of cream-color; hollow. Common in the spring of the year on sticks and decayed wood in fields or in old clearings. It is quite generally distributed. Edible but tough.

POLYPORUS ELEGANS. FR.

The pileus is fleshy, soon becoming woody; expanded, even, smooth, pallid.

Pores are plane, minute, nearly round, pallid, yellowish-white.

The stem is eccentric, even, smooth, pallid; base from the first abruptly black. This is quite common on rotten wood in the forests. It resembles P. picipes both in appearance and habitat.

POLYPORUS MEDULLA-PANIS. FR.

Effused, determinate, subundulate, firm, smooth, white, circumference naked, submarginate, wholly composed of middle sized, rather long, entire pores, the whole becoming yellowish in age.

I found this species on an elm log along Ralston's Run.

POLYPORUS ALBELLUS. PK.

The pileus is thick, sessile, convex or subungulate, subsolitary, two to four inches broad, one to one and a half thick, fleshy, rather soft; the adnate cuticle rather thin, smooth or sometimes slightly roughened by a slight strigose tomentum, especially toward the margin; whitish, tinged more or less with fuscus; flesh pure white, odor acidulous.

The pores are nearly plane, minute, subrotund, about two lines long; white, inclining to yellowish, the dissepiments thin, acute.

The spores are minute, cylindrical, curved, white, .00016 to .0002 inch long. Peck.

This species is quite common here and is very widely distributed in the United States.

POLYPORUS EPILEUCUS. FR.

This is quite a large and beautiful plant. It apparently grows without a stem, its color being an unequal gray. The pileus is somewhat coriaceous, firm, pulvinate, villous.

The pores are round, elongated, obtuse, entire, white.

This is not common with us, but I have met it a few times and always on elm logs or stumps.

POLYPORUS BETULINUS. FR.

THE BIRCH POLYPORUS. EDIBLE.

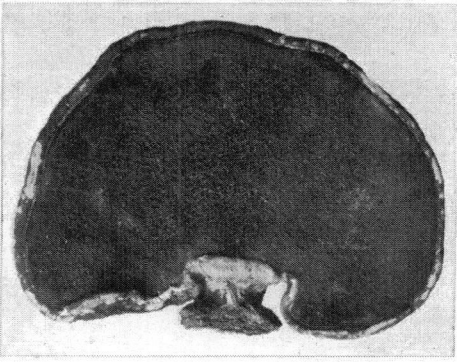

Figure 337   Polyporus betulinus.

Betulinus is from betulina, birch.

The pileus is from four to ten inches across, fleshy, soon corky, ungulate, obtuse, smooth, pale reddish-brown when mature, often mottled, roundish, or somewhat reniform, zoneless, the oblique vertex in the form of an umbo; pellicle thin, separating; flesh white, very thick.

The pores are short, round, minute, unequal, separable from the pileus when fresh, but really concrete with it; white or tinged with brown, developing slowly; when mature there are peculiar hair-like scales attached to the pore-surface, making the plant look like a Hydnum when viewed from the side. It is found wherever the birch tree grows. When young and fresh it is edible, but with a strong flavor unpleasant to many. In this state the deer eat it. The specimen in Figure 337 was found in Wisconsin, and photographed by Dr. Kellerman. This species is the Piptoporus suberosus (L.) of Merrill.

POLYPORUS CINNABARINUS. SCHW.

CINNABAR POLYPORUS.

Figure 338   Polyporus cinnabarinus. One-third natural size.

Cinnabarinus like cinnabar (vermilion). The pileus is dry, more or less spongy, pliant, rather thick, fibrous on top; flesh light or yellowish-red, shelving.

The pores are carmine, quite small, round, entire.

This species is quite common in the woods about Chillicothe. It is easily identified by the beautiful carmine color of the pileus and the pore surface, the latter being a shade darker than the former, as will be seen in Figure 338.

The specimens photographed were found in December. They grow on dead logs and branches, commonly on the oak and wild cherry, sometimes on maple. It is called by some authors Trametes cinnabarina.

POLYPORUS VULGARIS. FR.

COMMON EFFUSED POLYPORUS.

Vulgaris, common. Quite broadly effused, very thin, adheres closely to its host; even, white, dry. Circumference soon smooth and the whole surface composed of firm, crowded, small, round, nearly equal pores.

Effused on dead wood, fallen branches, and frequently on moist boards.

POLYPORUS LACTEUS. FR.

The pileus is white, or whitish, fleshy, somewhat fibrous, fragile, triangular in form, pubescent, azonate, margin somewhat inflexed, acute.

The pores are thin, acute, dentate, finally lacerate and labyrinthiform.

This species is found in the woods, on beech logs. It is small and thin, not much more than an inch in width but sometimes elongated. Steep and gibbous behind, becoming at length smooth and equal. It is not abundant in our woods, but I have found it often. August and September.

POLYPORUS CÆSIUS. SCHRAD.

The pileus is white, with a bluish tinge occasionally upon its surface, soft, tenacious, unequal, silky.

The pores are small, unequal, long, flexuous, dentate, lacerate.

It is found in woods on partially decayed sticks. I have only occasionally found a specimen in our woods.

POLYPORUS PUBESCENS. SCHW.

Figure 339 Polyporus pubescens. White without and within, pubescent and shiny.

Pubescens means downy; so called from the satiny finish of its pileus, which is fleshy, quite tough and corky, soft, convex, subzonate, pubescent and shiny; white without and within; the margin acute, becoming at length yellowish and hard, with a shiny lustre.

The pores are short, minute, nearly round and plane.

The pileus is from one to two inches in width, laterally confluent and usually very much imbricated. Quite plentiful in woods on beech logs. July to November.

POLYPORUS VOLVATUS. PK.

Volvatus, bearing a volva. This is a most interesting species. The pileus seems to be prolonged, making a volva-like protection of the spore surface. When this volva is ruptured small heaps of spores will often be seen on the volva, having been protected from the wind.

Figure 340   Polyporus volvatus. Natural size.

The plant is small, somewhat round, and before the volva is ruptured it is very like a puffball; fleshy, smooth, attached by a small point, whitish, slightly tinged with yellow, red or reddish-brown; the cuticle of the pileus enveloping the entire pore-surface, thick and firm. The pores are rather long, small, the mouths yellowish, with a tinge of brown. The spores are elliptical and flesh-colored, .0003 to .00035 inch long and about .0002 broad.

This plant has a wide distribution, being found in the New England and Eastern States, and the States of the Pacific slope. I presume it will be found wherever the spruce tree is a native.

The specimens in Figure 340 were found near Boston and were sent me about the first of May by Mrs. Blackford. The first package I took, before examining them, to be a new puffball, which they seemed to resemble in their undeveloped state.

POLYSTICTUS BIFORMIS. FR.

Figure 341   Polystictus biformis. Natural size. Frequently covered with green lichen.

Biformis means two shapes or appearances; referring to the condition of the pores in the young and the old plant.

The pileus is two to three inches wide, projecting from one to three inches, often imbricated so as to cover a large surface; laterally confluent, coriaceous, flexible, tough, subzonate, with innate radiating fibres, the cortex fibrillose, concolorous.

The pores at first very large, simple, compound, or confluent, round, elon gated, flexuous; the dissepiments dentate, then lacerate, the hymenium finally resolved into teeth.

When I first found this plant the hymenium had resolved into teeth, and I supposed that I had found an Irpex. It is found in woods on logs and stumps. Very common with us. Frequently covered with a green lichen. July to November.

POLYSTICTUS HIRSUTUS. FR.

THE BRISTLY POLYSTICTUS.

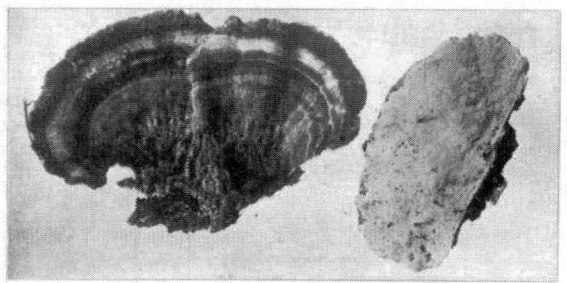

Figure 342   Polystictus hirsutus. Natural size.

Hirsutus means hairy or bristly. The pileus is corky, coriaceous, convex, then plane, hairy with rigid bristles, zoned with concentric furrows; of one color, whitish, sometimes these zones are quite marked as in Figure 342.

The pore surface is at first white, or whitish, becoming dark or brownish in age. The pores are round, the walls rather thick. It is found on logs and stumps in the woods. It is a very common plant and widely distributed.

POLYSTICTUS VERSICOLOR. FR.

THE COMMON ZONED POLYSTICTUS.

Figure 343   Polystictus versicolor. One-half natural size.

Versicolor means varying colors. The pileus is coriaceous, thin, rigid, plane, depressed behind; quite velvety, nearly even and shining, variegated with colored zones, sometimes entirely white or grayish-white, not unfrequently the whole surface is villous or woolly, and the zones mere depressions.

The pores are minute, round, acute, lacerated, white or cream-color.

It is very common, as well as very variable in form and color. It is frequently found on logs and is then densely imbricated. On our hillsides it frequently grows on a small bush as in Figure 343. It is one of the most beautiful plants in the woods.

POLYPORUS GILVUS. SCHW.

Gilvus means pale-yellow or deep-reddish flesh-color.

The pileus is corky, woody, hard, effuso-reflexed, imbricate, concrescent, subtomentose, then scabrous, uneven, reddish-yellow, then subferruginous, the margin acute.

The pores are minute, round, entire, brownish-ferruginous. Morgan.

It is very abundant throughout the state, being found on all kinds of logs and stumps.

POLYSTICTUS CINNAMONEUS. JACQ.

Figure 344   Polystictus cinnamoneus.

The pileus is an inch and a half, or less, broad, coriaceous, slightly depressed in the center; rather rough on the surface, but with a beautiful satiny lustre, and more or less zoned; caps often growing together, but with separate stems; shining, a light cinnamon-brown.

The spores are rather large, angular, torn with age; cinnamon-brown, growing darker in older plants.

The stem is one to two inches long, equal, or slightly tapering upward, cinnamon-brown, hollow or stuffed, tough, frequently sending forth branches from the side and base of the stem.

This is quite a beautiful plant, growing usually in patches of moss. The caps have quite a glossy cinnamon-brown surface, which will attract the attention of any one. They are very small and easily overlooked. Found in August and September.

This plant is called P. subsericeus by Dr. Peck.

POLYSTICTUS PERENNIS. FR.

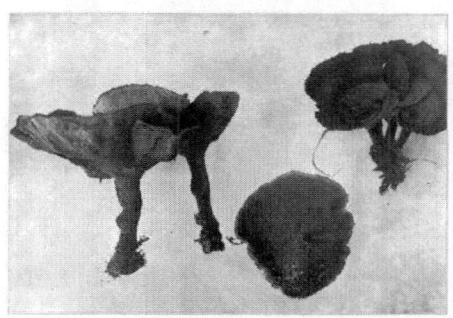

Plate XLVII. Figure 345   Polystictus perennis.

The pileus is thin, pliant when fresh but somewhat brittle when dry. It is minutely velvety on the upper surface, reddish-brown or cinnamon in color; expanded or umbilicate to nearly funnel-shaped. The surface is beautifully marked by radiations and fine concentric zones.

The stem is also velvety. The spore-tubes are minute, the walls thin and acute, and the mouths angular, and at last more or less torn. The margin of the cap is finely fimbriate, but in old specimens those hairs are apt to become rubbed off. Atkinson.

I found specimens by the roadside near Lone Tree Hill, near Chillicothe. It is the only place in which I have found this plant. I have found Polystictus subsericeus, or, as Prof. Atkinson calls it, P. cinnamomeus, in a number of localities.

POLYSTICTUS PERGAMENUS. FR.

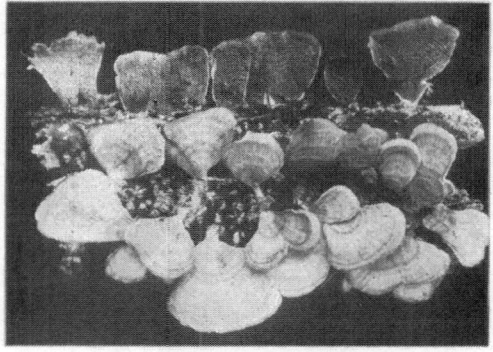

Figure 346   Polystictus pergamenus.

Pergamenus means parchment.

The pileus is coriaceous, thin, effused, reflexed, villous, zoned, cinereous-white, with colored zone; pliant when fresh.

The pores are unequal, torn, violaceous, then pallid. It is very common here on beech, maple, and wild cherry. The pores become torn so that they resemble the teeth of the Hydnum. This is one of the most common fungi in our woods.

The photograph is by Prof. J. D. Smith, of Akron, O.

FOMES LEUCOPHÆUS. MONT.

This has been called by many authors in America Fomes applanatus or Polyporus applanatus. It is very common in this country but very rare in Europe, while Fomes applanatus, which is common in Europe, is very scarce in the United States. In general appearance they are much alike, the applanatus having a softer tissue and echinulate spores, but our common species, leucophæus, has smooth spores.

The pileus is expanded, tuberculose, obsoletely zoned, pulverulent, or smooth; cinnamon, becoming whitish; cuticle crustaceous, rigid, at length fragile, very soft within; loosely floccose, margin tumid; white, then cinnamon. The pores are very small, slightly ferruginous, orifice whitish, brownish when bruised. The spore surface when fresh is soft and white.

This attractive plant is very common in our woods and furnishes an excellent stencil surface for drawing. Found all the year round.

FOMES FOMENTARIUS. FR.

THE BRACKET FOMES.

This species is very common in our woods. The brackets resemble a horse's hoof in shape. They are smoky, gray, and of various shades of brown. The upper surface of the bracket is quite strongly zoned and furrowed, so as to show each year's growth. The margin is thick and blunt, and the tube surface is concave; the openings of the tubes quite large, so that they can be readily seen by the naked eye. The tube surface is reddish-brown when mature. The inside was formerly used in making tinder-sticks, which were made by rolling the fungus wood until it was perfectly flexible and then dipping it into saltpetre.

FOMES RIMOSUS. BERKELEY.

CRACKED FOMES.

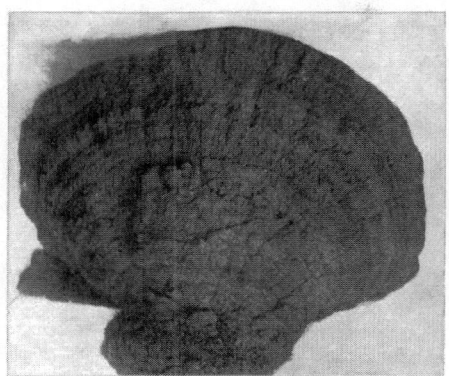

Figure 347　Fomes rimosus.

Rimosus means cracked. The fine checks in the pileus are clearly seen in the halftone.

The pileus is pulvinate-ungulate, much dilated, deeply sulcate; cinnamon, then brown or blackish; very much cracked or rimose. It is very hard, fibrous, tawny-ferruginous; the margin broad, pruinate-velvety, rather acute.

The pores are minute, indistinctly stratified, tawny-ferruginous, the mouths rhubarb-color. Morgan.

This plant is very common on the locust trees about Chillicothe. I have never found it on other wood.

FOMES PINICOLA. (SWARTZ.) FR.

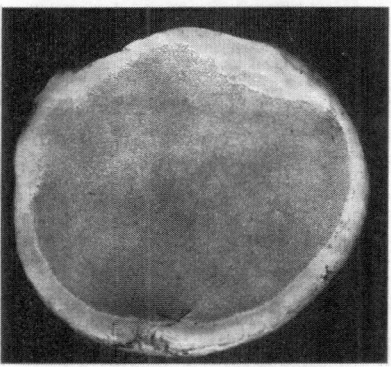

Figure 348　Fomes pinicola.

Pinicola means dwelling on pine. It is found on dead pine, spruce, balsam, and other conifers. It resembles Fomes leucophæus but is somewhat stouter and does not have as hard and firm a crust. The young growth is at the margin, and is whitish or tinged with yellow, while the old zones are reddish. The tube surface is whitish-yellow or yellowish. This is frequently called Polyporus pinicolus. (Swartz.) Fr.

FOMES IGNIARIUS. FR.

Figure 349   Fomes igniarius.

This is rather a common species in our state; black or brownish-black in color, somewhat triangular in shape, and frequently hoof-shaped. The zones indicating the yearly growth are plainly marked, and the tubes are quite long and of a dark brown color. Their growth is rather slow, and it requires years to produce some of the moderate sized specimens. Prof. Atkinson of Cornell University found a specimen which he believed to be over 80 years old.

This is called by many authors Polyporus igniarius (L.), Fr. Murrill calls it Pyropolyporus igniarius. This plant is widely distributed over the United States, and is met frequently in every wood in Ohio.

FOMES FRAXINOPHILUS. FR.

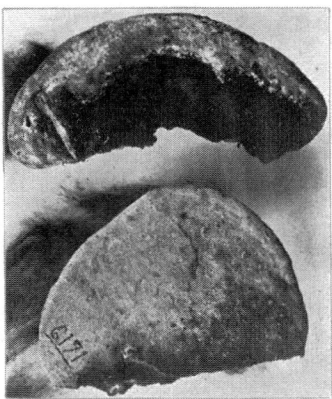

Figure 350   Fomes fraxinophilus.

Fraxinophilus means ash-loving; rathe: common in this country, but does not grow in Europe.

The pileus is between corky and woody, smooth, somewhat flattened, at first zoneless; white when young, then reddish-brown, white around the margin; at first even, then concentrically sulcate, pale within.

The tubes are short, pores minute, rusty-red but covered from the first with a white pubescence and continuous with the margin; the spores nearly round, 6–7μ.

The specimens in Figure 350 were found in Haynes' Hollow on a living ash, growing at intervals of five or six feet, one above another, to a height of thirty feet.

TRAMETES. FR.

In case of the genus Trametes the hymenophorum descends into the trama of the pores without any change, and is permanently concrete with the pileus. The pores are entire. There are, however, a few of the Polypori which are quite thin that have the trama of the same structure with the hymenophorum. These have been separated by Fries and have been called Polystictus. They are distinguished by the fact that the pores develop from the center out and are perpendicular to the fibrillose stratum above the hymenophorum while in the genus Trametes the hymenophorum is not distant from the rest of the pileus.

TRAMETES RUBESCENS. FR.

Figure 351   Trametes rubescens.

Figure 352   Trametes rubescens.

This is one of the neatest plants of this structure in our woods. It grows on the small branches and many times covers them quite well. It is resupinate, the cap being beautifully zoned as you see in Figure 351. Frequently they grow from the side of a small tree that has fallen to the ground and in this case they are shelving.

The pore surface is usually reddish or flesh-color, the pores being long and irregular and inclined to be labyrinthiform in older specimens as will be seen in Figure 352.

The whole plant is reddish or pale flesh-color. No one will fail to recognize it from these cuts.

TRAMETES SCUTELLATA. SCHW.

Scutellata means shield-bearing. It is frequently quite small, an inch or less; coriaceous, dimidiate, orbiculate or ungulate, fixed by the apex; the pilei quite hard: white, then brownish and blackish, becoming rugged and uneven, with white margin; hymenium disk-shaped, concave, white-pulverulent becoming dark; pores minute, long, with thick obtuse dissepiments. This is found on fence posts.

TRAMETES OHIENSIS. BERK.

The pilei are pulvinate, narrow, zoned, often laterally confluent; ochraceous-white, tomentose, then smooth, laccate. This plant resembles T. scutellata in many points, both in habit and in form.

TRAMETES SUAVEOLENS. (L.) FR.

Soft at first, pulvinate, white, villous, zoneless; pores rotund, rather large, obtuse, white, then darker; anise-scented. Found on willows.

MERULIUS. FR.

Merulius means a blackbird; from the color of the fungus.

Hymenophore covered with the soft waxy hymenium, which is incompletely porus, or arranged in reticulate, sinuous, dentate folds. This genus grows on wood, at first resupinate, expanded; the hymenophore springing from a mucous mycelium.

MERULIUS RUBELLUS. PK.

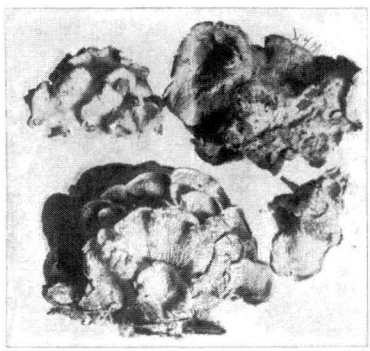

Figure 353   Merulius rubellus. Natural size.

Rubellus is the diminutive of ruber, reddish. The pileus grows in tufts, sessile, confluent and imbricated, repand, thin, convex, soft, dimidiate, quite tenacious; tomentose, evenly red, margin mostly undulately inflexed, growing pale in age. Hymenium whitish or reddish, folds much branched, forming anastomosing pores. The spores are elliptical, hyaline, minute, 4–5×2.5–3μ. The pileus is two to three inches long and an inch and a half broad.

It is found very frequently on decayed beech and sugar trees and I have found it growing on a live oak. The specimens in Figure 353 were collected near Columbus and photographed by Dr. Kellerman. It is probably the same as M. incarnati, Schw.

MERULIUS TREMELLOSUS. SCHRAD.

Figure 354   Merulius tremellosus

Tremellosus, trembling. Resupinate; margin becoming free and more or less reflexed, usually radiately-toothed, fleshy, tremelloid, tomentose, white; hymenium variously wrinkled and porus, whitish and subtranslucent-looking, becoming tinged with brown in the center. The spores are cylindrical, curved, about 4×1μ. From one to three inches across, remaining pale when growing in dark places. The margin is sometimes tinged with a rose-color, radiating when it is well developed. Massee.

This plant grows in woods on wood and is quite common in our woods—both the rose-colored and the translucent-brown. Captain McIlvaine calls Merulius tremellosus and M. rubellus emergency species. He says they are rather tasteless, tough, slightly woody in flavor. They are found in October and November.

MERULIUS CORIUM. FR.

Resupinate, effused, soft, papery, circumference at length free, reflexed, white, villous below. Hymenium netted, porus, pallid, tan-color.

Found on decaying-branches. Quite common.

MERULIUS LACRYMANS. FR.

Resupinate, fleshy, spongy, moist, tender, at first very light, cottony and white; when the veins appear they are of a fine yellow, orange or reddish-brown, forming irregular folds, so arranged to have the appearance of pores (but never anything like tubes), distilling when perfect drops of water which give rise to the specific name "weeping."

Dr. Charles W. Hoyt of Chillicothe, brought to my office two or three plants of this species that had grown on the under side of the floor in his wash-house. When he took up the floor the workmen discovered a number of pendant processes, some oval, some cone-shaped. Some were eight inches long, very white and beautiful but clearly illustrating the weeping process. The doctor called them white rats suspended by their tails.

DÆDALEA. PERS.

Dædalea is used with reference to the labyrinthiform pores; so named after Dædalos, the builder of the labyrinth of Crete.

The hymenophore descends into the trama without any change, pores firm, when fully grown sinuous and labyrinthiform, lacerated, and toothed. The habits of Dædalea are very much the same as Trametes, but they are inodorous. Care should be taken not to confound them with the species of Polyporus that have elongated curved pores.

DÆDALEA AMBIGUA. BERK.

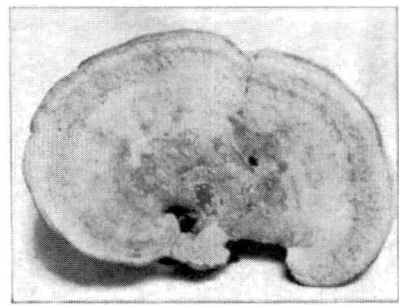

Figure 355   Dædalea ambigua. One-third natural size, showing upper surface.

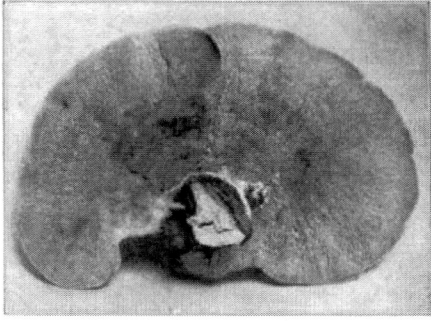

Figure 356   Dædalea ambigua. One-third natural size, showing the pore surface.

The pileus is white, corky, horizontal, explanate, reniform, subsessile, azonate, finely pubescent, becoming smooth.

Pores from round to linear and labyrinthiform, the dissepiments always obtuse and never lamellate.

It is a very common growth in Ohio, found on old logs of the sugar maple. You will see the beginning of the growth in the spring as a round white nodule which develops slowly. If the same plant is observed in the summer it will be found to be gibbous or convex in form. It finishes its growth in the fall when it has become explanate and horizontal, depressed above and with a thin margin. When fresh and growing it is of a rich cream-color and has a soft and velvety touch and a pleasant fragrance. In Figure 355, showing the surface of the cap, the growth of the plant shows in the form of the zones. Figure 356 shows the form of the dissepiments. In younger specimens these are frequently round, much like a Polyporus. There is one locality in Poke Hollow where the maple logs are white with this species, appearing, in the distance, to be oyster mushrooms.

DÆDALEA QUERCINA. PK.

THE OAK DÆDALEA.

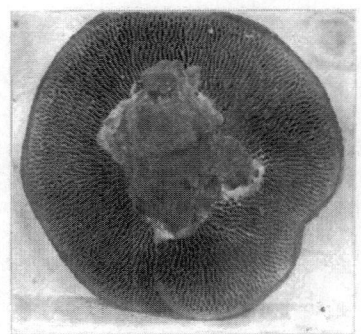

Figure 357   Dædalea quercina.

The pileus is a pallid wood color, corky, rugulose, uneven, without zones, becoming smooth; of the same color within as without; the margin in full-grown specimens thin, but in imperfectly developed specimens swollen and blunt.

The pores are at first round, then broken into contorted or gill-like labyrinthiform sinuses, with obtuse edges of the same color as the pileus, sometimes with a slight shade of pink.

They grow to be very large, from six to eight inches broad, being found on oak stumps and logs, though not as common in Ohio as D. ambigua. The specimen in Figure 357 were found in Massachusetts by Mrs. Blackford and photographed here.

DÆDALEA UNICOLOR. FR.

Villose-strigose, cinereous with concolorous zones; hymenium with flexuous, winding, intricate, acute dissepiments, at length torn and toothed. The pores are whitish cinereous, sometimes fuscous; variable in thickness, color, and character of hymenium; sometimes with white margin; often imbricated and fuliginous when moist. Widely distributed over the states and found on nearly all deciduous trees.

DÆDALEA CONFRAGOSA. BOTON.

THE WILLOW DÆDALEA.

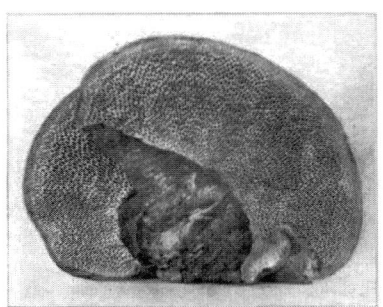

Figure 358   Dædalea confragosa.

Confragosa means broken, rough. The pileus is rather convex, corky, rough, slightly zonate, reddish-brown, unicolorous, somewhat of a rust-red within.

The pores are frequently round, like those of the Polyporus, but sometimes they are elongated into gills like the Lenzites; reddish-brown.

I have seen quite old specimens that were very difficult to distinguish from some of the forms of Lenzites. The young plants resemble very closely Trametes rubescens. It grows on Cratægus, willow and sometimes on other trees, and is widely distributed. The specimen in Figure 358 was found in Massachusetts by Mrs. Blackford, and photographed in my study.

FAVOLUS. FR.

Favolus is a diminutive of favus, honey-comb.

The hymenium is alveolate, radiating, formed of the densely irregularly uniting gills; elongated, diamond-shaped. Spores white. Semicircular in outline, somewhat stipitate.

Favolus canadensis. Klotsch.

Figure 359   Favolus Canadensis.

The pileus is fleshy, tough, thin, kidney-form, fibrillose, scaly, tawny, becoming pale and smooth.

The pores or alveoli are angular elongated, white at first, then straw-color.

The stem is eccentric, lateral, very short or lacking altogether.

This plant is very common around Chillicothe on fallen branches in the woods, especially on hickory. Found from September to frost. Not poisonous but too tough to eat. I do not believe there is any difference between F. canadensis and Favolus Europeus. I notice that our plant assumes different colors in different stages of its growth, and the form of the pores also changes.

CYCLOMYCES. KUNZ & FR.

Cyclomyces is from two Greek words, meaning a circle and fungus. This genus is very distinct from other tube-bearing genera. The pileus is fleshy, leathery or membranaceous, and usually cushion-formed. Upon the lower surface are the plate-like bodies resembling the gills of Agarics but which are composed of minute pores. These pore bodies are arranged in concentric circles around the stem.

CYCLOMYCES GREENII. BERK.

Figure 360   Cyclomyces Greenii

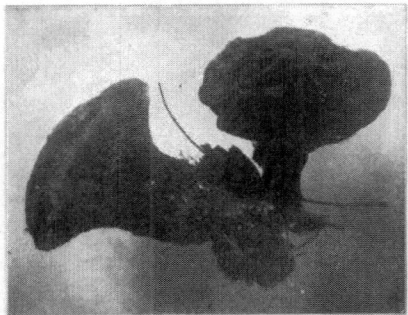

Figure 361   Cyclomyces Greenii. Old specimens.

The pileus is two to three inches broad, globose at first, convex, sometimes undulate, somewhat zoned, tomentose, dry, cushion-formed, cinnamon-brown, rather showy.

The gills are in concentric circles around the stem, growing larger and larger as they reach the margin of the cap. In the young plant the gills are divided into long divisions but in the older plant these division lines disappear as will be seen in Figure 361. The edges of the gills are white at first, as will be seen in Figure 361, but finally becoming cinnamon-brown.

The stem is central, tapering upward, quite large and swollen at times very much like Hydnum spongiosipes; the color is the same as the pileus.

This is a very interesting plant and quite rare in Ohio, however, I found several plants in the fall of 1905, on Ralston's Run. In the same locality I found Boletus badius, and when I first saw C. Greenii I came near mistaking it for the same plant and so neglecting it, the caps being at first glance so much alike.

GLŒOPORUS. MONT.

Glœoporus is from two Greek words, meaning gluten and pore. The plants of this genus resemble the polyporus and are frequently placed under that genus.

GLŒOPORUS CONCHOIDES. MONT.

Conchoides means like a shell.

The pileus is leathery or woody, at first fleshy, soft, effused, with upper margin reflexed; thin, silky, whitish, with edge of the margin often reddish. It has a trembling, gelatinous, spore-bearing surface, often somewhat elastic.

The pores are short, very small, round, cinnamon-brown.

There are several synonyms. Polyporus dichrous, Fr., and P. nigropurpurascens, Schw. Montgomery places it in the above genus because of its gelatinous hymenium.

CHAPTER VIII

HYDNACEAE—FUNGI WITH TEETH

There is, perhaps, no family in mycology that has a greater variety in form, size, and consistency than this. Some species are very large, some are small, some fleshy, and some are corky or woody. The fruiting surface is the special characteristic marking the family. This surface is covered with spines or teeth which nearly always point to the earth.

Many of the Hydnaceæ are shelving, growing on trees or logs; some grow on the ground on central, but usually eccentric, stems. The genera of Hydnaceæ are distinguished by the size, shape, and attachment of the teeth. The following genera are included:

- Hydnum—Spines discrete at the base.
- Irpex—Resupinate; with gill-like teeth concrete with the pileus.
- Mucronella—Plants with teeth only and no basal membrane.
- Radulum—Hymenium with thick, blunt, irregular spines.
- Sistotrema—Fleshy plants with caps and flattened teeth, on ground.
- Phlebia—Plants spread over the host with crowded folds or wrinkles.
- Grandinia—Covered with granules, more or less smooth, and excavated.
- Odontium—Covered with crested granules.

HYDNUM. LINN.

Hydnum is from a Greek word meaning an eatable fungus. The genus is characterized by awl-shaped spines which are distant at the base. These spines are at first papilliform, then elongated and round. They form the fruiting surface and take the place of the gills in the family Agaricaceæ and of the pores in the family of Polyporaceæ. The spines are simple or in some cases the tips are more or less branched.

This is the greatest genus in the family and it includes many important edible species. It may be divided into two groups: one, those species having a cap and a central or lateral stem; the other, the species growing with or without a distinct cap, in large imbricated masses. Some imitate coral in structure and some seem to be a mass of spines. Many of these plants grow to be very large and massive, frequently weighing over ten pounds.

HYDNUM REPANDUM. LINN.

THE SPREADING HYDNUM. EDIBLE.

Figure 362 Hydnum repandum. Two-thirds natural size.

Repandum, bent backward, referring to the position of the stem and the cap. The pileus is two to four inches broad, generally irregular, with the stem eccentric; fleshy, brittle, convex or nearly plane, compact, more or less repand, nearly smooth; color varying from a pale buff—the typical hue—to a distinct brick-red; flesh creamy-white, inclining to turn brown when bruised; taste slightly aromatic, margin often wavy.

The spines are beneath the cap, one-quarter to one-third of an inch long, irregular, entire, pointed, rather easily detached, leaving small cavities in the fleshy cap, soft, creamy, becoming darker in older specimens.

The stem is short, thick, solid in young specimens, hollow in older specimens; paler than the pileus, rather rough, often set eccentrically into the cap; one to three inches long, sometimes thickened at the base, sometimes at the top. The spores are globose or a broad oval, with a small papilla at one end.

The usual color of the cap is buff, sometimes very pale, almost white. The color and smoothness of the cap have given rise to the name of "doe-skin mushroom." I found this plant occasionally in the woods about Salem, Ohio. It is very variable in size and color, and is quite fragile, growing alone or in clusters. It is one of our best mushrooms if properly cooked, and may be dried and kept for winter use. Found in woods and open places from July to October, sometimes earlier. Specimens in Figure 362 were found in Poke Hollow.

HYDNUM IMBRICATUM. LINN.

THE IMBRICATED HYDNUM. EDIBLE.

Imbricatum is from imbrex, a tile, referring to the surface of the cap being torn into triangular scales, seeming to overlap one another like shingles on a roof.

The pileus is fleshy, plane, slightly depressed, tessellated scaly, downy, not zoned, umber in color or brownish as if scorched, flesh dingy-white, taste slightly bitter when raw, margin round.

The spines are decurrent, entire, numerous, short, ashy-white, generally equal in length.

The stem is firm, short, thick, even, whitish. The spores are pale yellow-brown, rough.

The bitter taste entirely leaves the plant when well cooked. It seems to delight in pine or chestnut woods. I found it in Emmanuel Thomas' woods, east of Salem, Ohio. It is found from September to November.

HYDNUM ERINACEUM. BULL.

THE HEDGEHOG HYDNUM. EDIBLE.

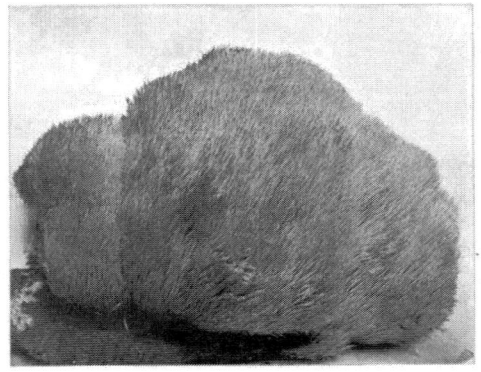

Plate XLVIII. Figure 363   Hydnum Erinaceum.
Two-thirds natural size. The plant is entirely white when fresh.

Erinaceum, a hedgehog. Two to eight inches or more across. Tufts pendulous. White and yellowish-white becoming yellowish-brown; fleshy, elastic, tough, sometimes emarginate (broadly attached as if tuft were cut in two or sliced off where attached), a mass of latticed branches and fibrils. Spines one and a half inches to four inches long, crowded, straight, equal, pendulous. The stem is sometimes rudimentary. The spores are subglobose, white, plain, 5–6μ. Peck, 22 N. Y. Report.

The spines when just starting are like small papillæ, as will be seen in Figure 364. Figure 363 represents a very fine specimen found on the end of a beech log, on the Huntington Hills, near Chillicothe. It made a meal for three families. I have found several basketfuls of this species on this same log, within the past few years. I have also found on the same log large specimens of Hydnum corralloides.

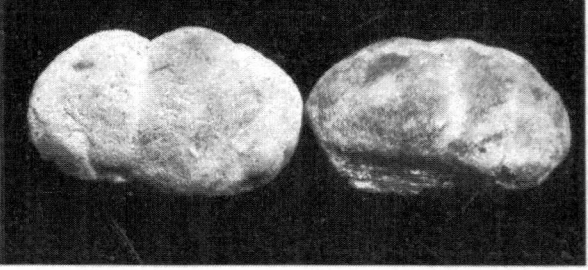

Figure 364   Hydnum erinaceum. Young state.

The photograph at the beginning of the book represents the largest specimen I ever saw of this species. It measured eighteen inches one way and thirteen the other, and was found on a maple tree on top of Mount Logan. It grew from a central stem, while the one in Figure 363 grew from a crack in a log, apparently without a stem. Plate I, Figure 1 was photographed after it was dried. The specimen can be seen in the Lloyd Library in Cincinnati. Found from July to October.

HYDNUM CAPUT-URSI. FR.

THE BEAR'S HEAD HYDNUM. EDIBLE.

Figure 365   Hydnum caput-ursi.

Caput-ursi means the head of a bear.

This is a very beautiful plant but not as common as some other species of Hydnum. It grows in very large pendulous tufts, as Figure 365 will indicate. It is found frequently on standing oak and maple trees, sometimes quite high up in the trees. It is more frequently found on logs and stumps, as are its kindred species. The plant arises out of the wood by a single stout stem which branches into many divisions, all of which are covered by long pendant spines. When it grows on top of a log or stump the spines are frequently erect. It is white, becoming in age yellow and brownish. It has a wide distribution through the states. As an esculent it is fine. The specimen in Figure 365 was found near Akron, Ohio, and was photographed by Mr. G. D. Smith. It is found from July to October.

HYDNUM CAPUT-MEDUSÆ. BULL.

THE MEDUSA'S HEAD HYDNUM. EDIBLE.

Caput-Medusæ, head of Medusa. This is a very striking plant when seen in the woods. The tufts are pendulous. The long wavy spines resemble the wavy locks of Medusa, hence the name. The long soft spines cover the entire surface of the fungus, which is divided into fleshy branches or divisions, each terminating in a crown of shorter drooping teeth.

The color at first is white, changing in age to a buff or a dark cream, which distinguishes it from H. caput-ursi. The taste is sweet and aromatic, sometimes slightly pungent. The stem is short and concealed beneath the growth.

Figure 366   Hydnum caput-Medusæ. One-third natural size.

I found this plant growing on a hickory log, on Lee's hill, near Chillicothe, from which came the specimen in Figure 366. I have also found it on elm and beech. Found from July to October.

It is both attractive and palatable.

HYDNUM CORALLOIDES. SCOP.

THE CORAL-LIKE HYDNUM. EDIBLE.

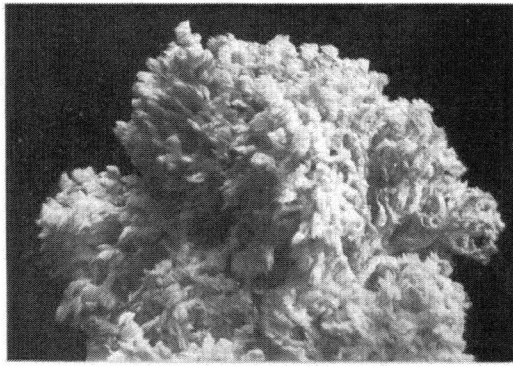

Figure 367   Hydnum coralloides. One-fourth natural size. Entire plant white.

This species grows in large, beautiful tufts on decaying logs, in damp woods. It grows from a common stem, dividing into many branches and then sub-dividing into many long and coral-like shoots, composed wholly of attenuated interlacing branches tapering to a point. The spines grow from one side of the flattened branches. It only needs to be seen once to be recognized as a coral-like mushroom. It is pure white at first, becoming creamy or dingy-white with age. It seems to delight in damp, hilly places, yet I found it to be abundant at Sidney, and to some extent about Bowling Green, Ohio, where it was very level. It is plentiful around Chillicothe. One hickory log, from which the specimen in the figure was taken, furnished me several basketfuls of this plant during three seasons, but at the end of the third season the log crumbled away, mycelium having literally consumed it. It is one of the most beautiful fungi that Dame Nature has been able to fashion. It is said that Elias Fries, when a mere boy, was so impressed with the sight of this beautiful fungus, which grew abundantly in his native woods in Sweden, that he resolved when he grew up to pursue the study of Mycology, which he did; and became one of the greatest authorities of the world in that part of Botany. In fact, he laid the foundation for the study of Basidiomycetes, and this beautiful little coral-like fungus was his inspiration.

It is found principally on beech, maple and hickory in damp woods, from July to frost. I have eaten it for years and esteem it among the best.

HYDNUM SEPTENTRIONALE. FR.

THE NORTHERN HYDNUM.

Plate XLIX. Figure 368   Hydnum septentrionale.
Grew from a small opening in a living beech tree.

Septentrionale, northern. This is a very large, fleshy, fibrous plant, growing usually upon logs and stumps.

There are many pilei growing one above the other, plane, margin straight, whole. The spines are crowded, slender and equal.

I have found a number of specimens about Chillicothe that would weigh from eight to ten pounds each. The plant is too woody to eat. Besides, it seems to have but little flavor. I have always found it on beech logs, from September to October.

A very large plant grows every year on a living beech tree on Cemetery Hill.

HYDNUM SPONGIOSIPES. PK.

Figure 369   Hydnum spongiosipes. One-third natural size.

Spongiosipes means a sponge-like foot. Pileus convex, soft, spongy-tomentose, but tough in texture, rusty-brown, the lower stratum firmer and more fibrous, but concolorous.

The spines are slender, one to two lines long, rusty-brown, becoming darker with age.

The stem is hard and corky within, externally spongy-tomentose; colored like the pileus, the central substance often transversely zoned, especially near the top. Spores globose, nodulose, purplish-brown, 4–6 broad. Pileus one and a half to four inches broad. Stem one and a half to three inches long, and four to eight lines thick. Peck, 50th Rep.

It is found in the woods, quite plentifully, about Chillicothe. I referred it to H. ferrugineum for a long time, but not being satisfied, sent some specimens to Dr. Peck, who classified it as H. spongiosipes. It is edible but very tough. Found from July to October.

HYDNUM ZONATUM. BATSCH.

THE ZONED HYDNUM.

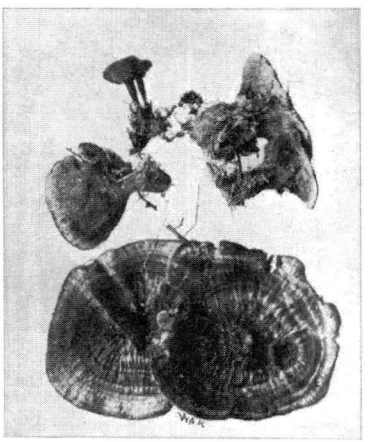

Figure 370   Hydnum zonatum.

Zonatum, zoned. Ferruginous; pileus equally coriaceous, thin, expanded, subinfundibuliform, zoned, becoming smooth; tough, almost leathery in texture, having a surface of beautiful brown, silky lustre, and with radiating striæ; margin paler; sterile.

The stem is slender, nearly equal, floccose, bulbous at the base.

The spines are slender, pallid, then of the same color as the pileus, equal. The spores are rough, globose, pale, 4μ.

The spore-bearing spines are shown in the upper plants in Figure 370. Two of them show coalesced caps, though the stems are separate. This is the case with H. scrobiculatum and H. spongiosipes. The plants in Figure 370 were collected by the roadside in woods on the State Farm, near Lancaster, and photographed by Dr. Kellerman.

HYDNUM SCROBICULATUM. FR.

Figure 371   Hydnum scrobiculatum. Two-thirds natural-size.

Scrobiculatum means marked with a ditch or trench; so called from the rough condition of the cap. The pileus is from one to three inches broad, corky, convex, then plane, sometimes slightly depressed; tough in texture, rusty-brown; the surface of the cap usually quite rough, marked with ridges or trenches, flesh ferruginous.

The spines are short, rusty-brown, becoming dark with age.

The stem is firm, one to two inches long, unequal, rusty-brown, often covered with a dense tomentum.

This species is very plentiful in our woods, among the leaves under beech trees. They grow in lines for some distance, the caps so close together that they are very frequently confluent. I found the plant at Salem, and in several other localities in the state, although I have never seen a description of it. Any one will be able to recognize it from Figure 371. It grows in the woods in August and September.

HYDNUM BLACKFORDÆ. PK.

The pileus is fleshy, convex, glabrous, grayish or greenish-gray, flesh whitish with reddish stains, slowly becoming darker on exposure; aculei subulate, 2–5 mm. long, yellowish-gray, becoming brown with age or drying; stem equal or stuffed, becoming hollow in drying; glabrous, colored like the pileus; spores brown, globose, verrucose, 8–10μ broad.

The pileus is 2.5–6 cm. broad; stem 2.5–4 cm. long, 3–4 mm. thick.

Mossy ground in low springy places in damp mixed woods. August. Peck.

This species was found at Ellis, Mass., and was sent to me through courtesy of the collector, Mrs. E. B. Blackford, Boston, for whom it was named.

HYDNUM FENNICUM. KARST.

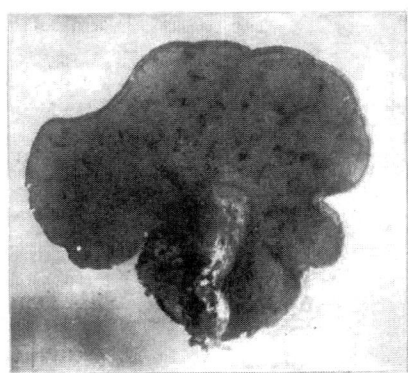

Figure 372   Hydnum fennicum. Natural size, showing the teeth.

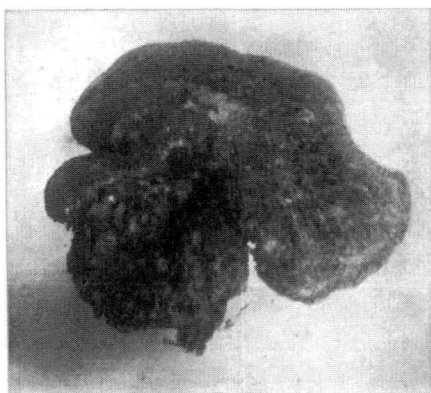

Figure 373   Hydnum fennicum. Natural size, showing the scaly cap.

Pileus fleshy, fragile, unequal; at first scaly, at length breaking up; reddish-brick color becoming darker; margin undulately lobed, two to four inches broad. Flesh white.

The teeth decurrent, equal, pointed, from white to dusky, about 4 mm. long.

The stem is sufficiently stout, unequal below, attenuated, flexuous or curved, smooth, of the same color as the cap, base acute, white tomentum outside, inside light pale-blue, or dark-gray.

The spores are ellipso-spheroidical or subspheroidical, rough, dusky, 4–6μ long, 3–5μ broad.

The plants in Figures 372 and 373 were found in Haynes' Hollow.

The plant is quite bitter and no amount of cooking will make it edible.

Found in woods from August to September.

HYDNUM ADUSTUM. FR.

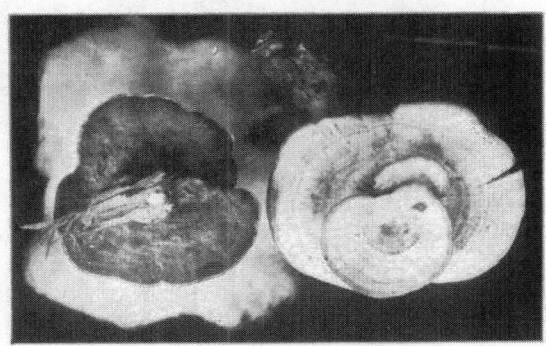

Figure 374   Hydnum adustum. Natural size.

Adustum means scorched, burned. The pileus is two to three inches broad, yellowish-white, blackish around the margin, coriaceous, slightly zoned; plane at first, then slightly depressed; tomen tose, thin; frequently a plant will be found growing on the top of another plant. The spines are at first white, adnate, short, turning flesh-color and when dried almost black.

The stem is short, solid, tapering upward.

The plant is found growing in the woods on trunks and sticks after a rain in July, August, and September. It is not as plentiful as Hydnum spongiosipes and H. scrobiculatum. It is an attractive plant when seen in the woods.

HYDNUM OCHRACEUM. P.

OCHREY HYDNUM.

Small, at first entirely resupinate, gradually reflexed, and somewhat repand, at first sparingly clothed with dirty-white down, at length rugose; one to three inches broad. The spines are short, entire, becoming pale. Fries.

It is occasionally found on decayed sticks in the woods.

HYDNUM PULCHERRIMUM. B. & C.

MOST BEAUTIFUL HYDNUM.

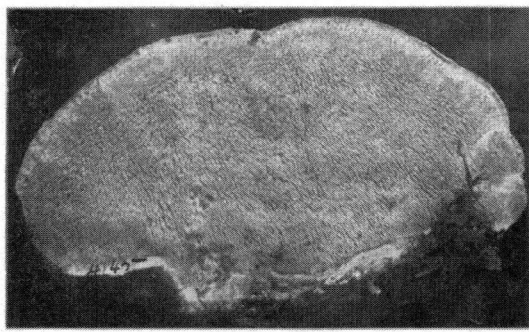

Figure 375 Hydnum pulcherrimum. Showing the under side of one of the pileoli.

Pulcherrimum is the superlative of pulcher, beautiful.

The pileus is fleshy, somewhat fibrous, alutaceus, hirsute; the margin thin, entire, incurved.

The aculei short, crowded, equal.

It is found on beech wood, frequently imbricated and laterally confluent; a single pileus two to five inches in breadth and projecting two to four inches. The spines are rather short, not exceeding a quarter of an inch.

The entire plant is quite fibrous and has a hirsute surface. The color varies from whitish to alutaceous and yellowish. It is not common with us. Figure 375 represents one of the pilei showing the spines.

HYDNUM GRAVEOLENS. DEL.

FRAGRANT HYDNUM.

Graveolens means sweet-scented.

The pileus is coriaceous, thin, soft, not zoned, rugose, dark-brown, brown within, margin becoming whitish. The stem is slender and the spines are decurrent. The spines are short, gray.

The whole plant smells of melilot; even after it has been dried and kept for years it does not lose this scent.

I found two specimens in Haynes's Hollow.

IRPEX. FR.

Irpex, a harrow, so called from a fancied resemblance of its teeth to the teeth of a harrow. It grows on wood; toothed from the first, the teeth are connected at the base, firm, somewhat coriaceous, concrete with the pileus, arranged in rows or like net-work. Irpex differs from Hydnum in having the spines connected at the base and more blunt.

IRPEX CARNEUS. FR.

This plant, as its specific name indicates, resembles the color of flesh. Reddish, effused, one to three inches long, cartilaginous-gelatinous, membranaceous, adnate. Teeth obtuse and awl-shaped, entire, united at the base. Fries.

Found on the tulip-tree, hickory, and elm. September and October.

IRPEX LACTEUS. FR.

Growing on wood, membranaceous, clothed with stiff hair, more or less furrowed, milk-white, as its specific name indicates.

The spines are compressed, radiate, margin porus. Found on hickory and beech logs and stumps.

IRPEX TULIPIFERA. SCHW.

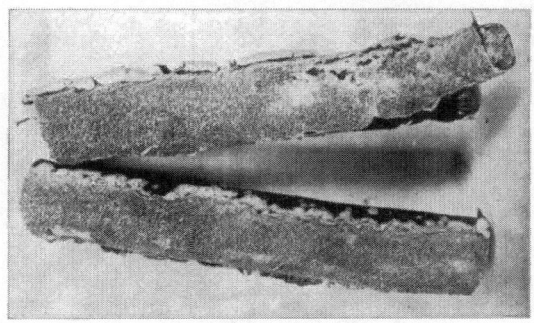

Figure 376 Irpex tulipifera.

Coriaceous-membranaceous, effused; hymenium inferior, at first toothed, teeth springing from a porus base, somewhat coriaceous, entirely concrete with the pileus, netted and connected at the base, white or whitish, turning yellowish with age.

This plant is very abundant here on fallen tulip trees. I have seen entire tree tops and trunks covered with this plant. The branches after they have been penetrated with the mycelial threads become very light and brittle.

PHLEBIA. FR.

Lignatile, resupinate, hymenium soft and waxy, covered with folds or wrinkles, edges entire or corrugated.

PHLEBIA RADIATA. FR.

Figure 377   Phlebia radiata.

Somewhat round, then dilated, confluent, fleshy and membranaceous, reddish or flesh-red, the circumference peculiarly radiately marked. The folds in rows radiating from the center.

The spores are cylindric-oblong, curved, $4–5 \times 1–1.5\mu$.

This is quite common on beech bark in the woods. Its bright color and mode of growth will attract attention.

GRANDINIA. FR.

Lignatile, effused, waxy, granulated, granules globular, entire, permanent.

GRANDINIA GRANULOSA. FR.

Effused, rather thin, waxy, somewhat ochraceous, circumference determinate, granules globular, equal, crowded.

Found on decayed wood. Quite common in our woods.

CHAPTER IX

THELEPHORACEAE

Thelephoraceæ is from two Greek words, a teat and to bear. The hymenium is even, coriaceous, or waxy, costate, or papillose. There are a number of genera under this family but I am acquainted with only the genus Craterellus.

CRATERELLUS. FR.

Craterellus means a small bowl. Hymenium waxy-membranaceous, distinct but adnate to the hymenophore, inferior, continuous, smooth, even or wrinkled. Spores white. Fries.

CRATERELLUS CANTHARELLUS. (SCHW.) FR.

YELLOW CRATERELLUS. EDIBLE.

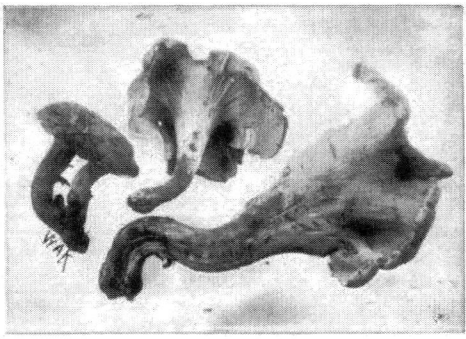

Figure 378   Craterellus cantharellus. Caps and stems yellow.

Cantharellus is a diminutive from a Greek word meaning a sort of drinking-cup.

The pileus is one to three inches broad, convex, often becoming depressed and funnel-shaped, glabrous, yellowish, or pinkish. Flesh white, tough, elastic.

Hymenium slightly wrinkled, yellow or faint salmon color.

The stem is one to three inches high, tapering downward, smooth, solid, yellow. The spores are yellowish or salmon color when caught on white paper, 7.5–10×5–6μ. Peck.

This plant resembles Cantharellus cibarius very closely. The color, form of growth, and the odor are very similar to the latter. It may be readily distinguished from C. cibarius by the absence of folds on the under or fruiting surface. The caps are often large and wavy, resembling yellow cauliflower. It is quite abundant about Chillicothe during the months of July and August. I have frequently gathered bushels of it for my mushroom-friends. It will be easily recognized from Figure 378, bearing in mind that the caps and stems are yellow.

CRATERELLUS CORNUCOPIOIDES FR.

THE HORN OF PLENTY CRATERELLUS. EDIBLE.

Figure 379   Craterellus cornucopioides. One-third natural size.

Cornucopioides is from cornu, a horn, and copia, plenty.

The pileus is thin, flexible, tubiform, hollow to the base, blackish-brown, sometimes a little scaly, the hymenium even or somewhat wrinkled, cinereous.

The stem is hollow, smooth, black, short, almost wanting. The spores are elliptical, whitish, $11-12 \times 7-8\mu$.

No one will have any trouble in recognizing this species, having once seen its picture and read its description. Its elongated or trumpet-shaped cap, and its dingy-gray or sooty-brown hue, will at once distinguish it. The spore-bearing surface is often a little paler than the upper surface. The cup is often three to four inches long. I have found it in quite large clusters in the woods near Bowling Green, and Londonderry, though it is found rather sparingly on the hillsides about Chillicothe. It has a wide distribution in other states. It does not look inviting, on account of its color, but it proves a favorite whenever tested, and may be dried and kept for future use. It is found from July to September.

CRATERELLUS DUBIUS. PK.

Dubius means uncertain, from its close resemblance to C. cornucopoides.

The pileus is one to two inches broad, infundibuliform, subfibrillose, lurid-brown, pervious to the base, the margin generally wavy, lobed. Hymenium dark cinereous, rugose when moist, the minute crowded irregular folds abundantly anastomosing; nearly even when dry. The stem is short. The spores are broadly elliptical or subglobose, 6–7.5μ long. Peck.

Figure 380   Craterellus dubius. Natural size.

It differs from C. cornucopioides in manner of growth, paler color, and smaller spores.

It is distinguished from Craterellus sinuosus by its pervious stem, while very similar in color to Cantharellus cinereus.

This plant, like C. cornucopoides, dries readily, and when moistened expands and becomes quite as good as when fresh. It needs to be stewed slowly till tender, when it makes a delightful dish.

The plants in Figure 380 were collected near Columbus by R. H. Young and photographed by Dr. Kellerman. They are found from July to October.

CORTICIUM. FR.

Entirely resupinate, hymenium soft and fleshy when moist, collapsing when dry, often cracked.

CORTICIUM LACTEUM. FR.

This is a very small plant, resupinate, membranaceous, and it is so named because of the milk-white color underneath. The hymenium is waxy when moist, cracked when dry.

CORTICIUM OAKESII. B. & C.

The plant is small, waxy-pliant, somewhat coriaceous, cup-shaped, then explanate, confluent, marginate, externally white-tomentose.

The hymenium is even, contiguous, becoming pallid. Spores elliptical, appendiculate.

I found very fine specimens of this plant on the Iron-wood, Ostrya Virginica, which grows on the high school lawn in Chillicothe. In rainy weather in October and November the bark would be white with the plant. It resembles a small Peziza at first.

CORTICIUM INCARNATUM. FR.

Waxy when moist, becoming rigid when dry, confluent, agglutinate, radiating. Hymenium red or flesh-color, covered with a delicate flesh-colored bloom. Some fine specimens were found on dead chestnut trees in Poke Hollow.

CORTICIUM SAMBUCUM. PK.

Effused on elder bark, white, continuous when growing, when dry cracked or flocculose and collapsing. It grows on the bark or the wood of the elder.

CORTICIUM CINEREUM. FR.

Waxy when moist, rigid when dry, agglutinate, lurid. The hymenium is cinereous, with a very delicate bloom. Common on sticks in the woods.

THELEPHORA. FR.

The pileus is without a cuticle, consisting of interwoven fibres. Hymenium ribbed, of a tough, fleshy substance, rather rigid, then collapsing and flocculent.

THELEPHORA SCHWEINITZII.

Figure 381   Thelephora Schweinitzii.

Schweinitzii is named in honor of the Rev. David Lewis de Schweinitz. Cæspitose, white or pallid. Pilei soft-corinaceous, much branched; the branches flattened, furrowed and somewhat dilated at the apex.

The stems are variable in length, often connate or fused together into a solid base.

The hymenium is even, becoming darker colored when older. Morgan.

This plant is known as T. pallida. It is very abundant on our hillsides in Ross County, and in fact throughout the state.

THELEPHORA LACINIATA. P.

The pileus is soft, somewhat coriaceous, incrusting, ferruginous-brown. The pilei are imbricated, fibrous, scaly, margin fimbriated, at first dirty white. The hymenium is inferior and papillose.

THELEPHORA PALMATA. FR.

The pileus is coriaceous, soft, erect, palmately branched from a common stalk; pubescent, purplish-brown; branches flat, even, tips fimbriated, whitish. The scent is very noticeable soon after it is picked. They grow on the ground in July and August.

THELEPHORA CRISTATA. FR.

The pileus is incrusting, rather tough, pallid, passing into branches, the apices compressed, expanded, and beautifully fringed. The plant is whitish, grayish, or purplish-brown. It is found on moss or stems of weeds. I found beautiful specimens at Bainbridge Caves.

THELEPHORA SEBACEA. FR.

The pileus is effused, fleshy, waxy, becoming hard, incrusting, variable, tuberculose or stalactitic, whitish, circumference similar; hymenium flocculose, pruinose, or evanescent.

It is found effused over grass. One meets with it often.

STEREUM. FR.

The hymenium is coriaceous, even, rather thick, concrete with the intermediate stratum of the pileus, which has a cuticle even and veinless, remaining unchanged and smooth.

STEREUM VERSICOLOR.

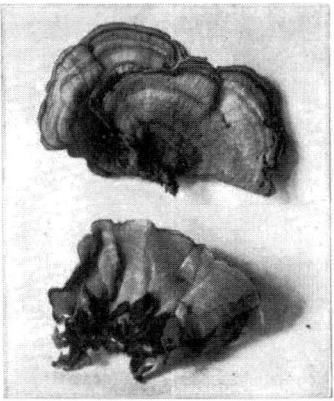

Figure 382   Stereum versicolor.

Versicolor means changing color, referring to the different bands of color. The pileus is effused, reflexed, having a number of different zones; in some plants the zones are more marked than in others, the zones appearing very much like those in Polyporus versicolor.

The hymenium is even, smooth and brown.

This is a very common plant, found everywhere on old logs and stumps. It is widely diffused and can be found at any time of the year.

STEREUM SPADICEUM. FR.

Pilei coriaceous and spreading, reflexed, villous, somewhat ferruginous; margin rather obtuse, whitish, even beneath; smooth, brownish, and bleeding when scratched or bruised.

STEREUM HIRSUTUM. FR.

Hirsutum means shaggy, hairy. The pilei are coriaceous and spreading, quite hairy, imbricated, more or less zoned, quite tough, often having a greenish tinge from the presence of a minute algæ; naked, juiceless, yellowish, unchanged when bruised or scratched. The hymenium is pale-yellow, smooth, margin entire, often lobed. I find it usually on hickory logs.

STEREUM FASCIATUM. SCHW.

Fasciatum means bands or fillets. Pileus is coriaceous, plane, villous, zonate, grayish; hymenium, smooth, pale-red. Growing on decayed trunks. Common in all of our woods.

STEREUM SERICEUM. SCHW.

Figure 383   Stereum sericeum.

Sericeum means silky or satiny; so called from its satin lustre. The plant is very small and easily overlooked, usually growing in a resupinate form; sessile, orbiculate, free, papyraceous, with a bright satin lustre, shining, smooth, pale-grayish color.

The plant grows on both sides of small twigs as is shown in the photograph. I do not find it on large trunks but it is quite common on branches. No one will fail to recognize it from its specific name.

When I first observed it I named it S. sericeum, not knowing that there was a species by that name. I afterwards sent it to Prof. Atkinson and was surprised to find that I had correctly named it.

STEREUM RUGOSUM. FR.

Rugosum means full of wrinkles.

Broadly effused, sometimes shortly reflexed; coriaceous, at length thick and rigid; pileus at length smooth, brownish.

The hymenium is a pale grayish-yellow, changing slightly to a red when bruised, pruinose. The spores are cylindrico-elliptical, straight, $11–12 \times 4–5\mu$. Massee.

This is quite variable in form, and agrees with S. sanguinolentum in becoming red when bruised; but it is thicker and more rigid in substance, its pores are straighter and larger.

STEREUM PURPUREUM. PERS.

Purpureum means purple, from the color of the plant.

Coriaceous but pliant, effuso-reflexed, more or less imbricated, tomentose, zoned, whitish or pallid.

The hymenium is naked, smooth, even; in color a pale clear purple, becoming dingy ochraceous, with only a tinge of purple, when dry. The spores are elliptical, $7–8 \times 4\mu$.

I found the plant to be very abundant in December and January, in 1906–7, on soft wood corded up at the paper mill in Chillicothe, the weather being mild and damp.

STEREUM COMPACTUM.

Broadly effused, coriaceous, often imbricated and often laterally joined, pileus thin, zoned, finely strigose, the zones grayish-white and cinnamon-brown.

The hymenium is smooth, cream-white.

This species is found on decayed limbs and trunks of trees.

HYMENOCHÆTE. LEV.

Hymenochæte is from two Greek words, hymen, a membrane; chæte, a bristle.

In this genus the cap or pileus may be attached to the host by a central stem, or at one side, but most frequently upon its back. The genus is known by the velvety or bristly appearance of the fruiting surface, due to smooth, projecting, thick-walled cells. I have found several species but have only been sure of three.

HYMENOCHÆTE RUBIGINOSA. (SCHR.) LEV.

Rubiginosa means full of rust, so called from the color of the plant.

The pileus is rigid, coriaceous, resupinate, effused, reflexed, the lower margin generally adhering firmly, somewhat fasciated; velvety, rubiginous or rusty in color, then becoming smooth and bright brown, the intermediate stratum tawny-ferruginous. The hymenium ferruginous and velvety. It is found here upon soft woods such as chestnut stumps and willow.

HYMENOCHÆTE CURTISII. BERK.

Curtisii is named in honor of Mr. Curtis.

The pileus is coriaceous, firm, resupinate, effused, reflexed, brown, slightly sulcate; the hymenium velvety with brown bristles. This is common on partially decayed oak branches in the woods.

HYMENOCHÆTE CORRUGATA. BERK.

Corrugata means bearing wrinkles or folds.

The pileus is coriaceous, effused, closely adnate, indeterminate, cinnamon colored, cracked and corrugated when dry, which gives rise to its name. The bristles are seen, under the microscope, to be joined. Found in the woods on partially decayed branches.

CHAPTER X

CLAVARIACEAE—CORAL FUNGI

Hymenium not distinct from the hymenophore, covering entire outer surface, somewhat fleshy, not coriaceous; vertical, simple or branched. Fries.

Most of the species grow on the ground or on well rotted logs. The following genera are included here:

- Sparassis—Fleshy, much branched, branches compressed, plate-like.
- Clavaria—Fleshy, simple or branched, typically round.
- Calocera—Gelatinous, then horn-like.
- Typhula—Simple or club-shaped, rigid when dry, usually small.

SPARASSIS. FR.

Sparassis, to tear in pieces. The species are fleshy, branched with plate-like branches, composed of two plates, fertile on both sides.

SPARASSIS HERBSTII. PK.

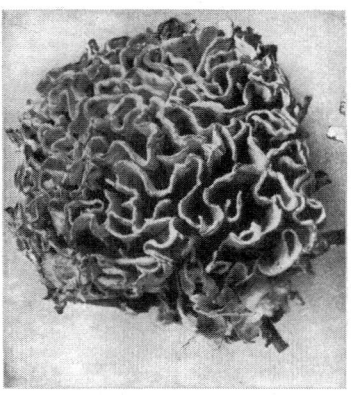

Figure 384   Sparassis Herbstii.

This is a plant very much branched, forming tufts four to five inches high, and five to six inches broad; whitish, inclining to creamy-yellow; tough, moist; the branches numerous, thin, flattened, concrescent, dilated above, spathulate or fan-shaped, often somewhat longitudinally curved or wavy; mostly uniformly colored, rarely with a few indistinct, nearly concolorous, transverse zones near the broad, entire apices.

The spores are globose, or broadly elliptical, .0002 to .00025 inch long, .00016 to .0002 broad.

This species was first found by the late Dr. William Herbst of Trexlertown, Pa., and was named by Dr. Peck in his honor. The specimen in Figure 384 was found at Trexlertown, Pa., and photographed by Mr. C. G. Lloyd. The plant delights in open oak woods, and is found through August and September. It is edible and quite good.

SPARASSIS CRISPA. FR.

Crispus, curly. This is a beautiful rosette-like plant, growing quite large at times, very much branched, whitish, oyster-colored, or pale yellow; branches intricate, flat and leaf-like, having a spore surface on both sides. The entire plant forms a large round mass with its leaf-like surface variously curled, folded, and lobed, with a crest-like margin, and springing from a well-marked root, most of which is buried in the ground.

No one will have any trouble to recognize it, having once seen its photograph. I found the plant quite frequently, in the woods about Bowling Green. It is not simply good, but very good.

CLAVARIA. LINN.

Clavaria is from clavus, a club. This is by far the largest genus in this family, and contains very many edible species, some of which are excellent.

The entire genus is fleshy, either branched or simple; gradually thickening toward the top, resembling a club.

In collecting clavaria special attention should be given to the character of the apices of the branches, color of the branches, color of spores, the taste of the plant, and the character of the place of its growth. This genus is readily recognized, and no one need to hesitate to eat any of the branching forms.

CLAVARIA FLAVA. SCHAEFF.

PALE-YELLOW CLAVARIA. EDIBLE.

Flava is from flavus, yellow. The plant is rather fragile, white and yellow, two to five inches high, the mass of branches from two to five inches wide, the trunk thick, much branched. The branches are round, even, smooth, crowded, nearly parallel, pointing upward, whitish or yellowish, with pale yellow tips of tooth-like points. When the plant is old, the yellow tips are likely to be faded, and the whole plant whitish in color. The flesh and the spores are white, and the taste is agreeable.

I have eaten this species since 1890, and I regard it as very good. It is found in woods and grassy open places. I have found it as early as June and as late as October.

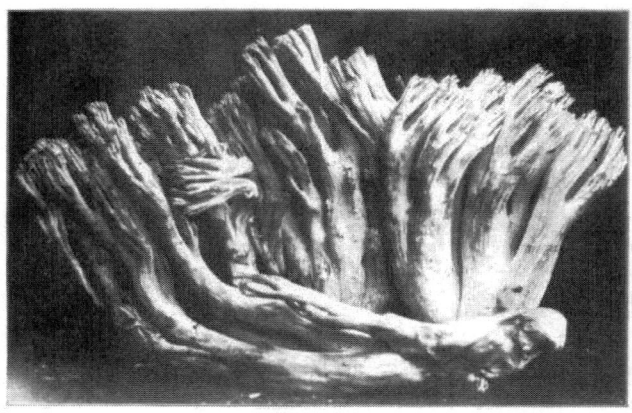

Figure 385   Clavaria flava. Natural size.

CLAVARIA AUREA. PERS.

THE GOLDEN CLAVARIA. EDIBLE.

This plant grows from three to four inches high. Its trunk is thick, elastic, and its branches are uniformly a deep golden yellow, often longitudinally wrinkled. The branches straight, regularly forked and round.

The stem is stout but thinner than in C. flava. The spores are yellowish and elliptical. It is found in woods during August and September.

CLAVARIA BOTRYTES. PERS.

THE RED-TIPPED CLAVARIA. EDIBLE.

Figure 386   Clavaria botrytes. One-half natural size.

Botrytes is from a Greek word meaning a cluster of grapes. This plant differs little from C. flava in size and structure, but it is easily recognized from the red tips of its branches. It is whitish, or yellowish, or pinkish, with its branches red-tipped.

The stem is short, thick, fleshy, whitish, unequal. The branches are often somewhat wrinkled, crowded, repeatedly branched. In older specimens the red tips will be somewhat faded. The spores are white and oblong-elliptical. It is found in woods and open places, during wet weather. I found this plant occasionally near Salem, from July to October, but it is not a common plant in Ohio.

CLAVARIA MUSCOIDES. LINN.

FORKED YELLOW CLAVARIA. EDIBLE.

Muscoides means moss-like. This plant is inclined to be tough, though graceful in growth; slender-stemmed, two or three time forked; smooth; base downy, bright yellow. The branchlets are thin, crescent-shape, acute. The spores are white and nearly round. The plant is usually solitary, not branching as much as some other species; quite dry, very smooth, except at the base, which is downy, in color resembling the yolk of an egg. It is frequently found in damp pastures, especially those skirting a wood.

CLAVARIA AMETHYSTINA. BULL.

THE AMETHYSTINE CLAVARIA. EDIBLE.

Figure 387 Clavaria amethystina.

Amethystina means amethyst in color. This is a remarkably attractive plant and easily recognized by its color. It is sometimes quite small yet often grows from three to five inches high. The color of the entire plant is violet; it is very much branched or almost simple; branches round, even, fragile, smooth, obtuse. The spores are elliptical, pale-ochraceous, sub-transparent, $10-12 \times 6-7\mu$.

This plant is quite common around Chillicothe, and it has a wide distribution over the United States. The specimens in Figure 387 were found in Poke Hollow.

CLAVARIA STRICTA. PERS.

THE STRAIGHT CLAVARIA. EDIBLE.

Figure 388   Clavaria stricta.

Stricta is a participle from stringo, to draw together. The plant is very much branched, pale, dull-yellow, becoming brownish when bruised; the stem somewhat thickened; branches very numerous and forked, straight, even, densely pressed, tips pointed. The spores are dark cinnamon. It is found on the Huntington hills near Chillicothe. Look for it in August and September.

CLAVARIA PYXIDATA. PERS.

THE CUP CLAVARIA. EDIBLE.

Figure 389   Clavaria pyxidata. Natural size.

Pyxidata is from pyxis, a small box. This plant is quite fragile, waxy, light-tan in color, with a thin main stem, whitish, smooth, variable in length, branching and rebranching, the branches ending in a cup. The spores are white.

It is found on rotten wood and is readily recognized by the cup-like tips. The specimen in Figure 389 was found near Columbus and photographed by Dr. Kellerman. Found from June to October.

CLAVARIA ABIETINA. SCHUM.

THE FIR-WOOD CLAVARIA.

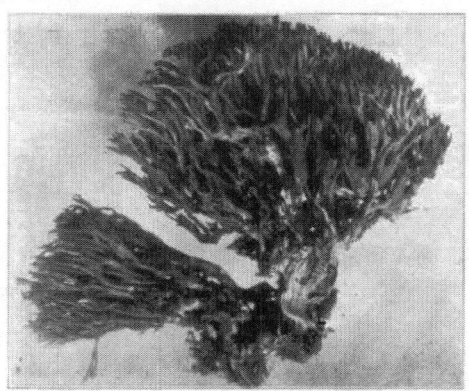

Figure 390   Clavaria abietina.

Abietina means fir-wood.

This plant grows in dense tufts, very much branched, ochraceous, trunk somewhat thickened, short, clothed with a white down; branches straight, crowded, longitudinally wrinkled when dry, branchlets straight.

The spores are oval and ochraceous.

It can be readily identified by its changing to green when bruised.

It is very common on our wooded hillsides. It is found from August to October.

CLAVARIA SPINULOSA. PERS.

Figure 391   Clavaria spinulosa.

Spinulosa means spiny or full of spines.

The trunk of this plant is rather short and thick, at least one-half to one inch thick, whitish. The branches are elongated, crowded, tense and straight; attenuated, tapering upward; color somewhat cinnamon-brown throughout.

The spores are elliptical, yellowish-brown, $11-13 \times 5\mu$.

It is usually given as found under pine trees, but I find it about Chillicothe in mixed woods, in which there are no pine trees at all. It is found after frequent rains, from August to October. As an edible, it is fairly good.

CLAVARIA FORMOSA. PERS.

BEAUTIFUL CLAVARIA. EDIBLE.

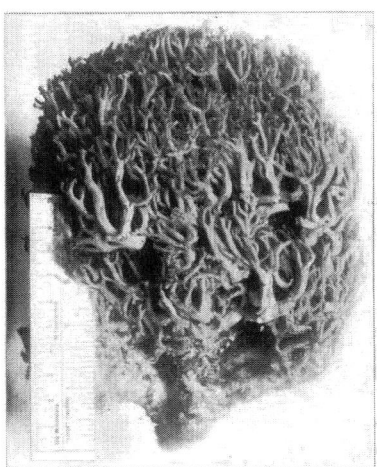

Figure 392 Clavaria formosa. Three-fourths its natural size.

Formosa is from formosus, meaning finely formed.

This plant is two to six inches high, trunk rather thick, often over an inch in thickness; whitish, or yellowish, elastic, the branches numerous, crowded, elongated, divided at the ends into yellow branchlets, which are thin, straight, obtuse or toothed.

The spores are elongated-oval, rough, buff-colored, $16 \times 8\mu$.

This is an extremely beautiful plant, very tender or brittle. When the plant is quite young, just coming through the ground, the tips of the branches are often of a bright red or pink. This bright color soon fades, leaving the entire plant a light yellow in color.

The plant has a wide distribution, and is found on the ground in the woods, frequently growing in rows. While the handsomest of the Clavarias, it is not the best, and only the tender parts of the plant should be used. It is found from July to October. The specimen in Figure 392 was found in Poke Hollow.

CLAVARIA CRISTATA. PERS.

THE CRESTED CLAVARIA. EDIBLE.

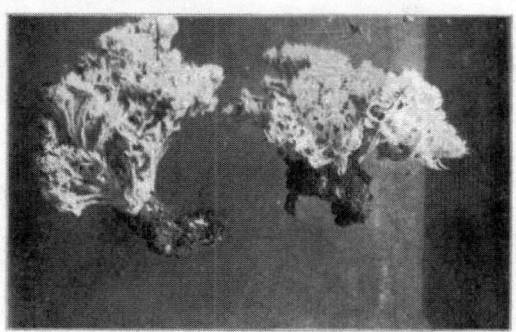

Figure 393　Clavaria cristata.

Cristata is from cristatus, crested. This is a smaller plant than the C. flava or C. botrytes. It is usually two to three inches high, white or whitish, the tufts of broad flattened branches, sometimes tinged with a dull pink or creamy-yellow. The branches are numerous, widened and flattened above, deeply cut into several finger-like points, sometimes so numerous as to give it a crested appearance. This peculiar characteristic distinguishes it from C. coralloides. When the plant is old the tips usually turn brown.

Sometimes a form will be found in which the crested appearance is wanting, and in that case the branches terminate in blunt points. The stem is short and inclined to be spongy.

It is found in the woods, in cool, moist, shady places. While it is tougher than some of the other species, if cut fine and well cooked it is very good. I have eaten it for years. It is found from June to October.

CLAVARIA CORONATA. SCHW.

THE CROWNED CLAVARIA. EDIBLE.

Figure 394　Clavaria coronata.

Pale yellow, then fawn color; divided immediately from the base and very much branched; the branches divergent and compressed or angulate, the final branchlets truncate-obtuse at apex and there encircled with a crown of minute processes. Morgan.

This plant is found on decayed wood. It is repeatedly branched in twos and forms clusters sometimes several inches in height. It resembles in form C. pyxidata, but it is quite a distinct species. In some localities it is found quite frequently. It is plentiful about Chillicothe. Found from July to October.

CLAVARIA VERMICULARIS. SCOP.

WHITE-TUFTED CLAVARIA.

Figure 395 Clavaria vermicularis.

Small, two to three inches high; cæspitose, fragile, white, club-shaped; clubs stuffed, simple, cylindrical, subulate.

Found on lawns, short pastures or in paths in woods. Someone has said they "look like a little bundle of candles." Edible, but too small to gather. June and July.

CLAVARIA CRISPULA. FR.

FLEXUOUS CLAVARIA. EDIBLE.

Very much branched, tan-colored, then ochraceous; trunk slender, villous, rooting; branches flexuous, having many divisions, branches of the same color, divaricating, fragile.

The spores are creamy-yellow, slightly elliptical. This plant is slightly acrid to the taste and retains a faint trace of acridity even after it is cooked. It is very plentiful in our woods. Found from July to October.

CLAVARIA KUNZEI. FR.

KUNZE'S CLAVARIA.

Rather fragile, very much branched from the slender cæspitose base; white; branches elongated, crowded, repeatedly forked, subfastigiate, even, equal; axils compressed. Specimens were found on Cemetery Hill under beech trees, and identified by Dr. Herbst. The spores are yellowish.

CLAVARIA CINEREA. BULL.

ASH-COLORED CLAVARIA. EDIBLE.

Cinerea, pertaining to ashes. This is a small plant, growing in groups, frequently in rows, under beech trees. The color is gray or ashy; it is quite fragile; stem thick, short, very much branched, with the branches thickened, somewhat wrinkled, rather obtuse. Its gray color will distinguish it from the other Clavaria.

CLAVARIA PISTILLARIS. L.

INDIAN-CLUB CLAVARIA. EDIBLE.

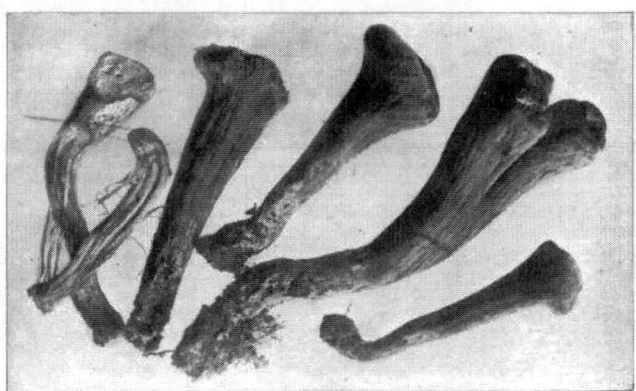

Figure 396 Clavaria pistillaris. One-half natural size.

Pistillaris is from pistillum, a pestle.

They are simple, large, stuffed, fleshy, everywhere smooth, three to ten inches high, attaining to one inch in thickness; light yellow, ochraceous, brownish, chocolate, club-shaped, ovate-rounded, puckered at the top; flesh white, spongy. The spores are white, $10 \times 5\mu$.

They are found in the leaf-mold of mixed woods, and you will sometimes find several growing together. They are found from July to frost.

The dark variety, which is frequently vertically wrinkled, is slightly acrid when raw, but this disappears upon cooking. The plant is widely distributed but abundant nowhere in our state. I found it occasionally in the woods near Chillicothe. The plants in Figure 396 were found near Columbus, and were photographed by Dr. Kellerman of Ohio State University.

CLAVARIA FUSIFORMIS. SOW.

SPINDLE-SHAPED CLAVARIA. EDIBLE.

Figure 397   Clavaria fusiformis. Natural size.

Fusiformis is from fusus, a spindle, and forma, a form.

The plant is yellow, smooth, rather firm, soon hollow, cæspitose; nearly erect, rather brittle, attenuated at each end; clubs somewhat spindle-shaped, simple, toothed, the apex somewhat darker; even, slightly firm, usually with several united at the base.

The spores are pale yellow, globose, $4-5\mu$.

They are found in woods and pastures. The plants in the figure were in the woods beside an untraveled road, on Ralston's Run.

They strongly resemble C. inæqualis. When found in sufficient quantities they are very tender and have an excellent flavor.

CLAVARIA INÆQUALIS. MULL.

THE UNEQUAL CLAVARIA. EDIBLE.

Inæqualis means unequal.

Somewhat tufted, quite fragile, from one to three inches high, often compressed, angular, often forked, ventricose; yellow, occasionally whitish, sometimes variously cut at the tip. The spores are colorless, elliptical, $9-10 \times 5\mu$.

One can readily distinguish it from C. fusiformis by the tips, these not being sharp pointed. It is found in clusters in woods and pastures from August to October. As delicious as C. fusiformis.

CLAVARIA MUCIDA. PERS.

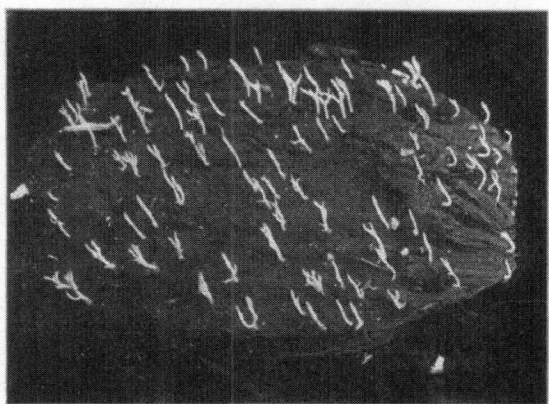

Figure 398   Clavaria mucida.

Mucida means slimy, so named from the soft and watery condition of the plants.

The plants are quite small, usually simple yet sometimes branched, club-shaped, one-eighth to an inch high, white, sometimes yellowish, frequently pinkish or rose-tinted.

These plants are very small and easily overlooked. It is found on decayed wood. I have found it late in the fall and early in the spring. You can look for it at any time of the year after warm rains or in damp places, on well decayed wood. The specimens in Figure 398 were photographed by Prof. G. D. Smith, Akron, Ohio.

CALOCERA. FR.

This plant is gelatinous, somewhat cartilaginous when moist, horny when dry, vertical, simple or branched, cæspitose or solitary.

The hymenium is universal; the basicia round and two-lobed, each lobe bearing a single one-spored sterigma. The spores are inclined to be oblong and curved.

This genus resembles Clavaria, but is identified by being somewhat gelatinous and viscid when moist and rather horn-like when dry, but especially by its two-lobed basidia.

CALOCERA CORNEA. FR.

Figure 399   Calocera cornea.

This is unbranched, cæspitose, rooting, even, viscid, orange-yellow or pale yellow; clubs short, subulate, connate at the base. The spores are round and oblong, $7-8 \times 5\mu$.

Found upon stumps and logs, especially upon oak where the timber is cracked, the plants springing from the cracks. When dry they are quite stiff and rigid.

CALOCERA STRICTA. FR.

These plants are unbranched, solitary, about one inch high, elongated, base somewhat blunt, even when dry, yellow.

Its habitat is very similar to C. cornea but more scattered. C. striata, Fr., is very similar to C. cornea, but is distinguished by its being solitary, and striate or rugose when dry.

TYPHULA. FR.

Epiphytal. Stem filiform, flaccid; clubs cylindrical, perfectly distinct from hymenium, sometimes springing from a sclerotium; hymenium thin and waxy.

This is distinguished from Clavaria and Pistillaria by having its stem distinct from the hymenium. It is a small plant resembling, in miniature, Typha, hence its generic name.

TYPHULA ERYTHROPUS. FR.

Simple; club cylindrical, slender, smooth, white; stem nearly straight, dark red, inclining to be black, springing usually from a blackish and somewhat wrinkled sclerotium. The spores are oblong, $5-6 \times 2-2.5\mu$.

This plant has a wide distribution, and is found in damp places upon the stems of herbaceous plants.

TYPHULA INCARNATA. FR.

Simple; club cylindrical, elongated, smooth; whitish, more or less tinged with pink above; one to two-inches high, base minutely strigose, springing from a compressed brownish sclerotium. The spores are nearly round, $5 \times 4\mu$.

This is a common and beautiful little plant and easily distinguished both by its color and the size and form of its spores. If the collector will watch the dead herbaceous stems in damp places, he will not only find the two just described, but another, differing in color, size, and form of spores, called T. phacorrhiza, Fr. It has a brownish color and its spores are quite oblong, $8-9 \times 4-5\mu$.

LACHNOCLADIUM. LEV.

Lachnocladium is from two Greek words meaning a fleece and a branch.

Pileus coriaceous, tough, repeatedly branched; the branches slender or filiform, tomentose. Hymenium amphigenous. Fungi slender and much branched, terrestrial, but sometimes growing on wood.

LACHNOCLADIUM SEMIVESTITUM. B. & C.

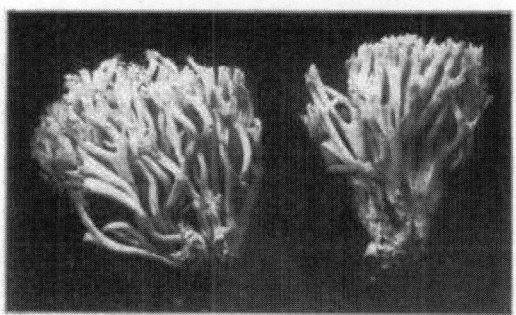

Figure 400 Lachnocladium semivestitum.

Pileus, much branched from a slender stem of variable length, expanded at the angles; the branches filiform, straight, somewhat fasciculate, smooth at the tips and paler in color.

This is quite a common specimen on our north hillsides. It is white and quite fragile. Found in damp places in August and September.

LACHNOCLADIUM MICHENERI. B. & C.

Figure 401 Lachnocladium Micheneri.

Coriaceous, tough, pale or whitish; stem well marked, branching from a point, branches numerous, tips pointed; white tomentum at the base of the stem.

This plant is very abundant here and is found very generally over the United States. It grows on fallen leaves in woods, after a rain, being found from July to October.

CHAPTER IX

TREMELLINI FR

Tremellini is from tremo, to tremble. The whole plant is gelatinous, with the exception, occasionally, of the nucleus. The sporophores are large, simple or divided. Spicules elongated into threads. Berk.

The following genera are included:

- Tremella—Immarginate. Hymenium universal.
- Exidia—Margined. Hymenium superior.
- Hirneola—Cartilaginous, ear-shaped, attached by a point.

TREMELLA. FR.

This plant is so called because the entire plant is gelatinous, tremulous, and without a definite margin, and also without nipple-like elevations.

TREMELLA LUTESCENS. FR.

YELLOWISH TREMELLA. EDIBLE.

This is a small gelatinous cluster, tremulous, convoluted, in wavy folds, pallid, then yellowish, with its lobes crowded and entire. Quite common over the state. It is found on decaying limbs and stumps from July to winter. It dries during absence of rain but revives and becomes tremulous during wet weather. It is called lutescens because of its yellowish color.

TREMELLA MESENTERICA. RETZ.

Mesenterica is from two Greek words meaning the mesentery. The plant varies in size and form, sometimes quite flat and thin but generally ascending and strongly lobed; plicated, and convoluted; gelatinous but firm; lobes short, smooth, covered with a frost-like bloom by the white spores at maturity. The spores are broadly elliptical. Common in the woods on decaying sticks and branches.

TREMELLA ALBIDA. HUD.

THE WHITISH TREMELLA. EDIBLE.

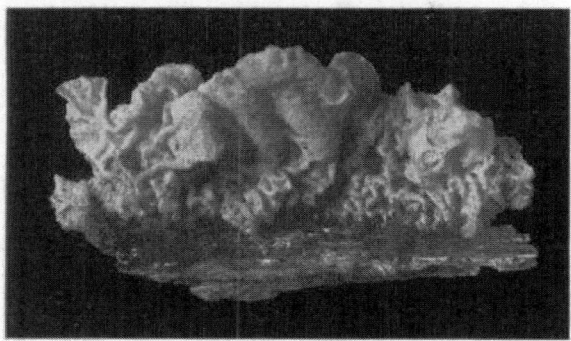

Figure 402   Tremella albida. Natural size.

Albida, whitish. This plant is very common in the woods about Chillicothe, and everywhere in the state where beech, sugar-maple, and hickory prevail.

It is whitish, becoming dingy-brown when dry; expanded, tough, undulated, even, more or less gyrose, pruinose. It breaks the bark and spreads in irregular and scalloped masses; when moist it has a gelatinous consistency, a soft and clammy touch, yielding like a mass of gelatine. Its spores are oblong, obtuse, curved, marked with tear-like spots, almost transparent, $12–14 \times 4–5\mu$. The specimen represented in Figure 402 was found near Sandusky and photographed by Dr. Kellerman.

TREMELLA MYCETOPHILA. PK.

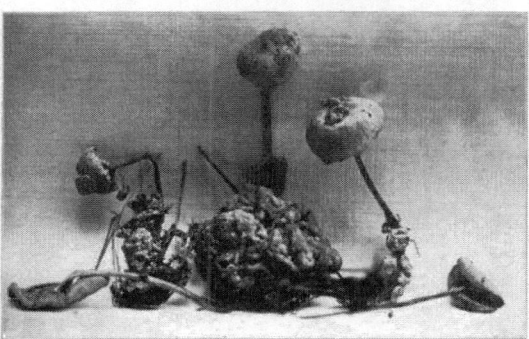

Figure 403   Tremella mycetophila.

Mycetophila is from two Greek words, mycetes, fungi; phila, fond of. The plant is so called because it is found growing upon other fungi.

Often nearly round, somewhat depressed, circling in folds, sometimes in quite large masses about the stems of the plant, as will be seen in Figure 403, tremelloid-fleshy, slightly pruinose, a dirty white or yellowish.

I have found it frequently growing on Collybia drophila, as is the case in Figure 403. Captain McIlvaine speaks in his book of finding this plant parasitic on Marasmius oreades in quite a large mass for this plant. I can verify the statement for I have found it on M. oreades during damp weather in August and September. It has a pleasant taste.

TREMELLA FIMBRIATA. PERS.

FIMBRIATA IS FROM FRIMBRIÆ, A FRINGE.

It is very soft and gelatinous, olivaceous inclining to black, tufted, two to three inches high, and quite as broad, erect, lobes flaccid, corrugated, cut at the margin, which gives rise to the name of species; spores are nearly pear shaped. Found on dead branches, stumps, and on fence-rails in damp weather. Easily known by its dark color.

TREMELLODON. PERS.

Tremellodon means trembling tooth.

These plants are gelatinous, with a cap or pileus; the hymenium covered with acute gelatinous spines, awl-shaped and equal. The basidia are nearly round with four rather stout, elongated sterigmata, spores very nearly round.

TREMELLODON GELATINOSUM. PERS.

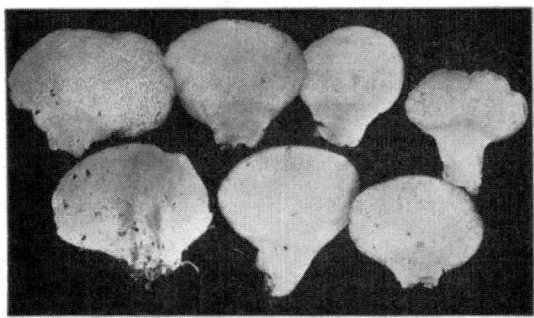

Figure 405   Tremellodon gelatinosum.

Gelatinosum means full of jelly or jelly-like, from gelatina, jelly.

The pileus is dimidiate, gelatinous, tremelloid, one to three inches broad, rather thick, extended behind into a lateral thick, stem-like base, pileus covered with a greenish-brown bloom, very minutely granular.

The hymenium is watery-gray, covered with hydnum-like teeth, stout, acute, equal, one to two inches long, whitish, soft, inclined to be glaucous. The spores are nearly round, $7–8\mu$.

These plants are found on pine and fir trunks and on sawdust heaps. They grow in groups and are very variable in form and size but easily determined, being the only tremelloid fungus with true spines. The plants in Figure 405 were photographed by Prof. G. D. Smith of Akron, Ohio. They are edible. Found from September to cold weather.

EXIDIA. FR.

Gelatinous, marginal, fertile above, barren below. Exidia may be known by its minute nipple-like elevations.

EXIDIA GRANDULOSA. FR.

Plate L. Figure 404   Excidia glandulosa.

This plant is called "Witches' Butter." It varies in color, from whitish to brown and deep cinereous, at length blackish; flattened, undulated, much wrinkled above, slightly plicated below; soft at first and when moist, becoming film-like when dry. Found on dead branches of oak.

HIRNEOLA. FR.

Hirneola is the diminutive of hirnea, a jug. Gelatinous, cup-shaped, horny when dry. Hymenium wrinkled, becoming cartilaginous when moistened. The hymenium is in the form of a hard skin which covers the cup-shaped cavities, and which can be peeled off after soaking in water, the interstices are without papillæ and the outer surface is velvety.

HIRNEOLA AURICULA-JUDÆ. BERK.

THE JEW'S EAR HIRNEOLA. EDIBLE.

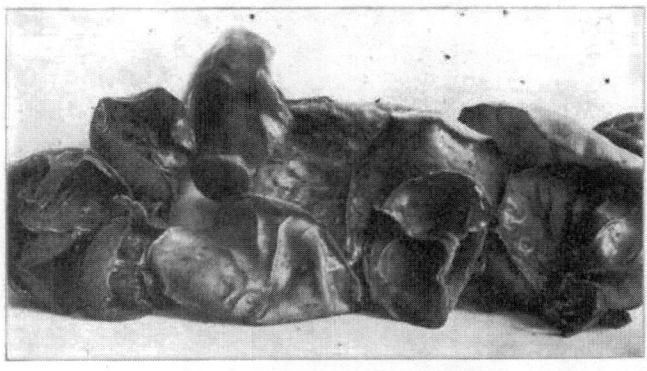

Figure 406   Hirneola auricula-Judæ.

Plate LI. Figure 407   Hirneola auricula-judae.

Auricula-Judæ, the ear of the Jew. The plant is gelatinous; one to four inches across; thin, concave, wavy, flexible when moist, hard when dry; blackish, fuzzy, hairy beneath; when covered with white spores it is cinereous. The hymenium by its corrugations forms depressions such as are found in the human ear. One will not fail to recognize it after seeing it once. It is not common in our woods, yet I have found it on several occasions. It is found on almost any timber but most frequently on the elm and elder. The plant in Figure 406 was found near Chillicothe. Its distribution is general.

GUEPINIA. FR.

Gelatinous, inclining to cartilaginous, free, different on the two sides, variable in form, substipitate. Hymenium confined to one side.

GUEPINIA SPATHULARIA.

Figure 408   Guepinia spathularia. Entire plant a light yellow.

Yellow, cartilaginous, especially when dry, spathulate, expanded above, hymenium slightly ribbed, contracted where it issues from a log.

It is quite common on beech and maple logs. I have seen beech logs, somewhat decayed, quite yellow with this interesting plant.

HYMENULA. FR.

Effused, very thin, maculæform, agglutinate, between wavy or gelatinous. Berk.

HYMENULA PUNCTIFORMIS. B. & BR.

POINT-LIKE HYMENULA.

Dirty white, quite pallid, gelatinous, punctiform, slightly undulated; consisting of erect simple threads; frequently there is a slight tinge of yellow. The spores are very minute. It looked very much like an undeveloped Peziza when I found it, in fact I thought it P. vulgaris until I had submitted a specimen to Prof. Atkinson.

CHAPTER XII

ASCOMYCETES—SPORE-SAC FUNGI.

Ascomycetes is from two Greek words: ascos, a sack; mycetes, a fungus or mushroom. All the fungi which belong to this class develop their spores in small membranous sacs. These asci are crowded together side by side, and with them are slender empty asci called paraphyses. The spores are inclosed in these sacs, usually eight in a sac. They are called sporidia to separate them from the Basidiomycetes. These sacs arise from a naked or inclosed stratum of fructifying cells, forming a hymenium or nucleus.

FAMILY—HELVELLACEAE

Hymenium at length more or less exposed, the substance soft. The genera are distinguished from the earth-tongues by the cup-like forms of the spore body, but especially by the character of the spore sacs which open by a small lid, instead of spores. The following are some of the genera:

- Morchella Pileus deeply folded and pitted.
- Gyromitra Pileus covered with rounded and variously contorted folds.
- Helvella Pileus drooping, irregularly waved and lobed.

MORCHELLA. DILL

Morchella is from a Greek word meaning a mushroom. This genus is easily recognized. It may be known by the deeply pitted, and often elongated, naked head, the depressions being usually regular but sometimes resembling mere furrows with wrinkled interspaces. The cap or head varies in form from rounded to ovate or cone shape. They are all marked by deep pits, covering the entire surface, separated by ridges forming a net-work. The spore-sacs are developed in both ridges and depressions. All the species when young are of a buff-yellow tinged with brown. The stems are stout and hollow, white, or whitish in color.

The common name is Morel, and they appear during wet weather early in the spring.

MORCHELLA ESCULENTA. PERS.

THE COMMON MOREL. EDIBLE.

The Common Morel has a cap a little longer than broad, so that it is almost oval in outline. Sometimes it is nearly round but again it is often slightly narrowed in its upper half, though not pointed or cone-like. The pits in its surface are more nearly round than in the other species. In this species the pits are irregularly arranged so that they do not form rows, as will be observed in Figure 409.

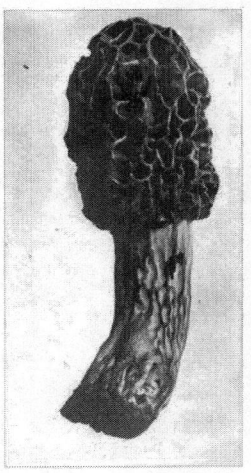

It grows from two to four inches high and is known by most people as the Sponge mushroom. It grows in woods and wood borders, especially beside wood streams. Old apple and peach orchards are favorite places for Morels. It makes no difference if the beginner cannot identify the species, as they are all equally good. I have seen collectors have for sale a bushel basketful, in which half a dozen species were represented. They dry very easily and can be kept for winter use. It is said to grow in great profusion over burnt districts. The German peasants were reputed to have burned forest tracts to insure an abundant crop. I find that more people know the Morels than any other mushroom. They are found through April and May, after warm rains.

Figure 409 Morchella esculenta. Two-thirds natural size.

MORCHELLA DELICIOSA. FR.

THE DELICIOUS MOREL. EDIBLE.

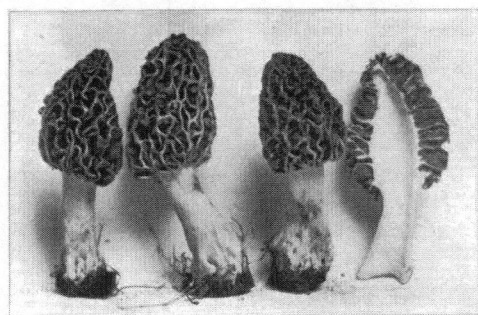

Figure 410 Morchella deliciosa. Two-thirds natural size.

This and the preceding species would indicate by their names that they have been held in high esteem for a long time, as Profs. Persoon and Fries, who named them, lived more than a hundred years ago. The Delicious Morel is recognized by the shape of its cap, which is generally cylindrical, sometimes pointed, and slightly curved. The stem is rather short and, like the stem of all Morels, is hollow from the top to the bottom.

It is found associated with other species of Morels, in woods and wood borders, also in old apple and peach orchards. They need to be cooked slowly and long. Coming early in the spring, they are not likely to be infested with worms. The flesh is rather fragile and not very watery. They are easily dried. Found through April and May.

MORCHELLA ESCULENTA VAR. CONICA. PERS.

THE CONICAL MOREL. EDIBLE.

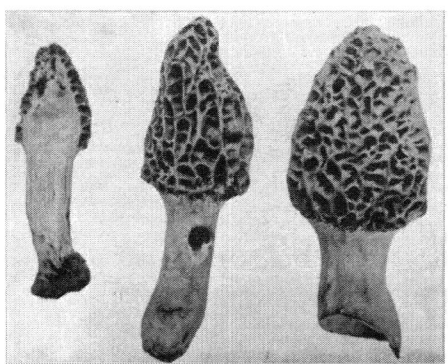

Plate LII. Figure 411   Morchella esculenta var. conica.

The Conical Morel is very closely related to M. esculenta and M. deliciosa, from which it differs in having the cap longer than it is wide, and more pointed, so that it is conical or oblong-conical. The plant, as a general thing, grows to be larger than the other species. It is, however, pretty hard to distinguish these three species. The Conical Morel is quite abundant about Chillicothe. I have found Morels especially plentiful about the reservoirs in Mercer County, and in Auglaize, Allen, Harden, Hancock, Wood and Henry Counties. I have known lovers of Morels to go on camping tours in the woods about the reservoirs for the purpose of hunting them, and to bring home large quantities of them.

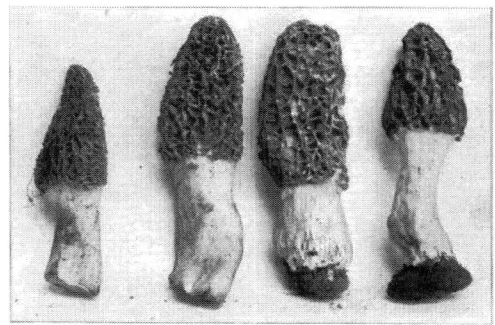

Figure 412   Morchella esculenta var. conica. Two-thirds natural size.

MORCHELLA ANGUSTICEPS. PK.

THE NARROW-CAP MOREL. EDIBLE.

Figure 413 Morchella angusticeps.

Angusticeps is from two Latin words: angustus, narrow; caput, head. This species and M. conica are so nearly alike that it is very difficult to identify them with any degree of satisfaction. In both species the cap is considerably longer than broad, but in angusticeps the cap is slimmer and more pointed. The pits, as a general thing are longer than in the other species. They are often found in orchards but are also frequently found in low woods under black ash trees. I have found some typical specimens about the reservoirs. The specimens in Figure 413 were collected in Michigan, and photographed by Prof. B. O. Longyear. They appear very early in the spring, even while we are still having frosts.

MORCHELLA SEMILIBERA. D. C.

THE HYBRID MOREL. EDIBLE.

Figure 414 Morchella semilibera. One-half natural size.

Semilibera means half free, and it is so called because the cap is bell-shaped and the lower half is free from the stem. The cap is rarely more than one inch long, and is usually much shorter than the stem, as is indicated in Figure 414. The pits on the cap are longer than broad. The stem is white or whitish and somewhat mealy or scurvy, hollow, and often swollen at the base. I found the specimens in Figure 414 about the last of May under elm trees, in James Dunlap's woods. They are quite plentiful there. I do not detect any difference in the flavor of these and other species.

MORCHELLA BISPORA. SOR.

THE TWO-SPORED MOREL. EDIBLE.

Figure 415   Morchella bispora. One-half natural size.

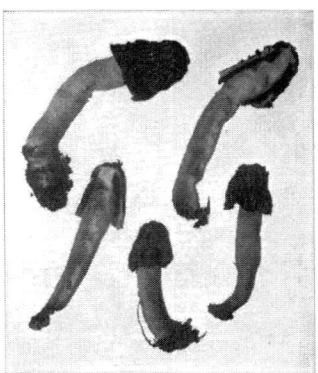

Plate LIII. Figure 416   Morchella bispora.

The two-spored Morel. Edible. Showing the cap free from the stem quite to the top.

Bispora, two-spored, differs from the other species in the fact that the cap is free from the stem quite to the top. The distinguishing characteristic, which gives name to the species, can be seen only by the aid of a strong microscope. In this species there are only two spores in each ascus or sac, and these are much larger than in the other species, which have eight spores in a sac or ascus. The ridges, as will be seen in Figure 415, run from the top to the bottom. The stem is much longer than the cap, hollow, and sometimes swollen at the base. The whole plant is fragile and very tender. The plants in Figure 415 were collected in Michigan by Prof. Longyear. Those in the full page display were found near Columbus and were photographed by Dr. Kellerman. It seems to have a wide range, but is nowhere very plentiful.

The spores can be readily obtained from morels by taking a mature specimen and placing it on white paper under a glass for a few hours.

The beginner will find much difficulty in identifying the species of Morels; but if he is collecting them for food he need not give the matter any thought, since none need be avoided, and they are so characteristic that no one need be afraid to gather them.

MORCHELLA CRASSIPES. PERS.

THE GIGANTIC MOREL. EDIBLE.

Crassipes is from crassus, thick; pes, foot.

The cap resembles the cap of M. esculenta in its form and irregular pitting, but it is quite a little larger. The stem is very stout, much longer than the pileus, often very much wrinkled and folded. I have found only a few specimens of this species. Found in April and May.

VERPA. SWARTZ.

Verpa means a rod. Ascospore smooth or slightly wrinkled, free from the sides of the stem, attached at the tip of the stem, bell-shaped, thin; hymenium covering the entire surface of the ascospore; asci cylindrical, 8-spored. The spores are elliptical, hyaline; paraphyses septate.

The stem is inflated, stuffed, rather long, tapering downward.

VERPA DIGITALIFORMIS. PERS.

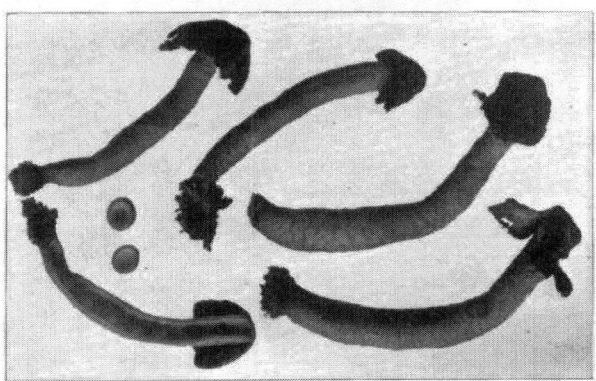

Figure 417   Verpa digitaliformis.

Digitaliformis is from digitus, a finger, and forma, a form.

The pileus is bell-shaped, attached to the tip of the stem, but otherwise free from it; olive-umber in color; smooth, thin, closely pressed to the stem, but always free; the edge sometimes inflexed.

The stem is three inches high, tapering downward, furnished at the base with reddish radicels; white, with a reddish tinge; apparently smooth, but under the glass quite scaly; loosely stuffed. The asci are large, 8-spored, the spores being elliptical. The paraphyses are slender and septate.

Figure 417 represents several plants, natural size. The one in the righthand corner is old, with a ragged pileus; the vertical section shows the pithy contents of the stem. The plants are found in cool, moist, and shady ravines from May to August. Edible, but not very good.

GYROMITRA. FR.

Gyromitra is from gyro, to turn; mitra, a hat or bonnet. This genus is so called because the plants look like a hood that is much wrinkled or plaited.

Ascophore stipitate; hymenophore subglobose, inflated and more or less hollow or cavernous, variously gyrose and convolute at the surface, which is everywhere covered with the hymenium; substance fleshy; asci cylindrical, 8-spored; spores uniseriate, elongated, hyaline or nearly so, continuous; paraphyses present. Massee.

GYROMITRA ESCULENTA. FR.

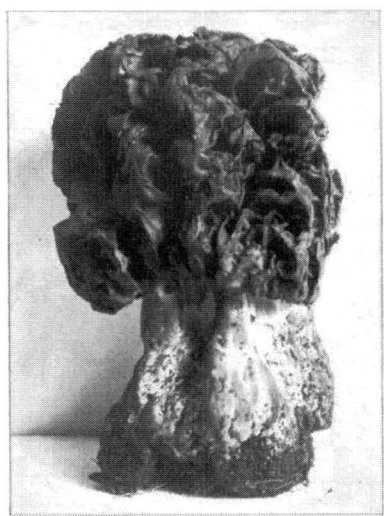

Plate LIV. Figure 418   Gyromitra esculenta.

Esculenta means edible. This is the largest spore-sac fungus. The original name was Helvella esculenta. It is bay-red, round, wrinkled or convoluted, attached to the stem, irregular, with brain-like convolutions.

The stem is hollow when mature, often very much deformed, whitish, scurvy, frequently enlarged or swollen at the base, sometimes lacunose, frequently attenuated upward, at first stuffed; asci cylindrical, apex obtuse, base attenuated, 8-spored; spores obliquely uniseriate, hyaline, smooth, continuous, elliptical, $17-25 \times 9-11\mu$; paraphases numerous.

This plant will be readily recognized from Figure 418, and its bay-red or chestnut-red cap with its brain-like convolutions. The books speak of its being found in pine regions, but I have found it frequently in the woods near Bowling Green, Sidney, and Chillicothe. Many authors give this plant a bad reputation, yet I have eaten it often and when it is well prepared it is good. I should advise caution in its use. It is found in damp sandy woods during May and June. The plant in Figure 418 was found near Chillicothe.

GYROMITRA BRUNNEA. UNDERWOOD.

THE BROWN GYROMITRA. EDIBLE.

Figure 419 Gyromitra brunnea.

Brunnea is from brunneus, brown. A stout, fleshy plant, stipitate, three to five inches high, bearing a broad, much contorted, brown ascoma. Stem is ¾ to 1.5 inch thick, more or less enlarged and spongy, solid at the base, hollow below, rarely slightly fluted, clear white; receptacle two to four inches across in the widest direction, the two diameters usually more or less unequal, irregularly lobed and plicate; in places faintly marked into areas by indistinct anastomosing ridges; closely cohering with the stem in the various parts; color a rich chocolate-brown or somewhat lighter if much covered with the leaves among which it grows; whitish underneath; asci 8-spored. Spores oval. This plant is found quite frequently about Bowling Green. The land is very rich there and produced both G. esculenta and G. brunnea in greater abundance than I have found elsewhere in the state. It is quite tender and fragile. The specimen in Figure 419 was found near Cincinnati and photographed by Mr. C. G. Lloyd.

HELVELLA ELASTICA. BULL.

THE PEZIZA-LIKE HELVELLA. EDIBLE.

Elastica means elastic, referring to its stem. The pileus is free from the stem, drooping, two to three lobed, center depressed, even, whitish, brownish, or sooty, almost smooth underneath, about 2 cm. broad.

Figure 420   Helvella elastica.

The stem is two to three and a half inches high, and three to five lines thick at the inflated base; tapering upward, elastic, smooth, or often more or less pitted; colored like the pileus, minutely velvety or furfuraceous; at first solid, then hollow. Spores hyaline, continuous, elliptical, ends obtuse, often 1-guttulate, 18–20×10–11; 1-serrate; paraphyses septate, clavate. Massee.

The plants in the figure were found near Columbus and photographed by Dr. Kellerman. I have not found the plant as far south as Chillicothe, though I found it frequently in the northern part of the state. It grows in the woods on leaf-mould.

HELVELLA LACUNOSA. AFZ.

THE CINEREOUS HELVELLA. EDIBLE.

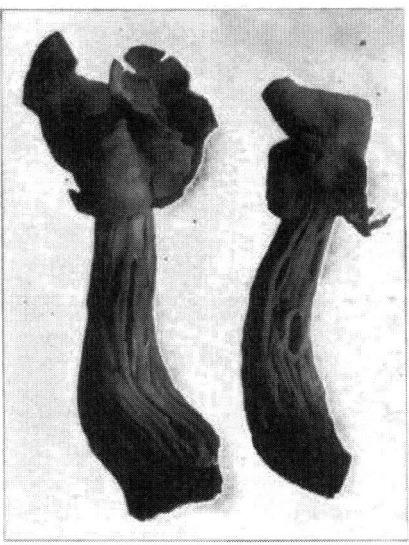

Figure 421   Helvella lacunosa.

Lacunosa, full of pits or pitted. This is a beautiful plant, very closely related to the Morchellas.

The pileus is inflated, lobed, cinereous black, lobes deflected, adnate.

The stem is hollow, white or dusky, exterior ribbed, forming intervening cavities.

The asci are cylindrical, and stemmed. The sporidia are ovate and hyaline.

The deep longitudinal grooves in the stem are characteristic of this species. The plants from which the halftone was made were collected near Sandusky and photographed by Dr. Kellerman. They grow in moist woods. I found the plants frequently in the woods near Bowling Green and occasionally about Chillicothe, growing about well-decayed stumps.

HYPOMYCES. TUL.

Hypomyces means upon a mushroom. It is parasitic on fungi. Mycelium byssoid; perithecia small; asci 8-spored.

HYPOMYCES LACTIFLUORUM. SCHW.

Figure 422 Hypomyces lactifluorum. The entire plant is a bright yellow. Natural size.

Lactifluorum means milk-flowing. It is parasitic on Lactarius, probably piperatus, as this species surrounded it. It seems to have the power to change the color into an orange-red mass, in many cases entirely obliterating the gills of the host-species, as will be seen in Figure 422.

The asci are long and slender. The sporidia are in one row, spindle-shaped, straight or slightly curved, rough, hyaline, uniseptic, cuspidate, pointed at the ends, $30–38 \times 6–8\mu$.

This very closely resembles Hypomyces aurantius, but the sporidia are larger, rough and warted and the felt-like mycelium at the base is wanting.

It occurs in various colors, orange, red, white, and purple. It is not plentiful, occurring only occasionally. Capt. McIlvaine says, "When it is well cooked in small pieces it is among the best." It is found from July to October.

LEPTOGLOSSUM LUTEUM. (PK.) SAC.

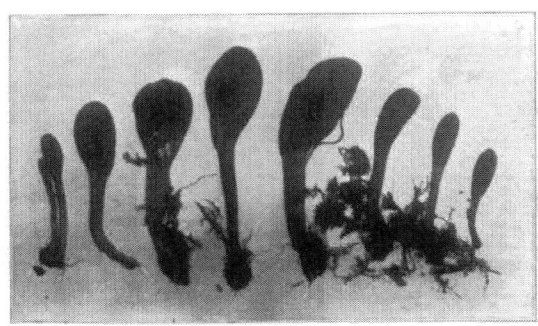

Figure 423   Leptoglossum luteum.

Leptoglossum is from two Greek words, meaning thin, delicate, and tongue; luteum means yellowish.

The club is distinct from the stem, smooth, compressed, generally with a groove on one side; luteous, often becoming brown at the tip or apex.

The stem is equal or slightly enlarged above, stuffed, luteous, minutely scaly.

The spores are oblong, slightly curved, in a double row, 1-1000 to 1-800 inch long. Peck.

These are found quite frequently among moss, or where an old log has rotted down, on the north hillsides about Chillicothe. The plants were first described by Dr. Peck as "Geoglossum luteum," but afterwards called by Saccardo "Leptoglossum luteum." The plants in Figure 423 were found in August or September, on Ralston's Run, near Chillicothe, and were photographed by Dr. Kellerman.

SPATHULARIA. PERS.

This is a very interesting genus, and one that will attract the attention of any one at first sight. It grows in the form of a spathula, from which it receives its generic name. The spore-body is flattened and grows down on both sides of the stem, tapering downward.

SPATHULARIA FLAVIDA. PERS.

THE YELLOW SPATHULARIA. EDIBLE.

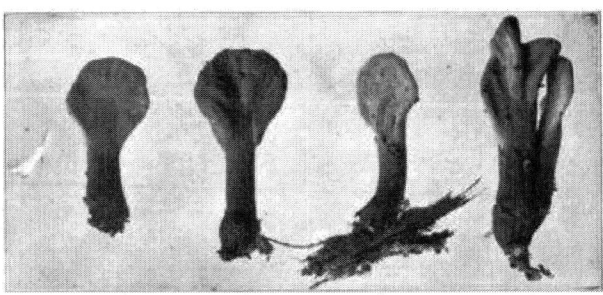

Figure 424   Spathularia flavida.

The spore body is a clear yellow, sometimes tinged with red, shaped like a spathula, the apex blunt, sometimes slightly cleft, the surface wavy, somewhat crisp, growing down the stem on opposite sides further than V. velutipes.

The stem is thick, hollow, white, then tinged with yellow, slightly compressed; asci clavate, apex somewhat pointed, 8-spored; spores arranged in parallel fascicles, hyaline, linear-clavate, usually very slightly bent, $50–60 \times 3.5–4\mu$; paraphyses filiform, septate, often branched, tips not thickened, wavy. While this is a beautiful plant it is not common. Found in August and September.

SPATHULARIA VELUTIPES. C. & F.

VELVET-FOOT SPATHULARIA. EDIBLE.

Velutipes is from velutum, velvet; pes, foot.

The spore body is flattened, shaped like a spathula, spore surface wavy, growing on the opposite sides of the upper part of the stem. tawny-yellow. The stem is hollow, minutely downy or velvety, dark brown tinged with yellow. It will dry quite as well as Morchella. It is found in damp woods on mossy logs. It is not a common plant. Found in August and September.

LEOTIA. HILL.

Receptacle pileate. Pileus orbicular, margin involute, free from the stem, smooth, hymenium covering upper surface.

The stem is hollow, central, rather long, continuous with pileus; the whole plant greenish-yellow.

Asci club-shaped, pointed, 8-spored. The spores are elliptical and hyaline. The paraphyses are present, usually slender and round.

LEOTIA LUBRICA. PERS.

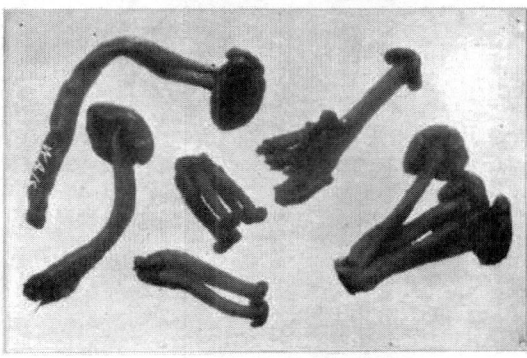

Figure 425   Leotia lubrica.

Lubrica means slippery; so called because the plants are usually slimy.

The pileus is irregularly hemispherical, somewhat wrinkled, inflated, wavy, margin obtuse, free from the stem, yellowish olive-green, tremelloid.

The stem is one to three inches long, nearly equal, hollow, and continuous with the cap; greenish-yellow, covered with small white granules.

The asci are cylindrical, slightly pointed at the apex, 8-spored. The spores are oblong, hyaline, smooth, sometimes slightly curved, 22–$25 \times 5\mu$. The paraphyses are slender, round, hyaline.

The plants are gregarious and grow among moss or among leaves in the woods. This species is quite plentiful about Chillicothe. It is distinguished from Leotia chlorocephala by the color of its stem and cap. The color of the latter is green or dark green. They are found from July to frost. They are edible but not choice.

LEOTIA CHLOROCEPHALA. SCHW.

Figure 426   Leotia chlorocephala.

Chlorocephala means green head. However, the entire plant is green.

They grow in clusters, pileus round, depressed, somewhat translucent, more or less waxy, margin incurved, dark-verdigris-green, sometimes rather dark-green.

The stem is rather short, almost equal; green, but often paler than the cap, covered with fine powdery dust, often twisted.

Asci cylindric-clavate, apex rather narrowed, 8-spored, spores smooth, hyaline, ends acute, often slightly curved, 17–$20 \times 5\mu$.

The specimens in Figure 426 were found in Purgatory Swamp, near Boston, by Mrs. Blackford. Both cap and stem were a deep verdigris-green. They were sent to me during the warm weather of August.

PEZIZA. LINN.

Peziza means stalkless mushroom. This is a large genus of discomycetous fungi in which the hymenium lines the cavity of a fleshy membranous or waxy cup. They are attached to the ground, decaying wood, or other substances, by the center, though sometimes they are distinctly stalked. They are often beautifully colored and are called fairy cups, blood cups, and cup fungi. They are all cup-or saucer-shaped; externally warted, scurvy or smooth; asci cylindrical, 8-spored. The genus is large. Prof. Peck reports 150 species. Found early in spring till early winter.

PEZIZA ACETABULUM. LINN.

RETICULATED PEZIZA. EDIBLE.

Acetabulum, a small cup or vinegar cup. The spore-bearing body stipitate, cup-shaped, dingy, ribbed externally with branching veins, which run up from the short, pitted and hollow stem; mouth somewhat contracted; light umber without and darker within. Found on the ground in the spring.

PEZIZA BADIA. PERS.

LARGE BROWN PEZIZA. EDIBLE.

Figure 427 Peziza badia.

Gregarious in its habits; sessile, or narrowed into a very short stout stem, more or less pitted; nearly round and closed at first, then expanded until cup-shaped; margin at first involute; externally covered with a frost-like bloom; disk darker than the external surface, very changeable in color; lobes more or less split and wavy, somewhat thick; spore-sacs cylindrical, apex truncante, sporidia oblong-ovate, epispore rough, 8-spored. Found on the ground in the grass or by the roadside in open woods. I found my first specimens in a clearing at Salem, but I have since found it at several points in the state. It should be fresh when eaten.

PEZIZA COCCINEA. JACQ.

THE CARMINE PEZIZA.

Coccinea means scarlet or crimson. Usually growing two or three on the same stick, the color is a very pure and beautiful scarlet, attractive to children; school children frequently bring me specimens, curious to know what they are. Specimens not large, disk clear and pure carmine within, externally white, as is the stem; tomentose, with short, adpressed down; sporidia oblong, 8-spored. It is readily recognized by the pure carmine disk and whitish tomentose exterior. It is found in damp woods on decayed sticks, being very common all over the state.

Figure 428   Peziza coccinea. One-third natural size.

PEZIZA ODORATA. PK.

THE ODOROUS PEZIZA. EDIBLE.

Gregarious in its habits. Cup yellowish, sessile, translucent, becoming dull brown when old, brittle when fresh, flesh moist and watery; the frame of the cup is separable into two layers; the outer one is rough, while the inner one is smooth. The disk is yellowish-brown. The asci are cylindrical, opening by a lid. On ground in cellars, about barns and outbuildings. A very beautiful cluster grew upon a water-bucket in my stable. The cups were quite large, two and a half to three inches across. Its odor is distinctive. It is very similar to Peziza Petersii from which it is distinguished by its larger spores and peculiar odor. Found in May and June.

PEZIZA STEVENSONI.

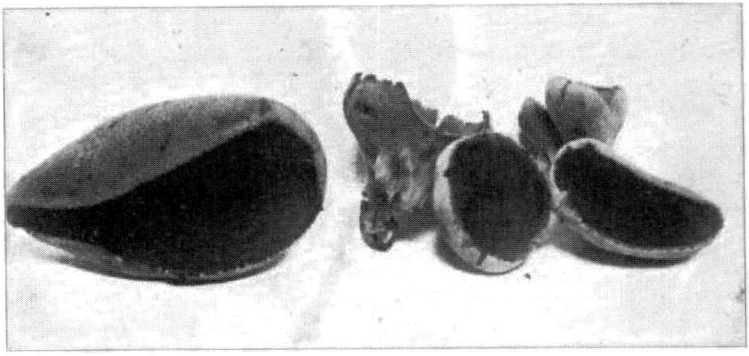

Figure 429   Peziza Stevensoni.

This plant is sessile or nearly so, growing on the ground in dense clusters. The specimens in Figure 429 grew in Dr. Chas. Miesse's cellar, in Chillicothe. They grow quite large at times; are ovate, externally grayish-white, covered with a minute down or tomentum, internally reddish-brown, the rim of the cup finely serrated, as will be seen in the figure below. They are found from May to July.

PEZIZA SEMITOSTA.

Figure 430 Peziza semitosta.

Semitosta, from its scorched appearance, or umber-like color.

The cup is one to one and a half inches across, hemispherical, hirsute-velvety without, date-brown within; margin indexed.

The stem is ribbed or wrinkled. Sporidia are subfusiform, .00117 inch long.

These plants are found on the ground in damp places. It was formerly called Peziza semitosta or Sarcoscypha semitosta. The plants in Figure 430 were found in August or September on the north side of the Edinger Hill, near Chillicothe, and were photographed by Dr. Kellerman. No doubt edible, but the writer has not tried them. This is called Macropodia semitosta.

PEZIZA AURANTIA. FR.

ORANGE-GROUND PEZIZA. EDIBLE.

Aurantia means orange color.

Subsessile, irregular, oblique, externally somewhat pruinose, whitish. The sporidia are elliptic, rough.

Found on the ground in damp woods. The cups are often quite large and very irregular. Found in August and September.

PEZIZA REPANDA. WAHL.

Repanda means bent backward. These plants are found in dark moist woods, growing on old, wet logs, or in well wooded earth. The cups are clustered or scattered, subsessile, contracted into a short, stout, stem-like base. When very small they appear like a tiny white knot on the surface of the log. This grows, so that soon a hollow sphere with an opening at the top is produced. The plant now begins to expand and flatten, producing an irregular, flattened disk with small upturned edges. The margin often becomes split and wavy, sometimes drooping and revolute; disk pale or dark brown, more or less wrinkled toward the center; externally the cup is a scurvy-white. The asci are 8-spored, quite large. The paraphyses are few, short, separate, clavate, and brownish at the tips. The spores are elliptical, thin-walled, hyaline, non-nucleate, $14 \times 9\mu$.

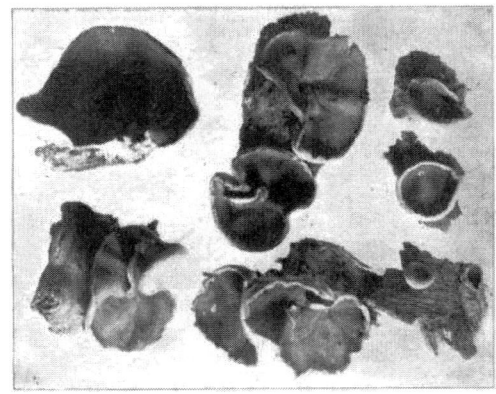

Figure 431   Peziza repanda.

Found from May to October. Edible.

PEZIZA VESICULOSA. BULL.

THE BLADDERY PEZIZA. EDIBLE.

Figure 432   Peziza vesiculosa.

Often in thick clusters. Those in the center are frequently distorted by mutual pressure; large, entire, sessile, at first globose; closed at first, then expanding; the margin of the cup more or less incurved, sometimes slightly notched; disk pallid-brown, externally; surface is covered with a coarsely granular or warty substance which plainly shows in the photograph. The hymenium is generally separable from the substance of the cap. The spores are smooth, transparent, continuous, elliptical, ends obtuse.

They are found on dung-hills, hot-beds or wherever the ground has been strongly fertilized and contains the necessary moisture. This is an interesting plant and often found in large numbers. Vesicolosa means full of bladders, as the picture will suggest.

I found a very nice cluster on the 25th of April, 1904, in my stable.

PEZIZA SCUTELLATA. LINN.

THE SHIELD-LIKE PEZIZA.

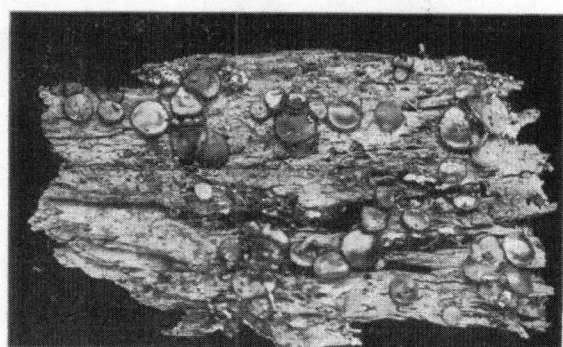

Figure 433   Peziza scutellata. Very small but will show form under the glass.

Becoming plane, vermillion-red, externally paler, hispid towards the margin with straight black hairs. Spores ellipsoid. Found on damp rotten logs from July to October. Very plentiful and very pretty under the magnifying glass.

PEZIZA TUBEROSA. BULL.

THE TUBEROUS PEZIZA.

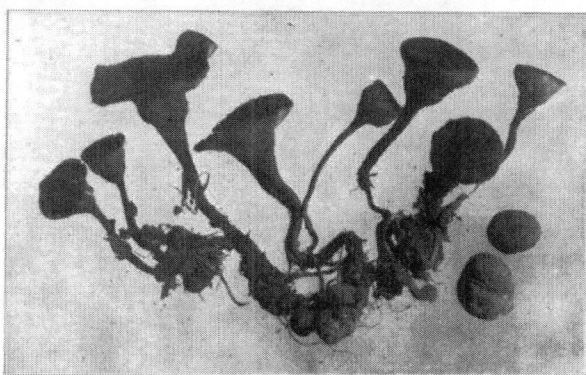

Figure 434   Peziza tuberosa. Natural size.

Tuberosa, furnished with a tuber or sclerotium. The cup is thin, infundibuliform, bright brown, turning pale.

The stem is elongated, springing from an irregular black tuber, called sclerotium. The stems run deep into the earth and are attached to a sclerotium, which will be seen in the halftone. Many fungus plants have learned to store up fungus starch for the new plant.

The sporidia are oblong-ellipsoid, simple. It is called by some authors Sclerotinia tuberosa. It grows on the ground in the spring and may be known by its bright brown color and its stem running deep into the earth and attached to a tuber.

PEZIZA HEMISPHERICA. WIGG.

Sessile, hemispherical, waxy, externally brownish, clothed with dense, fasciculate hairs; disk glaucous-white. This is called by Gillet Lachnea hemispherica. The cups are small, varying much in color and the sporidia are ellipsoidal. They are found on the ground in September and October. Found in Poke Hollow.

PEZIZA LEPORINA. BATSCH.

Substipitate, elongated on one side, ear-shaped, subferruginous externally, farinose internally; base even. It is sometimes cinereous or yellowish. Sporidia ellipsoidal. This is called frequently Otidea leporina, (Batsch.) Fckl. It is found on the ground in the woods during September and October. Found in Poke Hollow.

PEZIZA VENOSA. P.

This plant is saucer-shaped, sometimes many inches broad; sessile, somewhat twisted, dark umber, white beneath, wrinkled with rib-like veins. Odor often strong. Found growing on the ground in leaf mold. Found in the spring, about the last of April, in James Dunlap's woods, near Chillicothe. This is also called Discina venosa, Suec.

PEZIZA FLOCCOSA. SCHW.

Figure 435   Peziza floccosa. Natural size.

This is a beautiful plant growing upon partially decayed logs. I have always found it upon hickory logs. The cap is cup-shaped, very much like a beaker. The stem is long and slender, rather woolly; the rim of the cap is fringed with long, strigose hairs. The inner surface of the cup represents the spore-bearing portion.

The inside and the rim of the cup are very beautiful, being variegated with deep scarlet and white. Also called Sarcoscypha floccosa.

The plant is found from June to September.

PEZIZA OCCIDENTALIS.

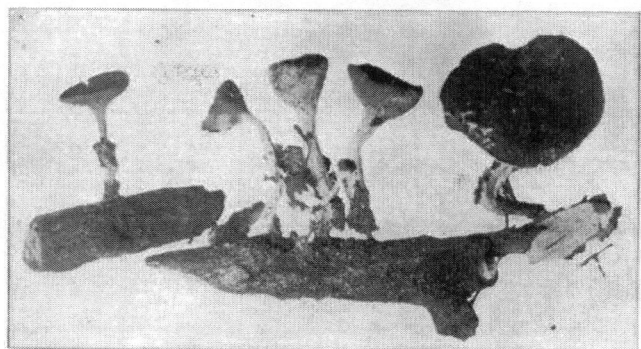

Figure 436   Peziza occidentalis. Natural size.

This is another very showy plant, quite equal in attractiveness to P. floccosa and P. coccinea.

The cup is infundibuliform, the outside as well as the stem whitish, and downy, the bowl or disk is reddish-orange. This is known by some authors as Sarcoscypha occidentalis. It grows on rotten sticks upon the ground. May and June.

PEZIZA NEBULOSA. COOKE.

Figure 437   Peziza nebulosa.

Nebulosa means cloudy or dark, from nebula, a cloud; from its color.

Ascophore stipitate, rather fleshy, closed at first, then cup-shaped, becoming somewhat plane, the margin slightly incurved, externally pilose or downy, pale gray or sometimes quite dark.

Asci are cylindrical; spores spindle-shaped, straight or bow-shaped, rough, 35–8; paraphyses thread-shaped.

These plants are found on decayed stumps or logs in the wood. The woods where I have found them have been rather dense and damp. The plants in Figure 437 were found in Haynes' Hollow and photographed by Dr. Kellerman.

URNULA CRATERIUM. (SCHW.) FR.

Urnula means burned; craterium means a small crater; hence the translation is a burned-out crater, which will appear to the student as a very appropriate name. It is a very common and conspicuous Ascomycetous, or cup fungus, growing in clusters on rotten sticks that lie in moist places. When the plants first appear they are small, black stems with scarcely any evidence of a cup. In a short time the end of the stem shows evidence of enlargement, showing lines of separation on the top. It soon opens and we have the cup as you see it in Figure 438. The hymenium, or spore bearing surface, is the interior wall of the cup. The cup is lined inside with a palisade of long cylindrical sacs, each containing eight spores with a small amount of liquid. These sacs are at right angles to the inner surface, and are provided with lids similar to that of a coffee-pot; at maturity the lid is forced open and the spores are shot out of these sacs, and, by jarring the fungus when it is ready to make the discharge, they can be seen as a little cloud an inch or two above the cup. Place a small slip of glass over the cup and you will see spores in groups of eight in very small drops of liquid on the glass. This species appears in April and May, and is certainly a very interesting plant. It is called by some Peziza craterium, Schw.

Figure 438　Urnula craterium. Two-thirds natural size.

HELOTIUM. FR.

Disc always open, at first punctiform, then dilated, convex or concave, naked. Excipulum waxy, free, marginate, externally naked.

HELOTIUM CITRINUM. FR.

LEMON-COLORED HELOTIUM.

Figure 439 Helotium citrinum. Disc-fungus, yellow growing on rotten logs. Slightly magnified.

This is a beautiful little Disc-fungus, yellow, growing upon rotten logs in damp woods. They often grow in dense clusters; a beautiful lemon-yellow, the head being plane or concave, with a short, thick, paler stem, forming an inverted cone. Asci elongated, narrowly cylindrical, attenuated at the base into a long, slender, crooked pedicel, 8-spored.

Sporidia oblong, elliptical, with two or three minute nuclei.

This is quite a common plant in our woods during wet weather or in damp places, growing upon old logs and stumps, in woods, in the fall. Figure 439 will give an idea of their appearance when in dense clusters. The plants photographed by Dr. Kellerman.

HELOTIUM LUTESCENS. FR.

YELLOWISH HELOTIUM.

Lutescens means yellowish. The plants are small, sessile, or attached by a very short stem; closed at first, then expanding until nearly plane; disk yellow, smooth; asci clavate, 8 spored; spores hyaline, smooth.

Gregarious or scattered. Found on half-decayed branches.

HELOTIUM ÆRUGINOSUM. FR.

THE GREEN HELOTIUM.

Æruginosum means verdigris-green. Gregarious or scattered, staining the wood on which they grow to a deep verdigris-green; ascophore at first turbinate and closed, then expanding, the margin usually wavy and more or less irregular; flexible, glabrous, even, somewhat contracted, and minutely wrinkled when dry; every part a deep verdigris-green, the disc often becoming paler with a tinge of tan color; 1–4 mm. across; stem 1–3 mm. long, expanding into the ascophore; hypothecium and excipulum formed of interlaced, hyaline hyphæ, 3–4μ. thick, these becoming stouter and colored green in the cortex; asci narrowly cylindric-clavate, apex slightly narrowed, 8-spored; spores irregularly 2-seriate, hyaline or with a slight tinge of green, very narrowly cylindric-fusiform, straight or curved, 10–14×2.5–3.5μ. 2-gutullate, or with several minute green oil globules; paraphyses slender, with a tinge of green at the tip. Massee.

Massee calls this Chlorosplenium æruginosum, De Not. It is quite common on oak branches, staining to a deep green the wood upon which it grows. It is widely distributed, specimens having been sent me from as far east as Massachusetts. The mycelium-stains in the wood are met more frequently than the fruit.

BULGARIA. FR.

Bulgaria—probably first found in that principality.

Receptacle orbicular, then truncate, glutinous within, at first closed; hymenium even, persistent, smooth.

BULGARIA INQUINANS. FR.

The Blackish Bulgaria.

Figure 440   Bulgaria inquinans. Two-thirds natural size.

Inquinans means befouling or polluting; so called because of the blackish, gelatinous coating of the cap.

Receptacle orbicular, closed at first, then opening, forming a cup, as shown on the right in Figure 440; disk or cup becoming plane; black, sometimes becoming lacunose; tough, elastic, gelatinous, dark-brown, or chocolate, almost black, wrinkled, and rough externally; stem very short, almost obsolete; cup light umber; sporidia large, elliptical, brown.

This plant is quite plentiful in some localities near Chillicothe. It is found in woods, on oak trunks or limbs partially decayed.

CHAPTER XIII

NIDULARIACEAE—BIRD'S NEST FUNGI

Spores produced on sporophores, compacted into one or more globose or disciform bodies, contained within a distinct peridium. Berkeley.

There are four genera included in this order.

- Cyathus—Peridium cup-shaped, composed of three different membranes.
- Crucibulum—Peridium of a uniform spongy membrane.
- Nidularia—Peridium globose, sporangia enveloped in mucus.
- Sphærobolus—Peridium double, sporangia ejected singly.

CYATHUS. PERS.

Cyathus is from a Greek word meaning a cup.

The peridium is composed of three membranes very closely related, closed at first by a white membrane, but finally bursting at the top. Sporangia plane, umbilicate, attached to the wall by an elastic cord.

CYATHUS STRIATUS. HOFFM.

STRIATE CYATHUS.

Figure 441   Cyathus striatus.

The plants are small, obconic, truncate, broadly open; externally ferruginous, with a hairy tomentum, internally lead-color, smooth, striated.

The sporangia are somewhat trigonous, whitish, broadly umbilicate; covering of the cup thin, evanescent, somewhat thicker underneath, and cottony, often covered with down-like meal.

The spores are thick and oblong.

This is a very interesting little plant. It is quite widely distributed. I have had it from several states, including New England. It is easily identified by the striations, or lines, on the inside of the cup, being the only species thus marked by internal striæ. The peridioles of the species fill only the lower part of the cup, below the striations.

CYATHUS VERNICOSUS. D. C.

VARNISHED CYATHUS.

Figure 442   Cyathus vernicosus.

Vernicosus means varnished. It is bell-shaped, base narrowly subsessile, broadly open above, somewhat wavy; externally rusty-brown, silky tomentose, finally becoming smooth, internally lead-colored.

The sporangia are blackish, frequently somewhat pale, even; covering rather thick, sprinkled with a grayish meal. Spores elliptical, colorless, $12–14 \times 10\mu$. I have frequently seen the ground in gardens and stubble-fields covered with these beautiful little plants. The quite firm, thick, and flaring cup will easily distinguish the species. The eggs or peridioles are black and quite large, appearing white because covered with a thin white membrane. Found in late summer and fall. The plants in Figure 442 were photographed by Prof. G. D. Smith.

CYATHUS STERCOREUS.

Figure 443   Cyathus stercoreus.

Stercoreus is from stercus, dung. This species, as the name suggests, is found on manure or manured grounds. Mr. Lloyd gives the following description: "The cups are even inside, and with shaggy hairs outside. When old they become smoother, and are sometimes mistaken for Cyathus vernicosus. However when once learned, the plants can be readily distinguished by the cups. Cyathus stercoreus varies considerably, however, as to shape and size of cups, according to habitat. If growing on a cake of manure, they are shorter and more cylindrical; if in loose manured ground, especially in grass, they are more slender and inclined to a stalk at the base." The peridioles or eggs are blacker than other species. They are found in late summer and fall.

CRUCIBULUM. TUL.

The peridium consists of a uniform, spongy, fibrous felt, closed by a flat scale-like covering of the same color.

The sporangia are plane, attached by a cord, springing from a small nipple-like tubercle.

This genus is distinguished from Cyathus, its nearest ally, by the peridial wall, consisting of two layers only.

CRUCIBULUM VULGARE. TUL.

Figure 444   Crucibulum vulgare.

The peridium is tan-colored, thick externally nearly even, internally quite even, smooth, shining; mouths of young plants are covered with a thin yellowish membrane called the epiphragm. When old the cups bleach out and lose their yellow color. The peridioles or eggs are white, that is they are covered with a white membrane. Their yellowish color and white eggs will readily distinguish this species.

They are found on decayed weeds, sticks, and pieces of wood. The specimens in the halftone grew on an old mat and were photographed by Mr. C. G. Lloyd.

NIDULARIA. TUL.

The peridium is uniform, consisting of a single membrane; globose, at first closed, finally ruptured or opening with a circular mouth.

The sporangia are quite small and numerous, not attached by a funiculus to the peridium, enveloped in mucus.

NIDULARIA PISIFORMIS. TUL.

PEA-SHAPED NIDULARIA.

Figure 445 Nidularia pisiformis.

Pisiformis is from two Latin words meaning pea and form.

The plant is gregarious, nearly round, sessile, rootless, hairy, brown or brownish, splitting irregularly.

The sporangia are subrotund or discoidal in form, dark brown, smooth, shining.

The spores are colorless, round or elliptical or pear-shaped, produced on sterigmata, $7–8 \times 8–9\mu$. Sometimes found on the ground and on leaves, but their favorite home is an old log. Found from July to September.

CHAPTER XIV

SUB-CLASS BASIDIOMYCETES

GROUP GASTROMYCETES

Gastromycetes is from two Greek words: gaster, stomach; mycetes, fungus. We have already seen that, in the group, Hymenomycetes, the spore-bearing surface is exposed as in the common mushroom or in the pore-bearing varieties, but in the Gastromycetes the hymenium is inclosed in the rind or peridium. The word peridium comes from peridio (I wrap around); because the peridium entirely envelops the spore-bearing portion, which, in due time, sheds the inclosed spores that have been formed inside the basidia and spicules, as will be seen in Figure 2. The cavity within the peridium consists of two parts: the threaded part, called the capillitium, which can be seen in any dried puffball, and a cellular part, called the gleba, which is the spore-bearing tissue, composed of minute chambers lined with the hymenium. The peridium breaks in various ways to permit the spores to escape. When children pinch a puffball to "see the smoke," as they say, issue from it, little do they know that they are doing just what the puffball would have them do, in order that its seeds may be scattered to the winds.

In case of the Phalloides, the hymenium deliquesces, instead of drying up.

Berkeley, in his "Outlines," gives the following characterization of this family: "Hymenium more or less permanently concealed, consisting in most cases of closely packed cells, of which the fertile ones bear naked spores in distinct spicules, exposed only by the rupture or decay of the investing coat or peridium."

The following families will be treated here:

- ◉ Phalloideæ—Terrestrial. Hymenium deliquescent.
- ◉ Lycoperdaceæ—Cellular at first. Hymenium drying up in a mass of threads and spores.
- ◉ Sclerodermaceæ—Peridium inclosing sporangia.

PHALLOIDEÆ. FR.

Volva universal, the intermediate stratum gelatinous. Hymenium deliquescent. Berkeley's Outlines.

The following genera will be represented:

⊙ Phallus—Pileus free around the stem.

⊙ Mutinus—Pileus attached to the stem.

PHALLUS DUPLICATUS. BOSC

LACED STINKHORN

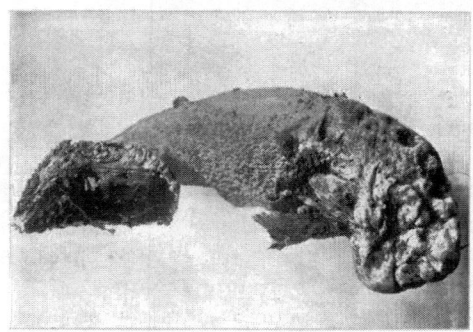

Plate LV. Figure 446 Phallus duplicatus.

Natural size, showing veil.

Volva egg-shaped, thick, whitish, frequently having a pinkish tinge.

The stem is cylindrical, cellulose, tapering upward. The veil is reticulate, frequently surrounding the whole of the stem from the pileus to the volva, often torn. The pileus is pitted, deliquescent, six to eight inches high, apex acute. Spores elliptic-oblong.

I am sure I never saw finer lace-work than I have seen on this plant. A few years ago one of these plants insisted upon growing near my house, where a fence post had formerly been, with the effect of almost driving the family from home. One can hardly imagine so beautiful a plant giving off such an odor. It is not a common plant in our state.

PHALLUS RAVENELII. B. & C

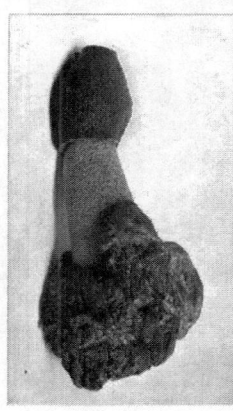

Figure 447 Phallus Ravenelii. Natural size, showing volva at base, receptacle and cap.

This plant is extremely abundant about Chillicothe. I have seen hundreds of fully developed plants on a few square yards of old sawdust; and one might easily think that all the bad smells in the world had been turned loose at that place. The eggs in the sawdust can be gathered by the bushel. In Figure 449 is represented a cluster, of these eggs. The section of an egg in the center of the cluster shows the outline of the volva, the pileus, and the embryo stem. Inside of the volva, in the middle, is the short undeveloped stem; covering the upper part and sides of the stem is the pileus; the fruit-bearing part, which is divided into small chambers, lies on the outside of the pileus. The spores are borne on club-shaped basidia as shown in Figure 448, within the chamber of the fruit-bearing part, and when the spores mature, the stem begins to elongate and force the gleba and pileus through the volva, leaving it at the base of the stem, as will be seen in Figure 448. The large egg on the left in the background of Figure 449 is nearly ready to break the volva. I brought in a large egg one evening and placed it on the mantle. Later in the evening, the room being warm, while we were reading my wife noticed this egg beginning to move and it developed in a few minutes to the shape you see in Figure 447. The development was so rapid that the motion was very perceptible. The pileus is conical in shape, and after the disappearance of the gleba the surface of the pileus is merely granular. The plants are four to six inches high. The stem is hollow and tapers from the middle to each end. This plant is also known as Dictyophora Ravenelii, Burt.

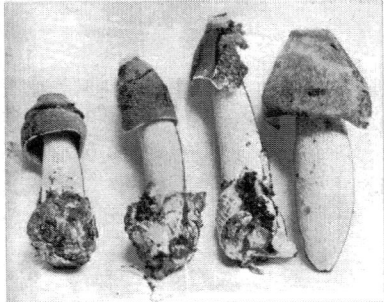

Figure 448   Phallus Ravenelii. Two-thirds natural size.

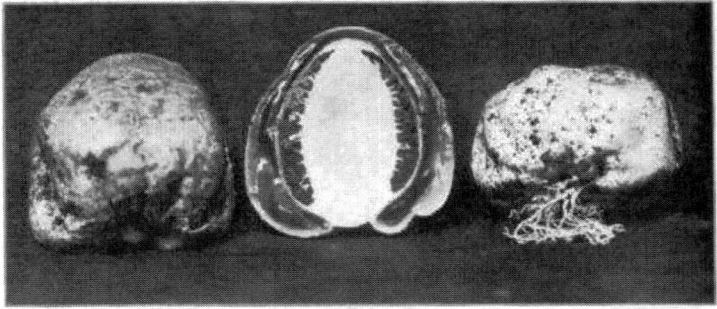

Figure 449   Phallus Ravenelii. Two-thirds natural size, showing the egg stage.

LYSURUS BOREALIS. BURT.

Figure 450   Lysurus borealis.

The receptacle is borne on a stalk, hollow, attenuated toward the base, divided above into arms, which do not join at their apices, and which bear the spore mass in their inner surfaces and sides, inclosing the spore mass when young, but later diverging.

The stem of the phalloid is white, hollow, attenuated downward; the arms are narrow, lance-shaped, with pale flesh-colored backs, traversed their entire length by a shallow furrow.

The egg in the center is about ready to break the volva and develop to a full grown plant. The plants in Figure 450 were found near Akron, Ohio, and photographed by G. D. Smith.

MUTINUS. FR.

The gleba is borne directly on the upper portion of the stem, which is hollow and composed of a single layer of tissue; and the plant has no separate pileus, by which characteristic the genus differs from Phallus.

MUTINUS CANINUS. FR.

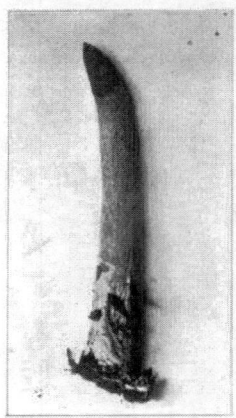

Figure 451   Mutinus caninus.

The gleba-bearing portion is short, red or flesh-colored, subacute, wrinkled, the cap or gleba forming the spore-bearing mass which is usually conical, some times oblong or ovoid, covering one-fourth to one-sixth the total length of the stem.

The stem is elongated, spindle-shaped, hollow, cylindrical, cellular, white, sometimes rosy. The spores are elliptical, involved in a green mucus, $6 \times 4\mu$. The plant comes from an egg, which is about the size of a quail's egg. You can find them in the ground if you will mark the place where you have seen them growing. They are found in gardens and in old woods and thickets. I have found this species in several localities about Chillicothe, but always in damp thickets. Mr. Lloyd thought this more nearly resembled the European species than any he had seen in this country. Found in July, August, and September.

MUTINUS ELEGANS. MONTAGNE.

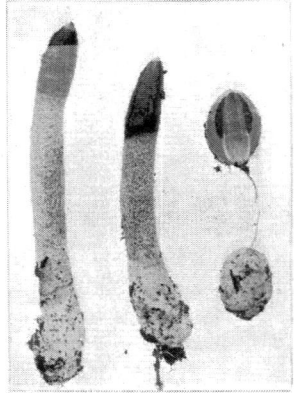

Plate LVI. Figure 452   Mutinus elegans.

Natural size, showing an egg and a section of an egg.

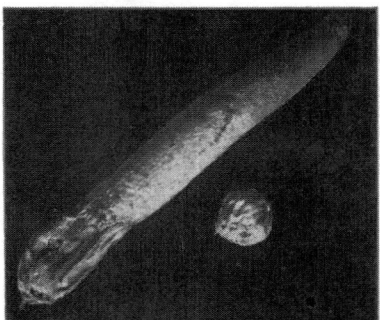

Figure 453 Mutinus elegans. One-third natural size, showing volva, white receptacle and red cap.

The pileus is acuminate, perforated at apex. The stem is cylindrical, tapering gradually to the apex, whitish or pinkish below, pileus bright red.

The volva is oblong-ovoid, pinkish, segments two or three. The spores are elliptical-oblong. Morgan.

The odor of this plant is not as strong as in some of the Phalloids. The eggs of Phallus and Mutinus are said to be very good when fried properly, but my recollection of the odor of the plant has been too vivid for me to try them. It is usually found in mixed woods, but sometimes in richly cultivated fields. I have found them frequently about Chillicothe six to seven inches high. In Figure 452 on the right is shown an egg and above it is a section of an egg containing the embryonic plant. This plant is called by Prof. Morgan Mutinus bovinus. After seeing this picture the collector will not fail to recognize it. It is one of the curious growths in nature. Found in July and August.

CHAPTER XV

LYCOPERDACEAE—PUFF-BALLS

This family includes all fungi which have their spores in closed chambers until maturity. The chambers are called the gleba and this is surrounded by the peridium or rind, which in different puffballs exhibits various characteristic ways of opening to let the spores escape. The peridium is composed of two distinct layers, one called the cortex, the other the peridium proper. The plant is generally sessile, sometimes more or less stemmed, at maturity filled with a dusty mass of spores and thread.

It affords many of our most delicious fungus food products. The following genera are considered here:

- ◎ Calvatia—The large puffball.
- ◎ Lycoperdon—The small puffball.
- ◎ Bovista—The tumbling puffball.
- ◎ Geaster—Earth Star.
- ◎ Scleroderma—The hard puffball.

CALVATIA. FR.

This genus represents the largest sized puffballs. They have a thick cord-like mycelium rooting from the base. The peridium is very large, breaking away in fragments when ripe and exposing the gleba. The cortex is thin, adherent, often soft and smooth like kid leather, sometimes covered with minute squamules; the inner peridium is thin and fragile, at maturity cracking into areas. The capillitium is a net-work of fine threads through the tissues of spore-bearing portion; tissue, snow white at first, turning greenish-yellow, then brown; the mass of spores and the dense net-work of threads (capillitium) attached to the peridium and to the subgleba or sterile base which is cellulose; limited and concave above. Spores small, round, usually sessile.

CALVATIA GIGANTEA. BATSCH.

THE GIANT PUFFBALL. EDIBLE.

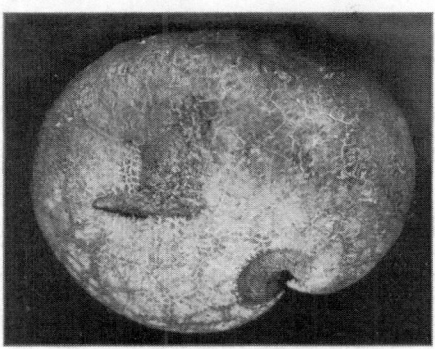

Plate LVII. Figure 454 Calvatia gigantea.

This species grows to an immense size (often twenty inches in diameter); round or obovoid, with a thick mycelial cord rooting it to the ground, sessile, cortex white and glossy, sometimes slightly roughened by minute floccose warts, becoming yellowish or brown. The inner peridium is thin and fragile, after maturity breaking up into fragments, apparently without any subgleba; capillitium and spores yellowish-green to dingy-olive. The spores are round, sometimes minutely warted.

Not common about Chillicothe, but in the northwestern part of the state they are very plentiful in their season, and very large. Standing in Mr. Joseph's wood-pasture, east of Bowling Green, I have counted fifteen giant puffballs whose diameters would average ten inches, and whose cortex was as white and glossy as a new kid glove. A friend of mine, living in Bowling Green, and driving home from Deshler, saw in a wood-pasture twenty-five of these giant puffballs. Being impressed with the sight and having some grain sacks in his wagon he filled them and brought them home. He at once telephoned for me to come to his house, as the mountain was too big to take to Mohammed. He was surprised to learn that he had found that proverbial calf which is all sweet-breads. That evening we supplied twenty-five families with slices of these puffballs.

They can be kept for two or three days on ice. The photograph, taken by Prof. Shaffner of Ohio State University, will show how they look growing in the grass. They seem to delight to nestle in the tall bluegrass. This species has been classed heretofore as Lycoperdon giganteum. Found from August to October.

Figure 455 Calvatia gigantia. One-fifth natural size, showing how they grow in the grass.

CALVATIA LILACINA. BERK.

LILAC PUFFBALL. EDIBLE.

Plate LVIII. Figure 456   Calvatia lilacina.

Natural size in a growing state.

The peridium is three to six inches in diameter; globose or depressed globose; smooth or minutely floccose or scaly; whitish, cinereous-brown or pinkish-brown, often cracking into areas in the upper part; commonly with a short, thick, stemless base; capillitium and spores purple-brown, these and the upper part of the peridium falling away and disappearing when old, leaving a cup-shaped base with a ragged margin. Spores globose, rough, purple-brown, 5–6.5 broad. Peck, 48th Rep. N. Y. State Bot.

It is very common all over the state. I have seen pastures in Shelby and Defiance counties dotted all over with this species. When the inside is white, they are very good and meaty. No puffball is poisonous, so far as is known, but if the inside has turned yellowish at all it is apt to be quite bitter. It will often be seen in pastures and open woods in the form of a cup, the upper portion having broken away and the wind having scooped out the purple spore-mass, leaving only the cup-shaped base. The specimens in Figure 457 are just beginning to crack open and to show purplish stains. They represent less than one-fourth of the natural size. They look very much like the smaller sized C. gigantea, but the purple spores and the subgleba at once distinguish the species. This species, found from July to October, is sometimes classed as Lycoperdon cyathiforme. The photograph was taken by Prof. Longyear.

Figure 457   Calvatia lilacina.

CALVATIA CÆLATA. BULL.

THE CARVED PUFFBALL. EDIBLE.

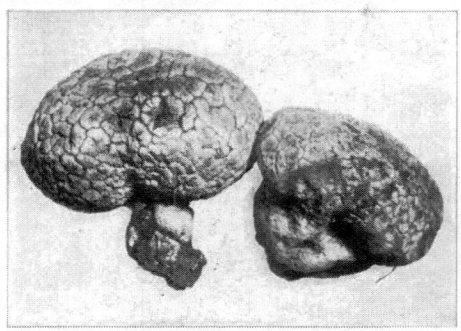

Plate LIX. Figure 458 Calvatia caelata.

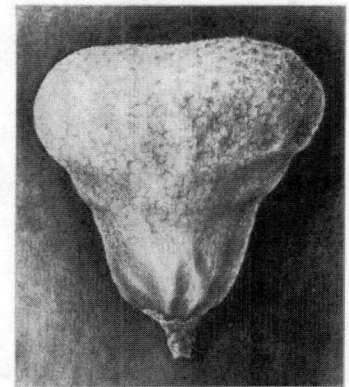

Figure 459 Calvatia cælata.

Cælata, carved. Peridium large, obovoid or top-shaped, depressed above, with a stout thick base and a cord-like root. Cortex a thickish floccose layer, with coarse warts or spines above, whitish then ochraceous or finally brown, at length breaking up into areola which are more or less persistent; inner peridium thick but fragile, thinner about the apex, where it finally ruptures, forming a large, irregular, torn opening. Subgleba occupying nearly half the peridium, cup-shaped above and for a long time persistent; the mass of spores and capillitium compact, farinaceous greenish-yellow or olivaceous, becoming pale to dark-brown; the threads are very much branched, the primary branches two or three times as thick as the spores, very brittle, soon breaking up into fragments. Spores globose, even, 4–4.5 in diameter, sessile or sometimes with a short or minute pedicel. Peridium is three to five inches in diameter. Morgan.

This species is much like the preceding but can be easily distinguished by the larger size and the yellowish-olive color of the mature spore-mass. The sterile base is often the larger part of the fungus and, as will be seen in Figure 459, it is anchored by a heavy root-like growth. It is found growing on the ground in fields and thin woods. When white through and through, sliced, rolled in egg and cracker crumbs, and nicely fried, you are glad you know a puffball. Found from August to October.

CALVATIA CRANIIFORMIS. SCHW.

THE BRAIN-SHAPED CALVATIA. EDIBLE.

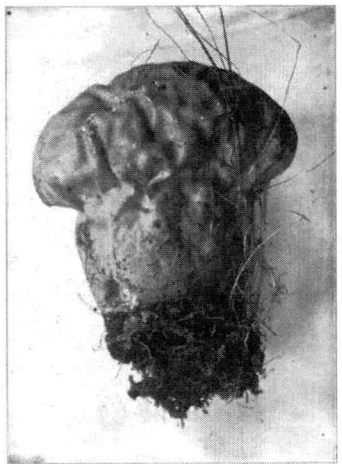

Plate LX. Figure 460   Calvatia craniiformis.

Figure 461   The sterile part of C. craniiformis.

Craniiformis is from Cranion, a skull; forma, a form.

The peridium is very large, obovoid or top-shaped, depressed above, the base thick and stout, with a cord-like root. The cortex is a smooth continuous layer, very thin and fragile, easily peeling off, pallid or grayish, sometimes with a reddish tinge, often becoming folded in areas; the inner peridium is thin, ochraceous to bright-brown, extremely fragile, the upper part, after maturity, breaking into fragments and falling away.

The subgleba occupies about one-half of the peridium, is cup-shaped above and for a long time persistent; the mass of spores and capillitium is greenish-yellow, then ochraceous or dirty olivaceous; the threads are very long, about as thick as the spores, branched. The spores are globose, even, 3–3.5μ in diameter, with minute pedicels. Morgan.

It is difficult to distinguish this from C. lilacina when fresh, but when ripe the color will tell the species. Figure 460 shows the plant as it appears on the ground, and figure 461 shows the subgleba or sterile base, which is frequently found on the ground after weathering the winter. This plant is very common on the hillsides under small oak shrubbery. I have gathered a basketful within a few feet. They grow very large, often five to six inches in diameter, seeming to delight in rather poor soil. When the spore-mass is white this is an excellent fungus, but exceedingly bitter after it has turned yellow. Found during October and November.

CALVATIA ELATA. MASSEE.

THE STEMMED CALVATIA. EDIBLE.

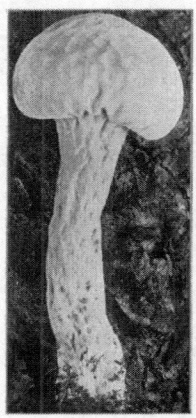

Figure 462   Calvatia elata.

Elata means tall; so called from its long stem.

The peridium is round, often slightly depressed above, plicate below, where it is abruptly contracted into a long stem-like base. The base is slender, round, and frequently pitted; mycelium rather plentiful, fibrous and thread-like. When in good condition it is a rich cream color. The cortex consists of a coat of minute per sistent granules or spinules. The inner peridium is white or cream-colored, becoming brown or olivaceous, quite thin and fragile, the upper part at maturity breaking up and falling away. The subgleba occupies the stem. The mass of spores and capillitium is usually brown or greenish-brown. The threads are very long, branched, branches slender. Spores round, even, sometimes slightly warted, 4–5μ, with a slight pedicel.

The plant grows on low mossy grounds among bushes, especially where it is inclined to be swampy. The plant in Figure 462 was found in a sphagnum swamp near Akron and was photographed by Prof. G. D. Smith. I am inclined to think it the same as Calvatia saccata, Fr.

LYCOPERDON. TOURN.

Mycelium fibrous, rooting from the base. Peridium small, globose, obovoid or turbinate, with a more or less thickened base; cortex a subpersistent coat of soft spines, scales, warts or granules; inner peridium thin, membranaceous, becoming papyraceous, dehiscent by a regular apical mouth. Morgan.

This genus includes puffballs with apical openings and is divided into two series, a purple-spored and an olive-spored series. The microscope shows that the gleba is composed of a great number of spores mixed with simple or branched threads. There are two sets of threads; one set arises from the peridial wall and the other from the subgleba or columella.

PURPLE-SPORED SERIES

LYCOPERDON PULCHERRIMUM. B. & C.

THE MOST BEAUTIFUL PUFFBALL. EDIBLE.

Figure 463   Lycoperdon pulcherrimum.

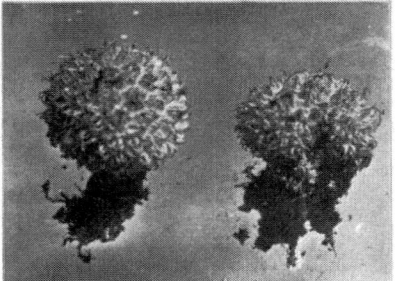

Figure 464   Lycoperdon pulcherrimum.

Pulcherrimum, most beautiful. The peridium is obovoid, with a short base, the mycelium forming a cord like a root. The cortex is covered with long white spines, converging at the apex, as will be seen in Figure 463. The spines soon fall from the upper part of the peridium, leaving the inner peridium with a smooth purplish-brown surface, often slightly scarred by the base of the spine. The subgleba occupies at least a third of the peridium. The spores and the capillitium are at first olivaceous, then brownish-purple, the spores rough and minutely warted. The plant is one to two inches in diameter. It is found in low, rich ground, in fields and wood margins. Only young and fresh plants are good.

The lower plant in Figure 463 shows where the spines have begun to fall, also the strong mycelial cord referred to in the description. I am indebted to Mr. Lloyd for the photograph. Found in September and October.

LYCOPERDON UMBRINUM. PERS.

THE SMOOTH PUFFBALL. EDIBLE.

Umbrinum, dingy umber. Peridium obovate, nearly sub-turbinate, with a soft, delicate, velvety bark; yellowish; inner peridium smooth and glossy, opening by a small aperture. The spores and capillitium, olivaceous, then purplish-brown. The capillitium with a central columella. A very attractive little plant, not frequently found. This plant is also called L. glabellum. In woods, September and October.

OLIVE-SPORED SERIES

LYCOPERDON GEMMATUM. BATSCH

THE GEMMED PUFFBALL. EDIBLE.

Plate LXI. Figure 465   Lycoperdon gemmatum.

Natural size. Entirely white when young. From the young to the matured dehiscing plant.

The peridium is turbinate, depressed above; the base short and obconic, or more elongated and tapering, or subcylindric, arising from a fibrous mycelium. The cortex consists of long, thick, erect spines or warts of irregular shape, with intervening smaller ones, whitish or gray in color, sometimes with a tinge of red or brown; the larger spines first fall away, leaving pale spots on the surface, and giving it a reticulate appearance. The subgleba is variable in amount, usually more than half the peridium; mass of spores and capillitium greenish-yellow, then pale-brown; threads simple or scarcely branched, about as thick as the spores. Spores globose, even, or very minutely warted. Morgan.

The species is readily recognized by the large erect spines which, because of their peculiar form and color, have given the notion of gems, whence the name of the species. These and the reticulations can be seen in Figure 465 by the aid of a glass. They are frequently found about Chillicothe.

LYCOPERDON SUBINCARNATUM. PK.

THE PINKISH PUFFBALL. EDIBLE.

Figure 466   Lycoperdon subincarnatum.

Subincarnatum means pale flesh-color. The peridium is globe-shaped, sessile, without a stem-like base. Not large, rarely over one inch in diameter. The subgleba is present but small. The outer peridium is pinkish-brown, with minute short, stout spinules, which fall away at maturity, leaving the inner ash-colored peridium neatly pitted by the falling off of the spinules of the outer coat, the pits not being surrounded by dotted lines. The capillitium and spores are first greenish-yellow, then brownish-olive. The threads are long, simple, and transparent. The columella is present and the spores are round and minutely warted.

They are often found in abundance on decayed logs, old stumps, and on the ground about stumps where the ground is especially full of decayed wood. They are found from August to October.

LYCOPERDON CRUCIATUM. ROTH.

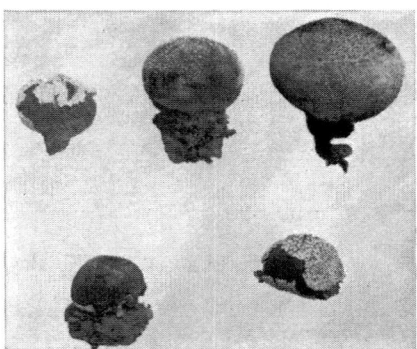

Figure 467   Lycoperdon cruciatum.

Peridium broadly ovate, often much depressed, plicate underneath, with a cord-like root; cortex a dense white coat of convergent spines, which at maturity peel off in flakes, as can be seen in the photograph, revealing a thin furfuraceous layer of minute yellowish scales covering the inner peridium. The subgleba broad, occupying about one-third of the cavity. The spores and capillitium are dark- brown. This species is very hard to distinguish from Wrightii. It was once called separans because of the fact that the outer coat separates, or peels off, so readily from the inner peridium. Found in open woods, or along paths in open woods or pastures. From July to October.

LYCOPERDON WRIGHTII. B. & C.

EDIBLE.

Figure 468   Lycoperdon Wrightii. Natural size.

The specific name is in honor of Charles Wright. The peridium is globe-like, sessile, white, minutely spinulose, often converging at the apex; when denuded, smooth or minutely velvety.

The spores and capillitium greenish-yellow, then brown-olive; the columella present, but very small. Spores small, smooth, 3–4μ.

The plants are very small, scarcely more than two cm. in diameter. They are generally cæspitose in short grass, along paths, and in sandy places.

I have frequently seen the ground white with them on Cemetery Hill where the specimens in Figure 468 were found. They were photographed by Dr. Kellerman. Found from July to the last of October.

LYCOPERDON PYRIFORME. SCHAEFF.

THE PEAR-SHAPED PUFFBALL. EDIBLE.

Plate LXII. Figure 469   Lycoperdon pyriforme.
Natural size when young as seen growing on decayed wood. The sections show
they are in the edible state.

Pyriforme means pear-shaped. The peridium is ovate or pear-shaped, with a profusion of mycelial threads, as will be seen in Figure 470.

The cortex is covered with a thin coat of minute brownish scales or granules, which are quite persistent. These can be seen in the photograph by the aid of a glass. They are sessile or have a short stem-like base; the subgleba is small and compact; the capillitium and spores are first white, then greenish-yellow, then dingy olivaceous; the inner coat is smooth, papery, whitish-gray or brownish, opening by an apical mouth; the spores are round, even, greenish-yellow to brownish-olive.

They grow in dense clusters, as will be seen in Figure 470. An entire log and stump, about four feet high, and the roots around it, were covered, as shown in Plate LXII. I gathered about three pecks, at this one place, to divide with my friends. It is one of the most common puffballs, and you may usually be sure of getting some, if you go into the woods where there are decayed logs and stumps. A friend of mine, who goes hunting with me occasionally, eats them as one would eat cherries. Found from July to November.

Figure 470   Lycoperdon pyriforme. Natural size.

LYCOPERDON PUSILLUM. PR.

THE SMALL LYCOPERDON. EDIBLE.

Pusillum means small.

Peridium is one-fourth to one inch broad, globose, scattered or cespitose, sessile, radicating, with but little cellular tissue at the base, white, or whitish, brownish when old, rimose-squamulose or slightly roughened with minute floccose or furfuraceous persistent warts; capillitium and spores greenish-yellow, then dingy olivaceous. Spores smooth 4μ in diameter. Peck.

These are found from June to cool weather in the fall, in pastures where the grass is eaten short. When mature they dehisce by a small opening, and when broken open will disclose the olive or greenish-yellow capillitium. The spores are of the same color, smooth and round.

LYCOPERDON ACUMINATUM. BOSC.

THE POINTED LYCOPERDON. EDIBLE.

Acuminatum means pointed.

The peridium is small, round, then egg-shaped; with a plentiful mass of mycelium in the moss in which the plants seem to delight. The plant is white and the outer rind is soft and delicate. There is no subgleba; the spores and capillitium are pale-greenish-yellow, then a dirty gray. The threads are simple, transparent, much thicker than the spores. The spores are round, smooth, 3μ in diameter.

I have found the plants frequently about Chillicothe on damp, moss-covered logs and sometimes at the base of beech trees, when covered with moss. They are very small, not exceeding one-half inch in diameter. The small ovoid form, with the white, soft, delicate cortex, will serve to distinguish the species. Found from September to October.

BOVISTA. DILL

The genus Bovista differs from Lycoperdon in several ways. When the Bovista ripens it breaks from its moorings and is blown about by the wind. It opens by an apical mouth, as does the genus Lycoperdon, but the species of Bovista have no sterile base. They are puffballs of small size. The outer coat is thin and fragile and at maturity peels off, leaving an inner coat firm, papery, and elastic, just such a coat as is suitable for the dispersion of its spores. Leaving its moorings at maturity, it is blown about the fields and woods, and with every tumble it makes it scatters some of its spores. It may take years to accomplish this perfectly. The species of the Lycoperdon do not leave their moorings naturally; their spores are dispersed through an apical mouth by a collapse of the walls of the peridium, after the fashion of a bellows, by which spores are driven out to the pleasure of the wind. In Bovista the threads are free or separate from the peridium, but in Lycoperdon they arise from the peridium and also from the columella.

BOVISTA PILA. B. & C

THE BALL-LIKE BOVISTA.

Plate LXIII. Figure 471   Bovista pila.
Natural size of matured specimens.

Pila means a round ball. The peridium is globe-like, sessile, with a stout mycelium, a cortex thin, white at first, then brown, forming a smooth continuous coat, breaking up at maturity and rapidly disappearing.

The inner peridium is tough, parchment-like, elastic, smooth, persistent, purplish-brown, fading to gray. The dispersion of spores takes place through an apical mouth. The capillitium is firm, compact, persistent, at first clay-colored, then purple-brown; threads small-branched, the ends being rigid, straight, pointed. There is something so noticeable about this little tumbler that you will know it when you see it, and if you often ramble over the fields you will soon meet it. However, I have as yet seen only the matured specimens.

BOVISTA PLUMBEA. PERS.

LEAD-COLORED BOVISTA. EDIBLE.

The plant is small, never growing to more than an inch and a fourth in diameter. The peridium is depressed globose, with a fibrous mycelium. The outer peridium is rather thick and when the plant is nearing maturity it breaks up readily unless handled very carefully; at maturity it scales off, except a small portion about the base. The outer peridium is white and comparatively smooth, the inner

is thin, tough, smooth, lead-colored, dehiscent at the apex by a round or oblong mouth. Mass of spores and capillitium not solid or hard; yellowish-brown, or olivaceous, then purplish-brown; the threads three to five times branched, the ends of the branches slender and tapering to a point. The spores are oval and smooth, with long transparent pedicels.

Figure 472   Bovista plumbea. Natural size. White when young.

This species grows on the ground in old pastures, being quite plentiful after warm rains, from the first of May till fall. It is one of the best of the puffballs, but should be eaten before the inner peridium begins to assume the tough form.

BOVISTELLA. MORGAN.

Bovistella, a diminutive of Bovista, though the plants are usually larger than the Bovistas.

The mycelium is cord-like; peridium nearly round, cortex a dense floccose coat; inner peridium thin, strong, elastic, opening by an apical mouth; subgleba present, cup-shaped; threads free and separate, branched; spores white. The genus Bovistella has the internal character of Bovista, and the habits of Lycoperdon.

BOVISTELLA OHIENSIS. MORGAN.

Peridium globe-like or broadly obovoid, sometimes much depressed, with small plications or wrinkles underneath, and a thick cord-like base or root, as will be seen in Figure 473. The outer coat is dense, floccose, or with soft warts or spines, white or grayish, drying to a buff color, and in time falling away; the inner coat is smooth, shining, with a pale brown or yellowish surface. The subgleba is large, occupying half of the peridium, extending up on the walls of the peridium, making it cup-shaped, and quite persistent. The spores and capillitium are rather loose, friable, clay-color to pale-brown. The threads, originating within the spore mass, and having no connection with the inner coat, are free, short, three to five times branching; branches tapering to the end. The spores are round to oval, with long translucent pedicels.

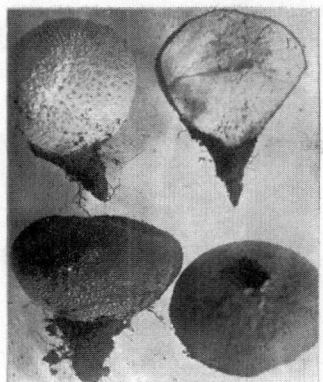

Figure 473   Bovistella Ohiensis. Natural size.

This can be readily distinguished from the species of Bovista because it has a sterile base; and from Lycoperdon because its threads are separate and free, while those of the Lycoperdon are attached both to the tissues of the inner peridium and to the columella or sterile base.

They are found growing on the ground in old pastures, or in open woods.

SCLERODERMA. PERS.

Scleroderma is from two Greek words: scleros, hard; derma, skin.

The peridium is firm, single, generally thick, usually bursting irregularly, and exposing the gleba, which is of uniform texture and consistency. There is no capillitium, but yellow flocci are found interspersed with the spores. The spores are globose, rough, usually mixed with the hyphæ tissue.

SCLERODERMA AURANTIUM. PERS.

THE COMMON SCLERODERMA. EDIBLE.

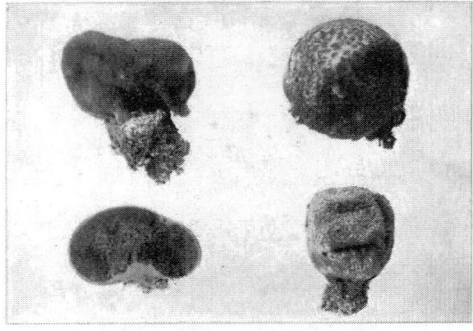

Plate LXIV. Figure 474   Scleroderma aurantium.
Natural size, showing a section of a young specimen.

Figure 475   Scleroderma aurantium.

Aurantium means colored like an orange. This is usually called S. vulgare. The peridium is rough, warty, depressed, globose, corky and hard, yellowish, opening by irregular fissures to scatter the spores; inner mass bluish-black, spores dingy. The plant remains solid until it is quite old. It is sessile, with a rooting base which is never sterile.

I have followed Mr. Lloyd's classification in separating the species, calling the rough-surfaced one S. aurantium, and the smooth-surfaced S. cepa.

In labeling it edible I wish only to indicate that it is not poisonous, as it is generally thought to be; however, it cannot be claimed as a very good article of food.

It has a wide distribution over the states. The plants in Figure 475 were found on Cemetery Hill, Chillicothe, and photographed by Dr. Kellerman. Found from August to November.

SCLERODERMA TENERUM. BERK.

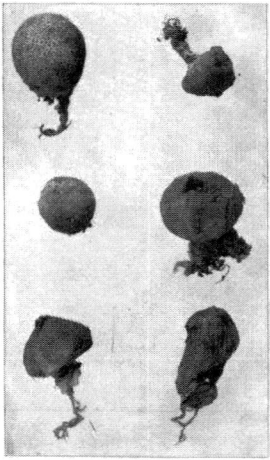

Figure 476   Scleroderma tenerum.

This species is often regarded as a small form of S. verrucosum, but it always seemed strange to me that this rather smooth plant should be called "verrucosum" when its frequently near neighbor, S. aurantium, is very verrucose.

S. tenerum is a very widely distributed species in the United States, somewhat constant as to form and quite frequent in occurrence. Mr. Lloyd, in his Mycological Notes, gives a very clear photograph of a plant that is quite local in this country and which he thinks should be called S. verrucosum of Europe.

The plant differs very widely from the one we find so commonly which by many authors has been called S. verrucosum. Some have even called it Scleroderma bovista.

The plant is nearly sessile, somewhat irregular, peridium thin, soft, yellowish, densely marked with small scales, dehiscence irregular, flocci yellow and spores dingy olive.

The species may be known by the thin and comparatively smooth peridium and yellow flocci. It is quite common in the United States, while the typical plant, S. verrucosum, is confined to a few localities along the Atlantic coast.

SCLERODERMA CEPA. PERS.

Cepa meaning an onion; having very much the appearance of an onion.

The peridium is thick, smooth, reddish-yellow to reddish-brown, opening by an irregular mouth. The plant is sessile and quite strongly rooted with fine rootlets. Its habitat, with us, is along the banks of small brooks in the woods. It has been classed heretofore as S. vulgare, smooth variety. I sent some to Prof. Peck, who quite agrees that they should be separated from S. vulgare. Found from August to November.

SCLERODERMA GEASTER. FR.

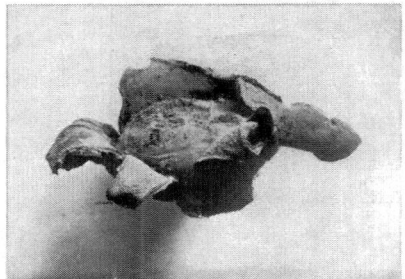

Plate LXV. Figure 477   Scleroderma geaster.

Geaster, so called because it has a star-like opening somewhat similar to the genus Geaster.

Peridium subglobose, thick, with a very short stem, or almost—sometimes entirely—sessile; hard, rough, splitting into irregular stellate limbs; frequently well buried in the ground. Inner mass dark-brown or blackish, sometimes with rather a purplish tinge. Some grow quite large with the peridium very thick. My attention was first attracted by some of the peridium shells upon the ground on Cemetery Hill. The plant is quite abundant there from September to December.

CATASTOMA. MORGAN

This is a small puffball-like plant, growing just beneath the ground and attached to its bed by very small threads which issue from every part of the cortex, which is quite thick. Breaking away at maturity in a circumscissile manner, the lower part is held fast to the ground, while the upper part remains attached to the inner peridium as a kind of cup. The inner peridium, with the top part of the outer peridium attached, becomes loose and tumbles over the ground, the mouth being in the base of the plant as it grew.

CATASTOMA CIRCUMSCISSUM. B. & C.

Figure 478 Catastoma circumscissum.

Circumscissum means divided into halves.

The peridium is usually round, more or less depressed, commonly rough because of the soil attached; the larger part of the plant remaining in the soil as a cup; the upper part with the inner peridium, depressed-globose, thin, pallid, becoming gray, with branny scales, with a small basal mouth. A thin spongy layer will frequently be seen between the outer and inner peridium. The mass of the spores is olivaceous, changing to pale-brown. The spores are round, minutely warted, 4–5μ. in diameter, often with very short pedicels.

The plants are usually found in pastures along paths. I have seen them in several parts of Ohio. They are found from Maine to the western mountains. This is called Bovista circumscissa by Berkeley.

There is a species of a western range called C. subterraneum. This differs mainly in having larger spores. It seems to be confined to the middle west. However, it does not grow under the ground, as its name would suggest.

There is also another species called C. pedicellatum. This species seems to be confined to the southern states and differs mainly in the spores having marked pedicels and closely warted.

PODAXINEÆ.

This tribe is characterized by having a stalk continuous with the apex of the peridium, forming an axis. Some of the plants are short stalked, some long stalked. The tribe forms a natural connecting link between the Gastromycetes and the Agarics. Thus: Podaxon is a true Gastromycetes, with capillitia mixed with spores; Caulogossum, with its permanent gleba chambers, is close to the Hymenogasters; Secotium is only a step from Caulogossum, the tramal plates being more sinuate-lamellate; and Montagnites, which is usually placed with the Agarics, is only a Gyrophragmium with the plates truly lamellate.

KEY TO THE GENERA.

Gleba with irregular, persistent chambers—

Peridium, elongated club-shaped	Cauloglossum.
Peridium, round or conical, and dehiscing by breaking away at the base	Secotium.
Gleba with sinuate-lamellate plates	Gyrophragmium.
Walls of gleba chambers not persistent	Podaxon.

SECOTIUM. KUNZ

This is a very interesting genus. When I found my first specimen I was much in doubt whether it was an Agaric or a puffball, as it seemed to be a sort of connecting link between the two classes. The genus is divided into smooth-spored and rough-spored species, both having a stalk continuing, as an axis, to the apex of the plant. The peridium is round or conical and it dehisces by breaking away at the base. Secotium is from a Greek word meaning chamber.

SECOTIUM ACUMINATUM. MONTAGNE.

Figure 479 Secotium acuminatum. Life size of small specimens.

This is an exceedingly variable species, as found about Chillicothe, yet the variability extends only to the outward appearance of the plant; some are almost round, slightly depressed, some (and a large majority) are inclined to be irregularly cone-shaped.

The peridium is light-colored, of a soft texture, not brittle; it slowly expels its spores by breaking away at the base; the stalk is usually short, but distinct and prolonged to the apex of the peridium, forming an axis for the gleba. The surface of the peridium is smooth, dingy-white or ash-colored, with minute white spots, due to scales. It is of various shapes; acute-ovate, sometimes obtuse, nearly spherical, sometimes slightly depressed and irregular cone-shaped. The gleba is composed of semi-persistent cells, plainly seen with a glass or even with the naked eye. It has no capillitium. The spores are globose and smooth, often apiculate. This plant is quite abundant about Chillicothe, and I have found it from the first of May to the last of October.

This species is widely distributed in America, and occurs in Northern Africa and Eastern Europe.

POLYSACCUM. DEC.

Polysaccum is from polus, many, and saccus, a sack. Peridium irregularly globose, thick, attenuated downward into a stem-like base, opening by disintegration of its upper portion; internal mass or gleba divided into distinct sac-like cells.

Allied to Scleroderma and distinguished by the cavities of the gleba containing distinct peridioles. Massee.

POLYSACCUM PISOCARPIUM. FR.

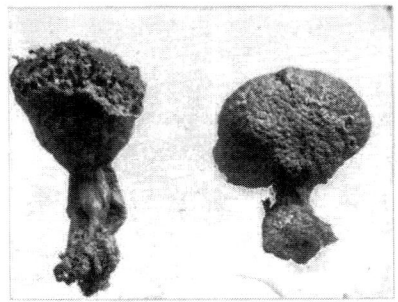

Figure 480   Polysaccum pisocarpium.

Pisocarpium is from two Greek words meaning pea and fruited.

Peridium irregularly globose, indistinctly nodulose, passing downward into a stout stem-like base, peridioles irregularly angular, 4–5×3μ, yellow. Spores globose, warted, coffee-color, 9–13μ. Massee.

I have found this plant only a few times about Chillicothe. Mr. Lloyd identified it for me. It has very much the shape of a pear. The skin is quite hard, smooth, olivaceous-black with yellow mottling patches not unlike the skin of a rattlesnake. The peridioles, which are small ovate sacs bearing the spores within, are very distinct. The interior of the plant when mature is dark, and it breaks and disintegrates from the upper part very like C. cyathiformis. This is a very interesting plant whose ovate sac-like cells will easily distinguish it. Found from August to October, it delights in sandy soil, in pine or mixed woods.

MITREMYCES. NEES.

Mitremyces is made up of two words: mitre, a cap; myces, a mushroom. It is a small genus, there being but three species found in this country. The spore-mass or gleba, in its young state, is surrounded by four layers. The outer layer is gelatinous and behaves itself somewhat differently in each species. This outer layer is known as the volva or volva-like peridium, which soon disappears. The next layer is called the exoperidium and is composed of two layers, the inner one quite thin and cartilaginous—in M. cinnabarinus it is a bright red; this is attached to a rather thick, gelatinous, outer layer which soon falls away, exposing the endoperidium, which is the layer seen in older specimens. Within the endoperidium are the spores, which are pale ochraceous or sulphur color, globose or elliptical in shape. They are contained in a separate membrane or sac; when they mature the sac contracts and forces the spores out into the air. The mycelium of this plant is especially peculiar, being composed of a bundle of root-like strands, translucent and jelly-like when young and fresh, but becoming tough and hard. This genus is called by some authors Calostoma, meaning a beautiful mouth, a very appropriate name, as the mouths of all American species are red and quite beautiful.

MITREMYCES CINNABARINUS. DESV.

Figure 481   Mitremyces cinnabarinus. Natural size.

The rooting strands are long, compact, dark when dry. Exporidium bright red, smooth internally; the outer layer thick, gelatinous when fresh, finally breaking into areas and curling inward. The separation is caused by the fact that the cells of the thick gelatinous portion expand by the absorption of water, while those of the inner layer do not, hence the rupture occurs. The endoperidium and rayed mouth are bright red when fresh, partially fading in old specimens.

The spores are elliptical-oblong, punctate-sculptured, varying much as to size in specimens from different localities; 6–8×10–14 in West Virginia specimens. Massachusetts specimens, 6–8×12–20. Lloyd.

I have seen these specimens growing in the mountains in West Virginia. They quickly arrest the attention because of their bright red caps. They seem not, as yet, to have crossed the Alleghenies—at least I have not found it in Ohio. It has a number of synonyms: Scleroderma calostoma, Calostoma cinnabarinum, Lycoperdon heterogeneum, L. calostoma.

The plants in Figure 481 were photographed by Dr. Kellerman. Mr. Geo. E. Morris of Waltham, Mass., sent me some specimens early in August, 1907.

GEASTER. MICH.

Geaster, an earth-star; so called because at maturity the outer coat breaks its connection with the mycelium in the ground and bursts open like the petals of a flower; then, becoming reflexed, those petals lift the inner ball from the ground and it remains in the center of the expanded, star-like coat. The coat of the inner ball is thin and papery, and opens by an apical mouth. The threads, or capillitium, which bear the spores proceed from the walls of the peridium and form the central columella. The threads are simple, long, slender, thickest in the middle and tapering towards the ends, fixed at one end and free at the other.

The Geaster is a picturesque little plant which will arrest the attention of the most careless observer. It is abundant and is frequently found in the late summer and fall in woods and pastures.

GEASTER MINIMUS. SCHW.

Figure 482   Geaster minimus. Natural size.

The outer coat or exoperidium recurved, segments acute at the apex, eight to twelve segments divided to about the middle. Mycelial layer usually attached, generally shaggy with fragments of leaves or grass, sometimes partly or entirely separating. Fleshy layer closely attached, very light in color, usually smooth on the limb of the exoperidium but cracked on the segments. Pedicel short but distinct. The inner peridium ovoid, one-fourth to one-half inch in diameter; white to pale-brown, sometimes almost black. Mouth lifted on a slight cone, lip bordered with a hair-like fringe; columella slender, as are also the threads. Spores brown, globe-shaped, and minutely warted. Found in the summer and early fall.

Nature seems to give it the power to lift up the spore-bearing body, the better to eject its spores to the wind. It is very frequently found in pastures all over the state. I have found it in many localities about Chillicothe. It is called "minimus" because it is the smallest Earth-star.

GEASTER HYGROMETRICUS. PERS.

WATER-MEASURING EARTH-STAR.

Figure 483   Geaster hygrometricus. Natural size.

The unexpanded plant is nearly spherical. The mycelial layer is thin, tearing away as the plant expands, the bark or skin falling with the mycelium. The outer coat is deeply parted, the segments, acute at the apex, four to twenty; strongly hygrometric, becoming reflexed when the plant is moist, strongly incurved when the plant is dry. The inner coating is nearly spherical, thin, sessile, opening by simply a torn aperture. There is no columella. The threads are transparent, much branched, and interwoven. The spores are large, globose, and rough.

The plant ripens in the fall and the thick outer peridium divides into segments, the number varying from four to twenty. When the weather is wet the lining of the points of the segments become gelatinous and recurve, and the points rest upon the ground, holding the inner ball from the ground. In dry weather the soft gelatinous lining becomes hard and the segments curve in and clasp the inner ball. Hence its name, "hygrometricus," a measurer of moisture. The plant is quite general.

GEASTER ARCHERI. BERK.

Figure 484   Geaster Archeri.

Young plant acute. Exoperidium cut beyond the middle into seven to nine acute segments. In herbarium specimens usually saccate but sometimes revolute. Mycelial layer closely adherent, compared to previous species relatively smooth. As in the previous species the mycelium covers the young plant but is not so strongly developed, so that the adhering dirt is not so evident on the mature plant. Fleshy layer when dry, thin and closely adherent. Endoperidium globose, sessile. Mouth sulcate, indefinite. Columella globose-clavate. Capillitium thicker than the spores. Spores small, 4 mc. almost smooth. Lloyd.

I first found the plant in the young state. The acute point, which will be seen in the photograph, puzzled me. I marked the place where it grew and in a few days found the developed Geaster. The plant is reddish-brown and it differs from other species "with sulcate mouths, in its closely sessile endoperidium." I have found the plant several times in Hayne's Hollow, near Chillicothe. I found it in the tracks of decayed logs.

The plant has been called Geaster Morganii in this country but had previously been named from Australia.

GEASTER ASPER. MICHELIUS.

Figure 485 Geaster asper. Natural size.

Exoperidium revolute, cut to about the middle in eight to ten segments. Both mycelial and fleshy layers are more closely adherent than in most species. Pedicel short and thick. Inner peridium subglobose, verrucose. Mouth conical, beaked, strongly sulcate, seated on a depressed zone. Columella prominent, persistent. Capillitium threads simple, long tapering. Spores globose, rough.

The characteristic of this plant is the verrucose inner peridium. Under a glass of low power it appears as though the peridium were densely covered with grains of sharp sand. This plant alone has this characteristic, to our knowledge; and although it is indicated in the figures of G. cornatus of both Schaeffer and Schmidel, we think that there it is only an exaggeration of the very minute granular appearance cornatus has. The word "asper" is the first descriptive adjective applied by Michelius. Fries included it in his complex striatus. Lloyd.

I have found the plant frequently about Chillicothe. The plants represented were photographed by Mr. Lloyd.

GEASTER TRIPLEX. JUNG.

Plate LXVI. Figure 486   Geaster triplex.

The unexpanded plant acute. Exoperidium recurved (or, when not fully expanded, somewhat saccate at base), cut to the middle (or usually two-thirds) in five to eight segments. Mycelial layer adnate. Fleshy layer generally peeling off from the segments of the fibrillose layer but usually remaining partially free, as a cup at base of inner peridium. Inner peridium subglobose, closely sessile. Mouth definite, fibrillose, broadly conical. Columella prominent, elongated. Threads thicker than spores. Spores globose, roughened, 3–6 mc. Lloyd, in Mycological Notes.

The color of Geaster triplex is reddish-brown. Notice the remains of a fleshy layer forming a cup at base of inner peridium, a point which distinguishes this species and which gives name to the species—triplex, three folds or apparently three layers. The photograph was made by Dr. Kellerman.

GEASTER SACCATUS. FR.

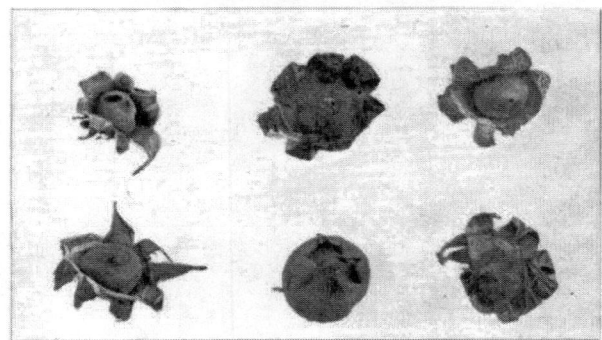

Figure 487   Geaster saccatus. Natural size.

The unexpanded plant is globose. Mycelium is universal. Exoperidium cut in six to ten segments about half way, the limb deeply saccate. Mycelial layer adnate to fibrillose. Fleshy layer, when dry, thin, adnate. Inner peridium sessile, globose, with a determinate fibrillose mouth.

The spores are globose, almost smooth. Lloyd.

Mr. Lloyd thinks this plant is practically the same as the G. fimbriatus of Europe, differing from it in being more deeply saccate and having a determinate mouth. This plant is very common on all the wooded hillsides about Chillicothe. I have seen the ground on the top of Mt. Logan almost completely covered with them. They are identified by Mr. Lloyd, Prof. Atkinson, and Dr. Peck. The plants in Figure 487 were photographed by Mr. Lloyd from typical specimens.

GEASTER MAMMOSUS. CHEV.

Exporidium thin, rigid, hygroscopic, smooth, divided almost to the base into about ten linear segments, often umbilicate at the base; inner peridium globose, smooth, sessile, furnished with a conical, even, protruding mouth, seated on a definite area.

Columella short, globose, evident (though distinct in mature plants).

Capillitium simple, tapering, hyaline, often flattened, slightly thinner than the spores. Spores globose, roughened, 3–7 mc. Lloyd.

Figure 488   Geaster mammosus.

This plant is found in the woods from July till late in the fall. It differs from G. hygrometricus by its even, conical mouth. I found specimens several times in Haynes's Hollow.

GEASTER VELUTINUS. MORG.

Figure 489   Geaster velutinus.

Unexpanded plants globose, sometimes slightly pointed at apex. Mycelium basal. Outer layer rigid, membranaceous, firm, light colored in the American plant. The surface is covered with short, dense, appressed velumen, so that to the eye the surface appears simply dull and rough, but its true nature is readily seen under a glass of low power.

The outer surface separates from the inner as the plant expands, and in mature specimens is usually partly free. The thickness and texture of the two layers are about the same. The fleshy layer is dark reddish-brown when dry, a thin adnate layer. Inner peridium sessile, dark colored, globose, with a broad base and pointed mouth. Mouth even, marked with a definite circular light-colored basal zone. Columella elongated, clavate. Spores globose, almost smooth, small, 2½—3½ mc. Lloyd.

MYRIOSTOMA COLIFORMIS. DICK.

Figure 490   Myriostoma coliformis. Natural size.

Exporidium usually recurved, cut to about the middle into six to ten lobes; if collected and dried when first open, rather firm and rigid; when exposed to weather becoming like parchment paper by the peeling off of the inner and outer layers. Inner peridium, subglobose, supported on several more or less confluent pedicels. Surface minutely roughened; mouths several, appressed fibrillose, round, plain or slightly elevated; columellæ several, filiform, probably the same in number as the pedicels; spores globose, roughened, 3–6 mc.; capillitium simple, unbranched, long, tapering, about half diameter of spores.

The inner peridium with its several mouths can be, not inaptly, compared to a "pepper-box." The specific name is derived from the Latin colum, a strainer, and the old English name we find in Berkeley "Cullender puffball" refers to a cullender (or colander more modern form) now almost obsolete in English, but meaning a kind of strainer. Lloyd.

Found in sandy soil. It is quite rare. Both the generic and specific names refer to its many mouths. The specimens in Figure 490 were found on Green Island, Lake Erie, one of the points where this rare species is found. It is found at Cedar Point, Ohio, also. The plant was photographed by Prof. Schaffner of the Ohio State University.

CHAPTER XVI

FAMILY—SPHAERIACEAE

Perithecia carbonaceous or membranaceous, sometimes confluent with the stroma, pierced at the apex, and mostly papillate; hymenium diffluent.—Berkeley Outlines.

There are four tribes in this family, viz:

- Nectriæi.
- Xylariæi.
- Valsei.
- Sphæriei.

Under Nectriæi we have the following genera:

Stipitate—

Clavate or capitate	Cordyceps.
Head globose, base sclerotioid	Claviceps.

Parasitic on grass—

Stroma myceloid	Epichlœ.

Variable—

Sporidia double, finally separating	Hypocrea.
Sporidia double, ejected in tendrils, parasitic on fungi	Hypomyces.
Stroma definite, perithecia free, clustered or scattered	Nectria.
Perithecia erect, in a polished and colored sac	Oomyces.

Under Xylariæi we have:

Stipitate—

Stroma corky, subelavate	Xylaria.
Stroma somewhat corky, discoid	Poronia.

CORDYCEPS. FR.

Cordyceps is from a Greek word meaning a club and a Latin word meaning a head. It is a genus of Pyrenomycetous fungi of which a few grow upon other fungi, but by far the greater number are parasitic upon insects or their larva, as will be seen in Figure 491.

The spores enter the breathing openings along the sides of the larva and the mycelium grows until it fills the interior of the larva and kills it.

In fructification a stalk rises from the body of the insect or larva and in the enlarged extremity of this the perithecia are grouped. The stroma is vertical and fleshy, head distinct, hyaline or colored; sporidia repeatedly divided and sub-moniliform.

CORDYCEPS HERCULEA. (SCHW.) SACC.

Figure 491 Cordyceps herculea. Showing the grub upon which this species grows.

Herculea is so called from its large size. The halftone will readily identify this species. The plant is quite large, clavate in form, the head oblong, round, slightly tapering upward with a decided protuberance at the apex, as will be seen in Figure 491. The head is a light yellow in all specimens I found, not alutaceous as Schw. states, nor is the head obtuse. I found several specimens on a sidehill in Haynes's Hollow in August and September, all growing from bodies of the large white grubs which are found about rotten wood. They were found during wet weather. They were identified by both Dr. Peck and Dr. Herbst.

CORDYCEPS MILITARIS. FR.

This is much smaller and more common than C. Herculea. Conidia—Subcæspitose, white; stem distinct, simple, becoming smooth; clubs incrassated, mealy; Conidia globose. Ascophore— Fleshy, orange-red; head clavate, tuberculose; stem equal; sporidia long, breaking up into joints. This is frequently called Torrubia militaris.

It is known as the caterpillar fungus. Its spores are cylindrical and are produced upon orange-red fruiting bodies in the fall. As soon as the spore falls on the caterpillar it sends out germ-threads which penetrate the caterpillar. Here the threads form long narrow spores which break off and form other spores until the body-cavity is entirely filled. The caterpillar soon becomes sluggish and dies. The fungus continues to grow until it has completely appropriated all of the insect's soft parts, externally a perfect caterpillar but internally completely filled with mycelial threads

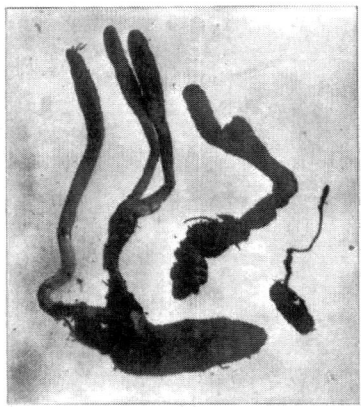

Figure 492   Cordyceps militaris.

Under favorable conditions this mycelial caterpillar, which has become a storage organ, will send up an orange-red club-shaped body, as will be seen in Figure 492, and will produce the kind of spores described above. Under some conditions this mycelial caterpillar may be made to produce a dense growth of threads from its entire surface, looking like a small white ball, and from these threads another kind of spore is formed. These spores are pinched off in great numbers and will germinate in the larva the same as the sac spore. The specimens were found by Mrs. E. B. Blackford near Boston, and photographed by Dr. Kellerman.

CORDYCEPS CAPITATA. FR.

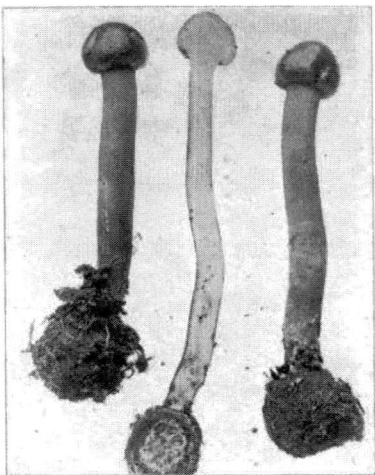

Figure 493   Cordyceps capitata. Natural size.

This plant is fleshy, capitate, head ovate, bay-brown, stem yellow, then blackish.

This plant is parasitic on Elaphomyces granulatus. It is shown at the base of the stem of the plant. It grows two or three inches under the surface and somewhat resembles a truffle in appearance.

Both are very interesting plants. The plant in Figure 493 was found near Boston, Mass. They are usually found in pine woods, often in tufts. The stems are from one to four inches long, nearly equal, smooth, lemon-colored, at length fibroso-strigose and blackish.

It is sometimes called Torrubia capitata.

CHAPTER XVII

MYXOMYCETES

The plants under this head belong to the slime-moulds and at first are wholly gelatinous. All the species and genera are small and easily overlooked, yet they are intensely interesting when carefully observed. In the morning you may see a mass of gelatinous matter and in the evening a beautiful net work of threads and spores, the transformation being so rapid. This gelatinous mass is known as protoplasm or plasmodium, and the motive power of the plasmodium has suggested to many that they should be placed in the animal kingdom, or called fungus animals. The same is true of Schizomycetes, to which all the bacteria, bacillus, spirillum, and vibrio, and a number of other groups belong. I have only a few Myxomycetes to present. I have watched the development of a number of plants of this group, but because of the scarcity of literature upon the subject I have been unable to identify them satisfactorily.

LYCOGALA EPIDENDRUM. FR.

Figure 494   Lycogala epidendrum.

This is called the Stump Lycogala. It is quite common, seeming in a certain stage to be a small puffball. The peridium has a double membrane, papery, per sistent, bursting irregularly at the apex; externally minutely warty, nearly round, blood-red or pinkish, then brownish; mouth irregular; spores becoming pale, or violet.

RETICULARIA MAXIMA. FR.

This is quite common on partially decayed logs. The peridium is very thin, tuberculose, effused, delicate, olivaceous-brown; spores olive, echinulate or spiny.

DIDYMIUS XANTHOPUS. FR.

These are very small yellow-stemmed plants, found on oak leaves in wet weather. The sporangium has an inner membranaceous peridium; the whole is round, brown, whitish. The stem is elongated, even, yellow. The columella is stipitate into the sporangia.

D. CINEREUM. FR.

Sporangia sessile, round, whitish, covered with an ashy-gray scurf. Spores black. Very small. On fallen oak leaves. Easily overlooked.

XYLARIA. SCHRANK.

Xylaria means pertaining to wood. It is usually vertical, more or less stipitate. The stroma is between fleshy and corky, covered with a black or rufous bark.

XYLARIA POLYMORPHA. GREV.

Figure 495   Xylaria polymorpha. Natural size.

Polymorpha means many forms. It is nearly fleshy, a number usually growing together, or gregarious; thickened as if swollen, irregular; dirty-white, then black; the receptacle bearing perithecia in every part.

This plant is quite common in our woods, growing about old stumps or on decayed sticks or pieces of wood. The spore-openings can be seen with an ordinary hand-glass.

XYLARIA POLYMORPHA, VAR. SPATHULARIA

Spathularia means in the form of a spathula or spatula. It is vertical and stipitate, the stem being more definite than in the X. polymorpha, the stroma being between fleshy and corky, frequently growing in numbers or gregarious, turgid, fairly regular, dirty-white, then brownish-red, finally black. An ordinary hand glass will show how it bears perithecia in all its parts. This will be clearly seen in the section on the right.

These plants are not as common as the X. polymorpha, but are found in habitats similar to those of the other plant, particularly around maple stumps or upon decayed maple branches

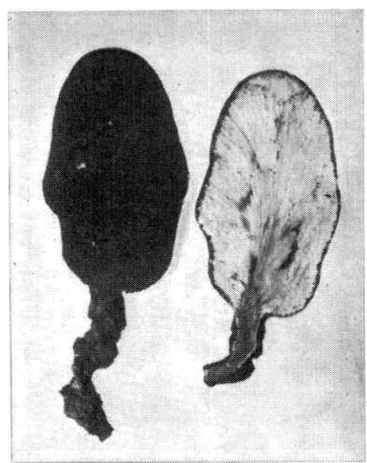

Figure 496 Xylaria polymorpha var. spathularia. Natural size.

STEMONITIS. GLED.

Stemonitis is from a Greek word which means stamen, one of the essential organs of a flower. This is a genus of myxomycetous fungi, giving name to the family Stemonitaceæ, which has a single sporangium or æthalium; without the peculiar deposits of lime carbonate which characterize the fructification of other orders, and the spores, capillitium, and columella are usually uniformly black, or brownish.

STEMONITIS FUSCA. ROTH.

Figure 497 Stemonitis fusca. Natural size.

Fusca means dark-brown, smoky. The sporangia are cylindrical and pointed at the apex, peridia fugacious, exposing the beautiful net-work of the capillitium. The reticulate capillitium springs from the dark, penetrating stem.

This is a very beautiful plant when studied with an ordinary hand-glass. I have frequently seen an entire log covered with this plant.

STEMONITIS FERRUGINEA. EHRB.

Ferruginea means rust color. The sporangia is very similar to that of S. fusca, cylindrical, peridium fugacious, exposing the reticulate capillitium, but instead of being dark-brown it is a yellowish or rusty-brown color.

CHAPTER XVIII

RECIPES FOR COOKING MUSHROOMS.

STEWED MUSHROOMS. NO. 1.

Choose them as nearly as possible of uniform size and free from insects. Drop them in salt water for five minutes to free them from any insects that may be hidden in the gills; drain them and wipe dry and clean with a rather rough cloth; cut off the stems close to the cap. Put them into a granite or porcelain saucepan, cover closely and stew gently fifteen minutes. Salt to taste. Rub a tablespoonful of butter into about a tablespoonful of flour, and stir this into the mushrooms, letting boil three or four minutes; stir in three tablespoonfuls of cream, mixed with a well-beaten egg, and stir the whole for two minutes without letting it boil, and serve either on toast or as a vegetable.

STEWED MUSHROOMS. NO. 2.

Clean mushrooms as directed above and stew in water ten minutes; then drain off part of the water and put in as much warm milk as you have poured off water; let this stew for five to ten minutes; then add some drawn butter, or veal or chicken gravy, and salt and pepper to taste. Thicken with a little corn starch wet in cold milk. Serve hot.

In cooking mushrooms they should always be kept as closely covered as possible in order the better to retain the flavor, and they should never be subjected to too great heat.

BAKED MUSHROOMS.

Be sure your mushrooms are fresh and free from insects; cut off the stems close to the caps and wipe the tops with a wet cloth. Arrange them in a pie dish with the gills uppermost, laying a little bit of butter on each; sprinkle pepper, salt, and a very little mace upon them. Put them into a hot oven and bake from fifteen minutes to half an hour, according to the tenderness of the mushrooms; if they are in danger of getting too dry baste them occasionally with butter and water. Pour over them some maitre d'hotel sauce and send to the table in the dish in which they were baked.

BROILED MUSHROOMS.

Select the finest and freshest you can get and prepare as for baking; put into a deep dish and pour over them some melted butter, turning them over and over in it. Salt and pepper and let them lie for an hour and a half in the butter. Put them, gills uppermost, on an oyster gridiron over a clear hot fire, turning them over as one side browns. Put them on a hot dish, having them well seasoned with butter, pepper, and salt and with a few drops of lemon juice squeezed upon each, if liked.

MUSHROOM AND VEAL RAGOUT.

Take equal quantities of cold veal steak or roast veal and small puffballs or other mushrooms, and mince all fine; mince a small onion and put with the mushrooms and meat into a pan with some cold veal gravy, if you have it, and water enough to cover the mixture. Add a tablespoonful of butter, pepper and salt well, and let the mixture cook until it is almost dry, stirring it frequently to keep it from scorching; it should cook fully half an hour. When almost done, add a large tablespoonful of good catsup, or Worcestershire sauce if preferred. Serve hot.

MUSHROOM PATÉS.

Wash mushrooms well, cut them into small pieces and drop them in salt water for five minutes. Have ready in a pan upon the stove about two ounces of butter to each pint of mushrooms, having pan and butter very hot but not scorching; dip the mushrooms from the salt water with a skimmer and drop them into the hot butter; cover them closely to retain the flavor, shaking the pan or stirring them over to keep them from scorching or sticking. Let them cook with moderate heat from fifteen to thirty minutes, according to the tenderness of the mushrooms. Remove the cover from the pan, draw the mushrooms to one side and lift the pan on one side so that the gravy will run down to the opposite side; stir into the gravy a level tablespoonful of sifted flour, and rub this smooth with the gravy; then add a half a pint of rich milk or cream; stir the mushrooms into this and allow it to boil for a minute. Have ready in the oven some paté shells, fill them with the mushrooms, seasoned to taste with salt and pepper, and set back in the oven for a few minutes to heat before serving. These are especially fine when made of Tricholoma personatum or Pleurotus ostreatus, but many other varieties will answer well.

BAKED BEEFSTEAK WITH MUSHROOM SAUCE.

Have your sirloin steak cut an inch or more thick, put into an exceedingly hot baking pan on top of the stove, in one minute turn steak over so that both sides will be seared. Put the pan into an exceedingly hot oven and allow it to remain for twenty minutes.

Have ready in a saucepan two tablespoonfuls of melted butter, heat well and add two cupfuls of fresh, clean mushrooms which have been allowed to stand in salt water for a period of five minutes; cover closely and cook briskly without burning for ten minutes; set on the back of the stove (after having seasoned them properly with salt and pepper) to keep hot until ready to use. Place the steak upon a hot dish, pour the mushrooms over it and send to the table at once. It is a dish fit for a king.

STUFFED MORELS.

Choose the freshest and best morels; cleanse them thoroughly by allowing the water from the faucet to run on them; open the stalk at the bottom; fill with veal stuffing, anchovy or any rich forcemeat you choose, securing the ends and dressing between slices of bacon; bake for a half an hour, basting with butter and water, and serve with the gravy which comes from them.

FRIED MORELS.

Wash a dozen morels carefully and cut off the ends of the stems. Split the mushrooms and put them into a pan in which two tablespoonfuls of butter have been melted. Cover closely and cook with a moderate heat for fifteen minutes. Mix two teaspoonfuls of corn starch in a half a pint of fresh milk and pour into the pan with the mushrooms, allowing it to boil for a minute or two; salt and pepper to taste and serve hot, upon toast if liked.

TO COOK BOLETI.

Cut off the stems, and remove the spore-tubes, after having wiped the caps clean with a damp cloth. They may be broiled in a hot buttered pan, turning them frequently until done, which will be about fifteen minutes. Dust with salt and pepper and put bits of butter over them as you would on broiled beefsteak.

They may be stewed in a little water in a covered saucepan, after being cut into pieces of equal size. Stew for twenty minutes and when done add pepper, salt, butter or cream.

Or they may be fried, after being sliced as you would egg plant, and dipped in batter or rolled in egg and cracker crumbs.

In preparing Boleti the spore tube should be removed unless very young, as they will make the dish slimy.

MUSHROOM CATSUP.

To two quarts of mushrooms allow a quarter of a pound of salt. The full grown mushrooms are better in making this as they afford more juice. Put a layer of mushrooms in the bottom of a stone jar, sprinkle with salt; then another layer of mushrooms till you have used all; let them lie thus for six hours, then break them into bits. Set in a cool place for three days, stirring thoroughly every morning. Strain the juice from them, and to every quart allow half an ounce of allspice, the same quantity of ginger, half a teaspoonful of powdered mace and half a teaspoonful of cayenne. Put it into a stone jar, cover it closely, set it in a saucepan of water over the fire, and boil hard for five hours. Take it off, empty it into a porcelain kettle and let it boil slowly for half an hour longer. Set it in a cool place and let it stand all night until settled and clear, then pour off carefully from the sediment, into small bottles, filling them to the mouth. Cork tightly and seal carefully. Keep in a dry, cool, dark closet.

MUSHROOMS WITH BACON.

Take some full-grown mushrooms, and, having cleaned them, procure a few rashers of nice streaky bacon and fry it in the usual manner. When nearly done add a dozen or so of mushrooms and fry them slowly until they are cooked. In the cooking they will absorb all the fat of the bacon, and with the addition of a little salt and pepper will form a most appetizing breakfast relish.

HYDNUM.

The Hydnums are sometimes slightly bitter and it is well to boil them for a few minutes and then throw away the water. Drain the mushrooms carefully; add pepper and salt, butter, and milk; cook in a covered saucepan slowly for twenty or twenty-five minutes; have ready some slices of toast, pour the mushrooms over these and serve at once.

OYSTER MUSHROOMS.

One of the best ways to cook an Oyster mushroom is to fry it as you fry an oyster. Use the tender part of the Oyster mushroom; clean thoroughly; add pepper and salt; dip in beaten egg and then bread crumbs and fry in fat or butter. Or parboil them for forty-five minutes, drain, roll in flour and fry.

The Oyster mushroom is also excellent when stewed.

LEPIOTA PROCERA.

Clean the caps with a damp cloth and cut off the stem close to the caps; broil lightly on both sides over a clear fire or in a very hot pan, turning the mushrooms carefully three or four times; have ready some freshly-made, well-buttered toast; arrange the mushrooms on the toast and put a small piece of butter on each and sprinkle with pepper and salt; set in the oven or before a brisk fire to melt the butter, then serve quickly.

Some persons think that slices of bacon toasted over the mushrooms improve the flavor.

BEEFSTEAK SMOTHERED IN MUSHROOMS.

Have ready a sufficient quantity of full-grown mushrooms, carefully cleaned; cut them in pieces and put into a baking pan with a tablespoonful of butter to two cupfuls of mushrooms, sprinkle with pepper and salt, and bake in a moderate oven forty-five minutes. Broil your steak until it is almost done; then put it into the pan with a part of the mushrooms under and the remainder over the steak; put it into the oven again and allow it to remain for ten minutes; turn out upon a hot dish and serve quickly.

Agaricus, Lepiota, Coprinus, Lactarius, Tricholoma, and Russula are especially fine for this method of preparation.

CHAPTER XIX

CULTIVATION OF THE MUSHROOM

GENERAL CONSIDERATIONS

Commercially, and in a restricted sense, the term "mushroom" is generally used indiscriminately to designate the species of fungi which are edible and susceptible of cultivation. The varieties which have been successfully cultivated for the market are nearly all derived from Agaricus campestris, Agaricus villaticus, and Agaricus Arvensis. They may be white, cream or creamy-white, or brown; but the color is not always a permanent characteristic, it is often influenced by surrounding conditions.

Mushrooms are grown for the market on a large scale in France and in England. It is estimated that nearly twelve million pounds of fresh mushrooms are sold every year at the Central Market of Paris. A large quantity of mushrooms are canned and exported from France to every civilized country. This industry has recently made remarkable progress in the United States, and fresh mushrooms are now regularly quoted on the markets of our large cities. They are sold at prices ranging from twenty-five cents to one dollar and fifty cents per pound, according to season, demand and supply.

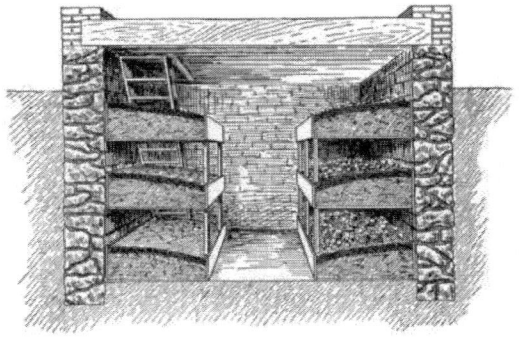

Figure 498   Mushroom Beds in a Cellar.

ESSENTIAL CONDITIONS

Mushrooms can be grown in any climate and in any season where the essential conditions may be found, obtained or controlled. These conditions are, first, a temperature ranging from 53°

to 60° F., with extremes of 50° to 63°; second, an atmosphere saturated (but not dripping) with moisture; third, proper ventilation; fourth, a suitable medium or bed; fifth, good spawn. It may be seen that in the open air, these conditions are rarely found together for any length of time. It is therefore necessary, in order to grow mushrooms on a commercial basis, that one or more of these elements be artificially supplied or controlled. This is usually done in cellars, caves, mines, greenhouses, or specially constructed mushroom houses. A convenient disposition of the shelves in a cellar is shown in Figure 498. A large installation for commercial purposes is shown in Figure 500, and a specially constructed cellar is shown in Figure 499. Where abandoned mines, natural or artificial caves are available, the required atmospheric conditions are often found combined and may be uniformly maintained throughout the year.

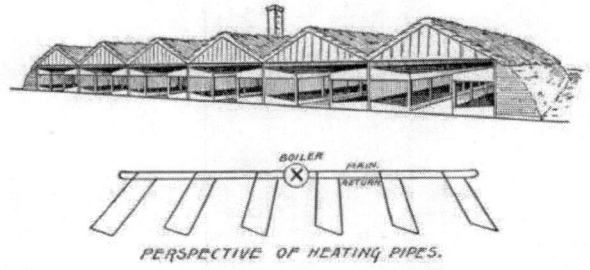

Figure 499   Specially Constructed Mushroom Houses.

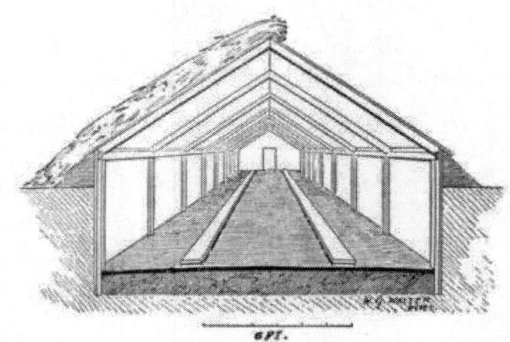

Figure 500   Mushroom Houses, Flat Beds.

TEMPERATURE

Within the limits prescribed, the temperature should be uniform throughout the growth of the crop. When too cold, the development of the spawn will be retarded or arrested. A high temperature will favor the development of molds and bacteria which will soon destroy the spawn or the growing crop. The cultivation of the mushroom, as a summer crop, is therefore greatly restricted. As a fall, winter or spring crop it may be grown wherever means are at hand to raise the temperature to about 58° F. Many florists are utilizing the waste space under the benches for that purpose; they have the advantage of being able to use the expended material of mushroom beds in growing flowers.

MOISTURE

Moisture is an important factor in the cultivation of the mushroom, and demands intelligent application. The mushroom requires an atmosphere nearly saturated with moisture, and yet the direct application of water on the beds is more or less injurious to the growing crop. It is therefore essential that the beds, when made, contain the requisite amount of moisture, and that this moisture be not lost by excessive evaporation. They should be protected from a dry atmosphere or strong draughts. Where watering becomes necessary, it should be applied in a fine spray around the beds with a view of restoring the moisture to the atmosphere, and on the beds after the mushrooms have been gathered.

VENTILATION

Pure air is essential to a healthy crop. Provision should therefore be made for a gradual renewal of the air in the mushroom house. However, draughts must be avoided as tending to a too rapid evaporation and cooling of the beds, an unfortunate condition which cannot thereafter be entirely remedied.

THE BEDS

The most common type of beds is known as the "flat bed." It is made on the floor or on shelves as shown in the illustrations. It is usually about 10 inches deep. Another type, principally used in France, is known as the "ridge bed," and requires more labor than the flat bed. The mushroom house and shelves, if used, should be frequently disinfected and whitewashed in order to avoid danger from insects and bacteria. The preparation of the beds and subsequent operations will be shown in connection with the other subjects.

PREPARATION OF THE MANURE

The best manure is obtained from horses fed with an abundance of dry and nitrogenous food. The manure of animals fed on greens is undesirable. Growers do not all follow the same method of fermenting or composting the manure. When first unloaded, the manure is left in its original state for a few days. It is then piled in heaps about three feet deep and well pressed down. In this operation the material should be carefully forked and well mixed, and wherever found too dry, it should be lightly sprinkled. It is allowed to remain in that condition for about six days when it is again well forked and turned. In the latter operation it receives an additional light sprinkling; the dry portions are turned inside in order that the whole mass may be homogenous and uniformly moist, and the heap is again raised to about three feet. About six days later the operation is repeated, and in about three days the manure should be ready for the beds. It is then of a dark brown color mixed with white, free from objectionable odor. It is unctuous, elastic and moist, though not wet, and should not leave any moisture in the hand.

Of course, the above rules are subject to modification according to the condition of the manure, its age and previous handling.

SPAWNING

The manure, having been properly composted, is spread evenly on the floor or shelves and firmly compressed in beds about ten inches in depth. The temperature of the bed is then too high for spawning and will usually rise still higher. It should be carefully watched with the aid of a special or mushroom thermometer. When the temperature of the beds has fallen to about 75° or 80°, they may be spawned. The beds must be spawned when the temperature falls, never when it rises. The bricks of spawn are broken into eight or ten pieces, and these pieces are inserted from one to two inches below the surface, about nine to twelve inches apart. The bed is then firmly compressed. An advantage is found in breaking and distributing the spawn over the surface of the bed a few days before spawning; this allows the mycelium to absorb some moisture and swell to some extent. If the bed is in proper condition it should not require watering for several weeks.

Figure 501   Brick Spawn, Pure Culture.

CASING THE BEDS

As soon as the spawn is observed to "run," or from eight days to two weeks, the beds are "cased" or covered with a layer of about one inch of light garden loam, well screened. The loam should be slightly moist, and free from organic matter. The beds should now be watched and should not be allowed to evaporate or dry out.

PICKING

Mushrooms should appear in from five to ten weeks after spawning, and the period of production of a good bed ranges from two to four months. In picking the mushrooms an intelligent hand will carefully twist it from the soil and fill the hole left in the bed with fresh soil. Pieces of roots or stems should never be allowed to remain in the beds, otherwise decay might set in and infect the surrounding plants. A good mushroom bed will yield a crop of from one-half to two pounds per square foot. Mushrooms should be picked every day or every other day; they should not be left after the veils begin to break.

For the market the mushrooms are sorted as to size and color, and packed in one, two or five-pound boxes or baskets. Since they are very perishable, they must reach the market in the shortest time.

OLD BEDS

It is not practicable to raise another crop of mushrooms in the material of an old bed, although this material is still valuable for garden purposes. The old material should be entirely removed, and the mushroom house thoroughly cleaned before the new beds are made. If this precaution be omitted the next crop may suffer from the diseases or enemies of the mushrooms.

Figure 502 A Cluster of 50 Mushrooms on One Root, Grown from "Lambert's Pure Culture Spawn" of the American Spawn Co., St. Paul, Minn.

SPAWN

The cultivated mushroom is propagated from "spawn," the commercial name applied to the mycelium; the term "spawn" includes both the mycelium and the medium in which it is carried and preserved. Spawn may be pro cured in the market in two forms, flake spawn and brick spawn. In both forms the mycelium growth is started on a prepared medium mainly consisting of manure and then arrested and dried. The flake spawn is short-lived by reason of its loose form, in which the mycelium is easily accessible to the air and destructive bacteria. It deteriorates rapidly in transportation and storage and can only be used to advantage when fresh. Growers, especially in the United States, have therefore discarded it in favor of brick spawn, which affords more protection to the mycelium and can be safely transported and stored for a reasonable period.

Until recently the manufacturer of spawn was compelled to rely entirely upon the caprice of nature for his supply. The only method known consisted in gathering the wild spawn wherever nature had deposited it and running the same into bricks or in loose material, without reference to variety. Neither the manufacturer nor the grower had any means of ascertaining the probable nature of the crop until the mushrooms appeared.

Figure 503 Agaricus villaticus.

PURE CULTURE SPAWN

The recent discovery of pure culture spawn in this country has made possible the selection and improvement of varieties of cultivated mushrooms with special reference to their hardiness, color, size, flavor and prolificness, and the elimination of inferior or undesirable fungi in the crop. The scope of this article precludes a description of the pure culture method of making spawn. It is now used by the large commercial growers and has in many sections entirely superseded the old English spawn and other forms of wild spawn. As now manufactured it resembles much in appearance the old English spawn (see Figure 501). Some remarkable results have been obtained by the use of pure culture spawn. We illustrate a cluster of fifty mushrooms on one root grown by Messrs. Miller & Rogers, of Mortonville, Pa., from "Lambert's Pure Culture Spawn" produced by the American Spawn Company, of St. Paul, Minn. (Figure 502). Several promising varieties have already been developed by the new method, and can now be reproduced at will. Figure 503 is a good illustration of Agaricus villaticus, a fleshy species in good demand. Figure 504 shows a bed of mushrooms grown from pure culture spawn in a sand rock cave, using the flat bed.

Figure 504 A Mushroom Cave, Showing One of the Test Beds of the American Spawn Co., St. Paul, Minn.

HOW TO COOK MUSHROOMS

To the true epicure there are but four ways of cooking mushrooms—broiling, roasting, frying them in sweet butter and stewing them in cream.

In preparing fresh mushrooms for cooking, wash them as little as possible, as washing robs them of their delicate flavor. Always bear in mind that the more simply mushrooms are cooked the better they are. Like all delicately flavored foods, they are spoiled by the addition of strongly flavored condiments.

BROILED MUSHROOMS

Select fine, large flat mushrooms, and be sure that they are fresh. If they are dusty just dip them in cold salt water. Then lay on cheese cloth and let them drain thoroughly. When they are dry cut off the stem quite close to the comb. Or, what is better, carefully break off the stem. Do not throw away the stems. Save them for stewing, for soup or for mushroom sauce. Having cut or broken off the stems, take a sharp silver knife and skin the mushrooms, commencing at the edge and finishing at the top. Put them on a gridiron that has been well rubbed with sweet butter. Lay the mushrooms on the broiling iron with the combs upward. Put a small quantity of butter, a little salt and pepper in the center of each comb from where the stem has been removed and let the mushrooms remain over the fire until the butter melts. Then serve them on thin slices of buttered and well browned toast, which should be cut round or diamond shape.

Serve the mushrooms just as quickly as possible after they are broiled, as they must be eaten when hot. So nourishing are broiled mushrooms that with a light salad they form a sufficient luncheon for anyone.

FRIED MUSHROOMS

Clean and prepare the mushrooms as for broiling. Put some sweet, unsalted butter in a frying pan—enough to swim the mushrooms in. Stand the frying pan on a quick fire, and when the butter is at boiling heat carefully drop the mushrooms in and let them fry three minutes, and serve them on thin slices of buttered toast.

Serve a sauce of lemon juice, a little melted butter, salt and red pepper with fried mushrooms.

STEWED MUSHROOMS

Stewed mushrooms after the following recipe make one of the most delicious of breakfast dishes: It is not necessary to use large mushrooms for stewing—small button ones will do. Take the mushrooms left in the basket after having selected those for broiling, and also use the stems cut from the mushrooms prepared for boiling. After cleaning and skinning them put them in cold water with a little vinegar, and let them stand half an hour. If you have a quart of mushrooms, put a tablespoonful of nice fresh butter in a stewpan and stand it on the stove. When the butter begins to bubble drop the mushrooms in the pan, and after they have cooked a minute season them well with salt and black pepper. Now take hold of the handle of the stewpan and, while the mushrooms are gently and slowly cooking, shake the pan almost constantly to keep the butter from getting brown and the mushrooms from sticking. After they have cooked eight minutes pour in enough rich, sweet cream to cover the mushrooms to the depth of half an inch, and let them cook about eight or ten minutes longer. Serve them in a very hot vegetable dish. Do not thicken the cream with flour or with anything. Just cook them in this simple way. You will find them perfect.

GLOSSARY

Abortive, imperfectly developed.

Aberrant, deviating from a type.

Acicular, needle-shaped.

Aculeate, slender pointed.

Acuminate, terminating in a point.

Acute, sharp pointed.

Adnate, gills squarely and firmly attached to the stem.

Adnexed, gills just reaching the stem.

Adhesion, union of different organs or tissues.

Adpressed, pressed into close contact, as applied to the gills.

Agglutinated, glued to the surface.

Alveolate, honey-combed.

Alutaceous, having the color of tanned leather.

Anastomosing, branching, joining of one vein with another.

Annual, completing growth in one year.

Annular, ring-shaped.

Annulate, having a ring.

Annulus, the ring around the stem of a mushroom.

Apex, in mushrooms the extremity of the stem next to the gills.

Apical, close to the apex.

Apiculate, terminating in a small point.

Appendiculate, hanging in small fragments.

Applanate, flattened out or horizontally expanded.

Arachnoid, cobweb-like.

Arculate, bow-shaped.

Areolate, pitted, net-like.

Ascus, spore case of certain mushrooms.

Ascomycetes, a group of fungi in which the spores are produced in sacs.

Ascospore, hymenium or sporophore bearing an ascus or asci.

Atomate, sprinkled with atoms or minute particles.

Atro (ater, black), in composition "black" or "dark."

Atropurpureous, dark purple (purpura, purple).

Aurantiaceous, orange-colored (aurantium, an orange).

Aureous, golden-yellow.

Auriculate, ear-shaped.

Azonate, without zones or circular bands.

Badious, bay, chestnut-color, or reddish-brown.

Basidium (pl. basidia), an enlarged cell on which spores are borne.

Basidiomycetes, the group of fungi that have spores borne on a basidium.

Bifid, cleft or divided into two parts.

Booted, applied to the stem of mushrooms when inclosed in a volva.

Boss, a knob or short rounded protuberance.

Bossed, furnished with a boss or knob, bulbate.

Byssus, a fine filamentous mass.

Cæspitose, growing in tufts.

Calyptra, applied to the portion of volva covering the pileus.

Campanulate, bell-shaped.

Cap, the expanded, umbrella-like receptacle of a common mushroom.

Capillitium, spore-bearing threads, often much branched, found in puffballs.

Carnose, flesh-color.

Cartilaginous, hard and tough.

Castaneous, chestnut-color.

Ceraceous, wax-like.

Cerebriform, brain-shaped.

Cespitose, growing in tufts.

Cilia, marginal hair-like processes.

Ciliate, fringed with hair-like processes.

Cinereous, light bluish gray or ash gray.

Circumscissile, breaking at or near the middle on equatorial line.

Circinate, rounded.

Clavate, club-shaped, gradually thickened upward.

Columella, a sterile tissue rising column-like in the midst of the Capillitium.

Concrete, grown together.

Continuous, without a break, one part running into another.

Cordate, heart-shaped.

Coriaceous, of a leathery or a cork-like texture.

Cortex, outer or rind-like layer.

Cortina, the web-like veil of the genus Cortinarius.

Cortinate, with a cortina.

Costate, with a ridge or ridges.

Crenate, notched, indented or escalloped at the edge.

Cryptogamia, applied to the division of non-flowering plants.

Cyathiform, cup-shaped.

Cyst, a bladder-like cell or cavity.

Cystidium (pl. cystidia), sterile cells of the hymenium, bladder-like.

Deciduous, of leaves falling off.

Decurrent, as when the gills of a mushroom are prolonged down the stem.

Dehiscent, a closed organ opening of itself at maturity.

Deliquescent, melting down, becoming liquid.

Dendroid, shaped like a tree.

Dentate, toothed.

Denticulate, with small teeth.

Dichotomous, paired, regularly forked.

Dimidiate, halved, applied to gills not entire.

Disc (disk), the hymenial surface, usually cup-shaped.

Discomycetes, Ascomycetes with the hymenium exposed.

Dissepiments, dividing walls.

Distant, applied to gills which are not close.

Discrete, distinct, not divided.

Echinate, furnished with stiff bristles.

Effused, spread over without regular form.

Emarginate, when the gills are notched or scooped out at junction with stem.

Ephemeral, lasting but a short time.

Epidermis, the external or outer layer of the plant.

Epiphytal, growing upon another plant.

Eccentric, out of the center; stem not attached to center of pileus.

Exoperidium, outer layer of the peridium.

Exotic, foreign.

Explanate, flattened or expanded.

Farinaceous, mealy.

Farinose, covered with a mealy powder.

Falcate, hooked or curved like a scythe.

Fasciculate, growing in bundles.

Fastigiate, bundled together with a sheath.

Ferruginous, rust-colored.

Fibrillose, clothed with small fibers.

Fibrous, composed of fibers.

Filiform, thread-like.

Fimbriated, fringed.

Fissile, capable of being split.

Fistular, fistulose, with the stem hollow or becoming hollow.

Flabelliform, fan-shaped.

Flaccid, soft and flabby.

Flavescent, turning yellow.

Flexuose, wavy.

Flocci, threads as of mold.

Floccose, downy.

Flocculose, covered with flocci.

Free, said of gills not attached to the stem.

Friable, easily crumbling.

Fugacious, disappearing quickly.

Fuliginous, sooty-brown or dark smoke-color.

Furcate, forked.

Furfuraceous, with bran-like scales or scurf.

Fuscous, dingy, brownish or brown tinged with gray.

Fusiform, spindle-shaped.

Gasteromyces, Basidiomycetes, in which the hymenium is inclosed.

Gelatinous, jelly-like.

Genus, a group of closely related species.

Gibbous, swollen at one point.

Gills, plates radiating from the stem on which the basidia are borne.

Glabrous, smooth.

Glaucous, with a white bloom.

Gleba, the spore-bearing tissue, as in puffballs and phalloids.

Globose, nearly round.

Granular, with a roughened surface.

Gregarious, growing in numbers in the same vicinity.

Habitat, the natural place of growth of a plant.

Hirsute, hairy.

Host, the plant or animal on which a parasitic fungus grows.

Hyaline, transparent, clear like glass.

Hygrophanous, looking watery when moist and opaque when dry.

Hygrometric, readily absorbing water.

Hymenium, the fruit-bearing surface.

Hymenophore, the portion which bears the hymenium.

Hypha, one of the elongated cells or threads of the fungus.

Imbricate, overlapping like shingles.

Immarginate, without a distinct border.

Incarnate, flesh-color.

Indehiscent, not opening.

Indigenous, native of a country or a place.

Indurated, hardened.

Indusium, a veil beneath the pileus.

Inferior, the ring low down on the stem of Agarics.

Infundibuliform, funnel-shaped.

Innate, adhering by growth.

Involute, edges rolled inward.

Isabelline, color of sole leather, brownish-yellow.

Laccate, varnished or coated with wax.

Lacerate, irregularly torn.

Laciniate, divided into lobes.

Lacunose, pitted or having cavities.

Lamella (lamellæ), gills of a mushroom.

Lanate, wooly.

Leucospore, white spore.

Livid, bluish-black.

Luteous, yellowish.

Maculate, spotted.

Marginate, having a distinct border.

Micaceous, covered with glistening scales, mica-like.

Micron, one-thousandth of a millimeter, nearly .00004 of an inch.

Mycelium, the delicate threads from germinating spores, called spawn.

Nigrescent, becoming black.

Obconic, inversely conical.

Obovate, inversely egg-shaped.

Obese, stout, plump.

Ochraceous, ochre-yellow, brownish-yellow.

Pallid, pale, undecided in color.

Papillate, covered with soft tubercles.

Paraphyses, sterile cells found among the reproductive cells of some plants.

Parasitic, growing on and deriving support from another plant.

Pectinate, toothed like a comb.

Peridium, the outer covering of a puffball, simple or double.

Perithecia, bottle-like receptacles containing asci.

Peronate, used when the stem has a distinct stocking-like coat.

Persistent, inclined to adhere firmly.

Pileate, having a cap or pileus.

Pileolus (pl. pileoli), a secondary pileus, arising from the primary one.

Pileus (pileus, a hat), the cap-like head of a fungus.

Pilose, covered with hairs, furry.

Pore, the opening of the tubes of a polyporus.

Pruinose, covered with a frost-like bloom.

Pubescent, downy.

Pulverulent, covered with dust.

Pulvinate, cushion-shaped.

Putrescent, soon decaying.

Punctate, dotted with points.

Reflexed, bent backwards.

Reniform, kidney-shaped.

Repand, bent or turned up or back.

Resupinate, attached to the matrix by the back.

Reticulate, marked with cross-lines, like the meshes of a net.

Revolute, rolled backward or upward.

Rimose, cracked or full of clefts.

Rimulose, covered with small cracks.

Ring, a part of the veil adhering to the stem of Agarics.

Rubescent, tending to a red-color.

Rubiginous, rust-color.

Rufescent, reddish in color.

Rugose, wrinkled.

Rufous, brownish-red.

Sapid, agreeable to the taste.

Saprophyte, a plant that lives on decaying animal or vegetable matter.

Scrobiculate, marked with little pits or depressions.

Serrate, saw-toothed.

Sinuate, wavy margin of gills or sinus where they reach the stem.

Spathulate, in the form of a spathula.

Spawn, the popular name for mycelium, used in growing mushrooms.

Spores, the reproductive bodies of mushrooms.

Sporophore, name given to the basidia.

Squamose, having scales.

Squamulose, covered with small scales.

Squarrose, rough with scales.

Stigmata, the slender supports of the spores.

Stipitate, having a stem.

Striate, streaked with lines.

Strigose, covered with lines sharp and rigid.

Strobiliform, pineapple-shaped.

Stuffed, stem filled with different material from the walls.

Sulcate, furrowed.

Tawny, nearly the color of tanned leather.

Terete, top-shaped.

Tesselated, arranged in small squares.

Tomentose, downy, with short hairs.

Trama, the substance between the plates of gills.

Truncate, cut squarely off.

Tubercle, a small wart-like excrescence.

Turbinate, top-shaped.

Umbillicate, having a central depression.

Umbo, the boss of a shield, applied to the central elevation of cap.

Umbonate, having a central boss-like elevation.

Uncinate, hooked.

Undulate, wavy.

Vaginate, sheathed.

Veil, a partial covering of stem or margin of pileus.

Veliform, a thin veil-like covering.

Venate or veined, intersected by swollen wrinkles below and on the sides.

Ventricose, swollen in the middle.

Vernicose, shining as if varnished.

Verrucose, covered with warts.

Villose, villous, covered with long, weak hairs.

Viscid, covered with a shiny liquid which adheres to the fingers; sticky.

Viscous, gluey.

Volute, rolled up in any direction.

Volva, a universal veil.

Zoned, zonate, marked with concentric bands of color.

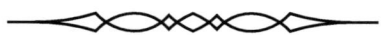

Made in the USA
Monee, IL
07 July 2026

56548174R00291